Ventilation Systems

Properly designed ventilation systems for cooling, heating and cleaning the outdoor air supplied to buildings are essential for maintaining good indoor air quality as well as for reducing a building's energy consumption. Modern developments in ventilation science need to be well understood and effectively applied to ensure that buildings are ventilated as efficiently as possible.

Ventilation Systems provides up-to-date knowledge based on the experience of internationally recognised experts to deal with current and future ventilation requirements in buildings. Presenting the most recent developments in ventilation research and its applications, this book offers a comprehensive reference to the subject, covering the fundamentals as well as more advanced topics. It is a unique publication as it covers the subject rigorously in a way needed by researchers but has a practical flavour that will be of value to a wide range of building professionals.

Hazim B. Awbi is Professor of Building Environmental Science at the University of Reading, UK, and founder of the Indoor Environment and Energy Research Group (IEERG).

Also available from Taylor & Francis

Ventilation of Buildings 2nd edition
H. Awbi
Hb: ISBN 978–0–415–27055–3
Pb: ISBN 978–0–415–27056–1

Building Services Engineering 5th edition
D. Chadderton
Hb: ISBN 978–0–415–41354–1
Pb: ISBN 978–0–415–41355–8

Tropical Urban Heat Islands
N. H. Wong et al.
Hb: ISBN 978–0–415–41104–2

Heat and Mass Transfer in Buildings
K. Moss
Hb: ISBN 978–0–415–40907–0
Pb: ISBN 978–0–415–40908–7

Energy Management and Operating Costs in Buildings
K. Moss
Hb: ISBN 978–0–415–35391–5
Pb: ISBN 978–0–415–35392–2

Renewable Energy Resources 2nd edition
J. Twidell and T. Weir
Hb: ISBN 978–0–419–25320–4
Pb: ISBN 978–0–419–25330–3

Mechanics of Fluids 8th edition
B. Massey and J. Ward Smith
Hb: ISBN 978–0–415–36205–4
Pb: ISBN 978–0–415–36206–1

Housing and Asthma
S. Howieson
Hb: ISBN 978–0–415–33645–1
Pb: ISBN 978–0–415–33646–8

Information and ordering details

For price availability and ordering visit our website **www.sponpress.com**
Alternatively our books are available from all good bookshops.

Ventilation Systems

Design and performance

Edited by Hazim B. Awbi

Routledge
Taylor & Francis Group

LONDON AND NEW YORK

First published 2008 by Taylor & Francis

2 Park Square, Milton Park, Abingdon, Oxfordshire OX14 4RN
52 Vanderbilt Avenue, New York, NY 10017

Routledge is an imprint of the Taylor & Francis Group, an informa business

First issued in paperback 2019

Typeset in Sabon by
Integra Software Services Pvt. Ltd, Pondicherry, India

British Library Cataloguing in Publication Data
A catalogue record for this book is available from the British Library

Library of Congress Cataloging-in-Publication Data
Ventilation systems: design and performance / edited by Hazim B. Awbi.
p. cm.
Includes bibliographical references and index.
ISBN-13: 978-0-419-21700-8 (hardback : alk. paper) 1. Ventilation.
I. Awbi, H. B. (Hazim B.), 1945–
TH7658.V455 2007
697.9′2–dc22
2006100008

ISBN13: 978-0-419-21700-8 (hbk)
ISBN13: 978-0-367-38792-1 (pbk)

Contents

Contributors

Hazim B. Awbi, Ph.D. is Professor of Building Environmental Science in the School of Construction Management and Engineering, University of Reading, UK, and founder of the Indoor Environment and Energy Research Group (IEERG). His research is in ventilation, room air movement, Computational Fluid Dynamics (CFD) and heat transfer in buildings.

James Axley, Ph.D. is Professor at the School of Architecture and the School of Forestry and Environmental Studies, Yale University, USA and his research is on the development of theory and computational tools for building thermal, airflow, and air quality simulation and design analysis.

Philomena Bluyssen, Ph.D. is a Research Scientist at TNO, Netherlands and has published extensively on indoor air quality.

Per Heiselberg, Ph.D. is Professor in the Department of Civil Engineering and Head of Hybrid Ventilation Centre at Aalborg University, Denmark.

Yuguo Li, Ph.D. is Professor in the Department of Mechanical Engineering, University of Hong Kong and his research is in natural ventilation, CFD, bio-aerosols and engineering control of respiratory infection.

Peter V. Nielsen, Ph.D., FASHRAE is Professor at Aalborg University, Denmark and Honorary Professor at the University of Hong Kong. His research is in room air movement and Computational Fluid Dynamics (CFD). He was awarded the John Rydberg Gold Medal in 2004.

Bjarne W. Olesen is Professor and Head of the International Centre for Indoor Environment and Energy, Department of Mechanical Engineering, Technical University of Denmark. The Centre is one of the world's leading research centres in indoor environment, people's health, comfort and productivity.

Claude-Alain Roulet, Ph.D., is Adjunct Professor at the EPFL (Swiss Federal Institute of Technology, Lausanne), Switzerland and private consultant in building physics and indoor environment quality.

Preface

This book is authored by eight distinguished researchers in ventilation and indoor air quality from five countries. It is a follow-on from the successful book *Ventilation of Buildings*, which is authored by the title editor. The new title draws from the vast experience of the eight authors in the field, includes their knowledge of the subject and presents the results from extensive international research programmes involving the authors as well as results from the work of other researchers.

The book deals with the applications of ventilation science in buildings. Buildings are responsible for a large proportion of a country's total energy consumption and a large part of this is used in ventilation, i.e. heating, cooling and cleaning of outdoor air supplied to buildings. Properly designed ventilation systems are essential for maintaining good indoor air quality, which is necessary for a productive building as well as for reducing a building's energy consumption. To achieve these aims, it is essential that modern development in ventilation science is well understood and effectively applied by those involved in building and system design and maintenance. This book aims to provide the building professionals with up-to-date knowledge based on the experience of internationally recognised experts to enable them implementing current and future ventilation requirements in buildings.

The book covers the fundamentals as well as the more advanced topics to cater for a wide range of readers. This unique publication covers the subject rigorously in a way needed by researchers and, at the same time, has a practical flavour and therefore should appeal to a wide range of building professionals. The book offers a comprehensive reference for researchers, designers, architects and specifiers of ventilation systems in buildings.

Chapter 1 presents the fundamental principles and physics of the airflow and heat transfer phenomena that occur within buildings. The basic fluid flow and heat transfer concepts and their analyses are presented with worked examples giving particular emphasis to the flow in enclosures. Chapter 2 presents the latest knowledge on human requirements for thermal comfort and air quality indoors and the impact of these on ventilation rates. The results and arguments presented in Chapter 2 show that there is a tendency for specifying higher fresh air supply rates in buildings than is recommended by most current ventilation guidelines. This will undoubtedly have a large impact on energy usage, which will require a proper assessment of the energy flow for ventilation to mitigate the impact. Chapter 3 describes methods used for assessing the energy flow in buildings and ventilation systems. It presents guidelines for improving the energy performance of buildings without compromising the indoor environment. Chapter 4 introduces the modeling of airflow into and within buildings by describing two categories of models that are commonly used nowadays: the macroscopic and the microscopic approaches. Whereas the macroscopic methods are based on modeling the air flow in buildings including their heating, ventilating and air-conditioning (HVAC) systems as collection of finite-sized control volumes, the microscopic methods, which are better known as computational fluid dynamics (CFD) models, on the other hand are based on the continuum approach that provides detailed descriptions of the flow, heat and mass transport processes within and outside the building. Chapter 5 deals with the characteristics of different types of air distribution systems, including new methods that have recently been developed, and the methods used for selecting and designing these for mechanically ventilated building enclosures. In Chapter 6, the types of HVAC systems are characterised, and the methods used for assessing the components of such systems are described, including the measurement techniques that are used to assess their performance. Chapter 7 describes the characteristics and performance of natural and hybrid ventilation systems and their components. Such systems are finding wider applications in modern buildings as, if properly designed, these can provide good indoor environment at lower energy consumption than conventional mechanical systems. Examples of buildings using hybrid systems are also presented. Finally, Chapter 8 describes various techniques that are applied in ventilation and room air movement measurements and airflow visualization. Such techniques are very useful for setting-up, commissioning and maintaining ventilation systems as well as estimating the airflow through the building envelop.

Hazim B. Awbi
Reading, UK, 2007

Acknowledgements

The following figures and tables are reproduced with permission:

Table 2.3 'Smoking free spaces in commercial buildings' according to ASHRAE 62.1, CR 1752, and EN15251. © American Society of Heating, Refrigerating and Air-Conditioning Engineers, Inc., www.ashrae.org

Figure 2.5 (a) de Dear, R. and Brager, G. S. (1998) 'Developing an adaptive model of thermal comfort and preference', *ASHRAE Transactions*, 104(1a): 145–167. © American Society of Heating, Refrigerating and Air-Conditioning Engineers, Inc., www.ashrae.org

Figure 5.25 Nielsen, P. V., Topp, C., Snnichsen, M. Andersen, H. (2005). 'Air distribution in rooms generated by a textile terminal – comparison with mixing ventilation and displacement ventilation'. *ASHRAE Transactions* 111 (Part 1): 733–739. © American Society of Heating, Refrigerating and Air-Conditioning Engineers, Inc., www.ashrae.org

Chapter 1

Airflow, heat and mass transfer in enclosures

Yuguo Li

1.1 Introduction

Airflow and transport phenomena play an important role in air quality, thermal comfort and energy consumption in buildings. Advances in airflow control in buildings in the past four decades have made it possible to design and evaluate building ventilation not only qualitatively but in many situations also quantitatively. In recent years, a broad range of practical ventilation problems have been investigated by the application of computational fluid dynamics (CFD) and advanced airflow measurement methods.

This chapter describes the fundamental principles of airflow, heat and mass transfer phenomena that take place in buildings. The need of emphasizing multi-disciplinary nature is noted here. Much of the basic theory and concepts on airflow, heat and mass transfer are described in classical textbooks of heat transfer and fluid mechanics, with new developments reported in journals such as *Journal of Heat Transfer, Journal of Fluid Mechanics* and *International Journal of Heat and Mass Transfer*. Historically, the concepts and technologies developed in other engineering disciplines have also been successfully applied and extended to ventilation application. Examples include the application of the residence time concept (Danckwerts, 1952) developed in chemical engineering to ventilation efficiency (Sandberg, 1981) and the application of CFD originally developed for the aerospace industry (Nielsen, 1974). Contribution of the ventilation community to fluid mechanics and heat and mass transfer has also been evident, such as the development of non-isothermal jets (Koestel, 1955).

The ultimate goal of an in-depth understanding of fluid mechanics in building airflow is to provide engineers effective and efficient design and analysis tools. Either experiments or numerical predictions (which may be considered as numerical experiments) can only provide data, but no conclusions. It is important that the fundamental principles can be applied in analysing the data from either experiments or CFD and also drawing conclusions from data. The quality of data is crucial for drawing any good or new conclusions. Effective and accurate methods for obtaining the required data are important parts of any airflow study, which again require good understanding of the fundamental principles. Airflow problems in buildings can be treated at various levels of theoretical rigour depending on the specific task and physical complexity of the problem.

1.2   Transport phenomena in buildings

1.2.1   Basic concepts

Fluid flow and transport phenomena are inherently associated with buildings as the primary function of buildings is to create an adequate indoor environment for the occupants or equipment therein. It is now established that the velocity, turbulence, temperature and humidity are all important thermo-fluid parameters for thermal comfort (Fanger, 1970; Fanger *et al.*, 1998), and some other quantities such as the composition of air, particle contents, odours etc. are all important parameters for indoor air quality (Spengler *et al.*, 2001). One of the basic fluid flow and transport problems in buildings is in the design of air distribution systems.

Figures 1.1–1.3 show sketches of three different ventilation systems, i.e. mixing ventilation (Figure 1.1), displacement ventilation (Figure 1.2) and kitchen local exhaust ventilation using a range hood (Figure 1.3). The three systems involve a broad range of airflow, heat and mass transfer phenomena, including wall jets (Figure 1.1), thermal plumes (Figures 1.2 and 1.3), gravity currents (Figure 1.2), natural convection along vertical walls (Figure 1.3), heat transfer and pollutant transport (all).

The airflow in these systems may be analysed using two different approaches. Let us take displacement ventilation as an example.

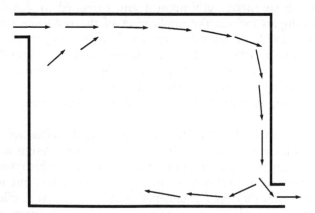

Figure 1.1 A sketch of air distribution pattern in a room ventilated by a mixing system.

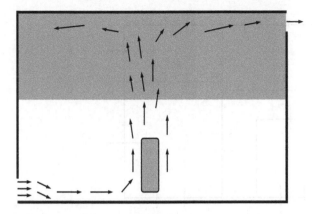

Figure 1.2 A sketch of air distribution pattern in a room ventilated by a displacement system.

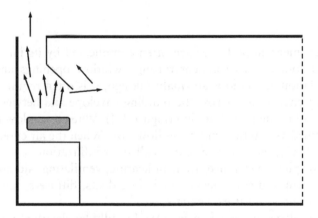

Figure 1.3 A sketch of airflow pattern through a kitchen local exhaust ventilation using a range hood.

- if our interest is to ensure that the location of the interface between the lower clean zone and the upper polluted zone, a macroscopic approach involving the use of the macroscopic mass balance equations can be used. The location of the interface is approximately taken where the total upward flow rate of the plume and the vertical boundary layer is equal to the supply airflow rate.
- however, if the detailed air velocity and turbulence level in the occupied region (particularly close to the supply register) is of interest, a microscopic approach involving the use of the differential governing equations of airflow will have to be solved.

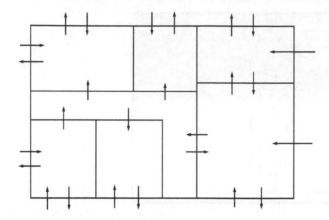

Figure 1.4 Illustration of the air exchange between zones in a building and between a building and outdoor environment.

The airflow phenomena in buildings are often complicated by its inter-action with the outdoor air environment through window openings and leakages. For ventilation and indoor air quality design, it is often required to determine the air-change rate across the building envelope and between the zones (rooms) within the same building (Figure 1.4). When a window is open, it may be termed as natural ventilation flow rate. When the air enters a building through background leakages, it is called air infiltration.

Various fluid flow phenomena also occur in heating, ventilating and air-conditioning equipment and components, including ducts, diffusers, fans, air-handling units, airflows, heat exchangers, etc.

The fluid flow problems discussed so far may broadly be classified into two categories:

- macroscopic fluid flow problems, in which the objective is to develop overall relationships between some lumped flow parameters. Examples include the required exhaust flow rate (fan power) for achieving a 100 per cent capture efficiency in kitchen range hoods (Li and Delsante, 1996); the dependence of air-change rate on the pressure difference across a building envelope, the heating and cooling load of a house because of air infiltration, the dependence of the clean zone interface in displacement ventilation on supply airflow rates, etc.
- microscopic fluid flow problems, in which the objective is to obtain detailed distribution of primitive airflow variables, such as the velocity distribution rather than the volumetric flow rate, the spatial tempera-ture distribution rather than the total heat gain or loss.

The macroscopic and microscopic fluid flow problems described above are related in nature. Generally, one could obtain the macroscopic flow quantities by integrating all the necessary microscopic quantities, but not vice versa.

1.2.2 Characteristic scales and non-dimensional parameters

One difficulty when attempting to analyse the fluid flow phenomena in buildings is that there are many geometrical and physical parameters that govern or influence the flow. These parameters can be the geometry, type and location of the air supply registers and exhaust openings, the supply air velocity, the heat generation in the room, etc. It is therefore helpful to estimate the order of length, time and velocity scales of airflow system. Such an analysis helps to identify the dominant parameters/factors in the flow and thus provides a reference point for further analysis.

The upper limit of the length scale may be the duct diameter, the dimension of the supply register and room or the dimension of furniture causing the flow disturbance. Let this typical length dimension be L. The characteristic velocity, U, which can be the supply air velocity or some kind of average velocity. The characteristic overall convection time scale, t_c, for a fluid element to be advected along the dimension L by the velocity U is

$$t_c = \frac{L}{U} \tag{1.1}$$

A second characteristic time scale because of the viscous diffusion may also be defined as:

$$t_d = \frac{\rho L^2}{\mu} \tag{1.2}$$

Almost in all airflow problems, a question can be asked about which of these two time scales dominate the flow. The Reynolds number, Re, is defined as the ratio between these two time scales.

$$Re = \frac{t_d}{t_c} = \frac{\rho U L}{\mu} \tag{1.3}$$

The Reynolds number plays an important role in airflow analysis. The airflow in a round ventilation duct is laminar when Re is small, and it is turbulent when Re is larger than 2300. It is noted that recent experiments showed that laminar flow existed at much higher Re. The rapidity at which mass and momentum can be transferred in a turbulent flow compared to that in a laminar flow is a very important feature of turbulence.

On the other hand, many building airflow problems are driven by thermal buoyancy forces. In a thermally driven flow, another non-dimensional parameter can be derived. Consider an airflow between two horizontal plates with the lower plate at temperature $T+\Delta T$ and the upper plate at T.

Because of the thermal expansion of the fluid, a fluid element will thus experience a buoyancy force per unit mass (i.e. acceleration) of

$$\frac{g\Delta\rho}{\rho} = -\beta g\Delta T \tag{1.4}$$

where g is the acceleration of gravity and β is the thermal expansion coefficient.

If the air element is allowed to accelerate freely from lower to upper surfaces, it would reach a convection velocity u_c:

$$u_c = \sqrt{g\beta\Delta TL} \tag{1.5}$$

where L is the vertical distance between the two surfaces. The time scale because of thermal convection for a length scale L, is

$$t_{tc} = \frac{L}{u_c} = \sqrt{\frac{L}{g\beta\Delta T}} \tag{1.6}$$

A thermal diffusion time scale can be obtained by analogy to Equation (1.2),

$$t_{td} = \frac{L^2}{\alpha} \tag{1.7}$$

where α is the thermal diffusivity, $\alpha = k/\rho c_p$, k is the thermal conductivity, and c_p is the specific heat capacity at constant pressure. The Rayleigh number is defined as the product of the ratio t_{td}/t_{tc} and the Reynolds number based on u_c and L.

$$Ra = \frac{\rho g\beta\Delta TL^3}{\alpha\mu} \tag{1.8}$$

The Rayleigh number is important in studying the convection in horizontal layers (so-called Bènard convection), see Tritton (1988). The convection flow driven by the buoyancy forces between the horizontal plates is laminar when Ra is small. The flow is turbulent at Ra values between 10^5 and 10^7.

In air-conditioned rooms, flows are often driven by both mechanical ventilation and temperature differences. The thermal convection velocity may still be represented by Equation (1.5). If U is the supply air velocity and L the characteristic length (e.g. room length), the time scale due to thermal convection is given by Equation (1.6), and the time scale because of advection of velocity U is given by Equation (1.1). The square of the ratio between t_{tc} and t_c defines the Archimedes number, which is widely used in the ventilation community (e.g. Croome and Roberts, 1980).

$$Ar = \left(\frac{t_c}{t_{tc}}\right)^2 = \frac{g\beta\Delta TL}{U^2} \tag{1.9}$$

In the mixed convection literature (Bejan, 2004), Ar is written as Gr/Re^2, where Grashof number is defined as:

$$Gr = \frac{\rho^2 g\beta\Delta TL^3}{\mu^2} \tag{1.10}$$

The Grashof number plays a special role in the study of many natural convection flows, for example flow in a room with two differentially heated vertical walls. It should be noted that $Ra = GrPr$, where Pr is the Prantl number, and $Pr = \nu/\alpha$.

In general, the Archimedes number is a measure of the relative importance of buoyant and inertia forces. The Archimedes number is important in building airflows because it combines two important air-conditioning design parameters – supply air velocity and room temperature difference. The relative roles of Re and Ar are evident from the following example.

Example 1.1. Two commonly used air distribution systems for ventilation and air conditioning are the mixing and displacement systems. Assuming the air temperature in the occupied zone of an office is 23°C and using the supply parameters below:

- mixing ventilation system – register length scale 0.1 m, supply air velocity 4 m s^{-1} and supply air temperature 14°C;
- displacement ventilation system – register length scale 0.5 m, supply air velocity 0.2 m s^{-1} and supply air temperature 19°C.

Estimate the Reynolds number and the Archimedes number for the two ventilation systems respectively.

Assume the following air properties: $\mu = 1.84 \times 10^{-5}$ Pa·s, $\rho = 1.189$ kg m^{-3}.

Solution:
For air, the thermal expansion coefficient $\beta = 1/T$.
For the mixing ventilation system,

$$Re = \frac{\rho UL}{\mu} = \frac{1.189 \times 4 \times 0.1}{1.84 \times 10^{-5}} = 2.585 \times 10^4$$

$$Ar = \frac{g\beta L \Delta T}{U^2} = \frac{9.8 \times 0.1 \times (23-14)}{(23+273.15) \times 4^2} = 0.0019$$

For the displacement ventilation system,

$$Re = \frac{\rho U L}{\mu} = \frac{1.189 \times 0.2 \times 0.5}{1.84 \times 10^{-5}} = 6.5 \times 10^3$$

$$Ar = \frac{g\beta L \Delta T}{U^2} = \frac{9.8 \times 0.5 \times (23-19)}{(23+273.15) \times 0.2^2} = 1.7$$

The apparent differences in the two non-dimensional parameters indicate that the airflow in the two systems will be different. The supply flow of a mixing system is often of a jet type (large Re), whereas the supply flow of displacement ventilation is often of a gravity current type (large Ar). In a jet-type flow, the motion is governed by the initial momentum of the supply air, which may be influenced by the gravity force. In contrast, in a gravity current, the motion is governed by the gravity force.

1.3   General governing equations of airflow, heat and mass transfer

Basic governing equations relate to the process variables, such as velocity, pressure, viscosity and density in airflows. They can be constructed using the principles of conservation of mass, momentum and energy. Development or selection of an appropriate model for a particular airflow process requires an understanding of the relative importance of the influencing factors.

There are generally two types of airflow equations that are used to describe the flow:

- integral equations of fluid flows, which are mostly used for solutions involving a macroscopic approach.
- differential equations of fluid flow, i.e. the Navier–Stokes equations, which are mostly used for solutions involving a microscopic approach.

The two types of equations can be developed in a more or less independent manner. In developing integral equations, the whole system or part of the system is taken as a control volume, and its boundary as the control surface. The overall balances of mass, momentum and energy are applied. Generally, only overall performance parameters are included, but the detailed flow structure in the control volume is not explicitly included. The overall performance parameters are sometimes a function of the detailed flow structure in the control volume. But the two types of equations are not independent of each other. The integral equations can be obtained by integrating the differential equations over a control volume (e.g. a room or a zone). Conversely, the latter can be obtained by reducing the control volume to an infinitesimal air element (Bird *et al.*, 2002).

1.3.1 Integral equations

The integral balance equations for mass, momentum and energy can be derived for an arbitrary control volume of finite size in Figure 1.5 (Bird *et al.*, 2002). In a vector form, these equations can be summarized as follows:

Integral mass balance

$$\frac{d}{dt}\int_V \rho dV = -\oint_A \rho v \cdot dA \tag{1.11}$$

Integral momentum balance

$$\frac{d}{dt}\int_V \rho v dV + \oint_A \rho v v \cdot dA = -\oint_A p dA - F_s + F_g \tag{1.12}$$

where the surface force F_s acts on the control surface, for example because of drag and the body forces F_g acts in the fluid, for example gravitational force.

Integral energy balance

$$\frac{d}{dt}\left[\int_V \rho\left(\hat{U}+\frac{1}{2}v^2+\hat{\Phi}\right)dV\right] = -\oint_A \left(\rho\hat{U}+\frac{1}{2}\rho v^2+\rho\hat{\Phi}\right)v\cdot dA$$

$$+Q-W-\oint_A pv\cdot dA \tag{1.13}$$

where $\hat{U}$ is the internal energy, $\frac{1}{2}\rho v^2$ is the kinetic energy and $\hat{\Phi}$ is the potential energy (gravitational). The net rate of heat added to the system

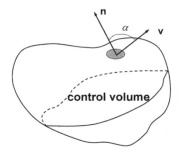

Figure 1.5 An arbitrary control volume of finite size.

from the surroundings is Q, which includes all thermal energy entering the fluid through the solid surfaces of the system and by conduction or radiation in the fluid at the inlet and outlet.

Integral mechanical energy balance

$$\frac{d}{dt}\left[\int_V \rho\left(\frac{1}{2}v^2+gz\right)dV\right]=-\oint_A \rho\left(\frac{1}{2}v^2+gz\right)v\cdot dA$$

$$-\oint \rho\left(\int_{p_1}^{p_2}\frac{1}{\rho}dp\right)v\cdot dA-W-E_v \qquad (1.14)$$

The integral mechanical energy balance Equation (1.14) is only applicable to isothermal flows.

For the simplified system in Figure 1.6, all inflow is normal to an area A_1 and all outflow is normal to an area A_2. There is no flow across other parts of the control volume. With these conditions, the above equations can be simplified.

Integral mass balance

$$\frac{d}{dt}\int_V \rho dV = \rho_1 v_{m1} A_1 - \rho_2 v_{m2} A_2 \qquad (1.15)$$

where v_m is the mean velocity given by

$$v_m = \frac{1}{A}\int_A v dA$$

Integral momentum balance

$$\frac{d}{dt}\int_{z_1}^{z_2} \rho v A dz = \rho_1 v_1^2 A_1 - \rho_2 v_2^2 A_2 + p_1 A_1 - p_2 A_2 - F_s + F_g \qquad (1.16)$$

We assumed that $dV = A dz$, and we also have $F_g = \left(\int_{z_1}^{z_2}\rho A dz\right)g$.

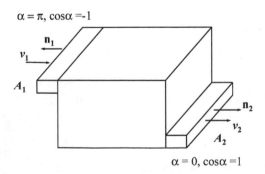

Figure 1.6 A room as a control volume.

Integral energy balance

$$\frac{d}{dt}\int_V EdV = \rho_1 \hat{U}_1 v_1 A_1 - \rho_2 \hat{U}_2 v_2 A_2 + \frac{1}{2}\rho_1 v_1^3 A_1 - \frac{1}{2}\rho_2 v_2^3 A_2 + \rho_1 \hat{\Phi}_1 v_1 A_1$$
$$- \rho_2 \hat{\Phi}_2 v_2 A_2 + Q - W + p_1 v_1 A_1 - p_2 v_2 A_2 \qquad (1.17)$$

where $E = \rho\hat{U} + \frac{1}{2}\rho v^2 + \rho\hat{\Phi}$ is the energy flux. For the simplified form in steady state, the integral mechanical energy balance becomes

$$\frac{1}{2}v_{m2}^2 - \frac{1}{2}v_{m1}^2 + g(z_2 - z_1) + \int_{p_1}^{p_2} \frac{1}{\rho}dp + W' + E_v' = 0 \qquad (1.18)$$

where $W' = W/q$, $E_v' = E_v/q$ and q is the mass flow rate $q = \rho v A$. Equation (1.18) is the Bernoulli's equation.

It is useful to further simplify the above set of integral equations. For example, the integral mass balance equation becomes at steady state

$$\rho_1 v_{m1} A_1 = \rho_2 v_{m2} A_2 \qquad (1.19)$$

If the density is constant, $\rho_1 = \rho_2$, it becomes

$$v_{m1} A_1 = v_{m2} A_2 \qquad (1.20)$$

In ventilation airflows, we often deal with contaminants in air. The word concentration c is used for either partial density (kg m^{-3}) or volume fraction (m^3 m^{-3}). Similar to Equation (1.11), for the control volume in Figure 1.5, we have

$$\frac{d}{dt}\int_V cdV = -\oint_A cv \cdot dA \qquad (1.21)$$

For the simplified control volume of Figure 1.6,

$$\frac{d}{dt}\int_V cdV = c_1 v_{m1} A_1 - c_2 v_{m2} A_2 \qquad (1.22)$$

If concentration c is constant within the control volume V, which is also a constant,

$$V\frac{dc}{dt} = c_1 v_{m1} A_1 - c_2 v_{m2} A_2 \qquad (1.23)$$

For the integral energy balance equation, we introduce the mass flow rate $q = \rho v A$, we have for the unsteady-state integral energy balance.

$$\frac{d}{dt}\int_V EdV = -\Delta\left[q\left(\hat{U} + \frac{1}{2}v^2 + \hat{\Phi} + \frac{p}{\rho}\right)\right] + Q - W \qquad (1.24)$$

This is just a statement of the first law of thermodynamics as applied to a flow system. At steady state,

$$\Delta\left[\hat{U}+\frac{1}{2}v^2+\hat{\Phi}+\frac{p}{\rho}\right]=\hat{Q}-\hat{W} \tag{1.25}$$

where $\hat{Q}$ is the heat added per unit mass of fluid flowing through the system and $\hat{W}$ is the amount of work done by a unit mass of fluid in traversing the system.

The quantity $\hat{U}+p/\rho$ is the enthalpy $\hat{H}$. For air

$$\Delta\hat{H}=\int_{T_1}^{T_2}c_p dT \tag{1.26}$$

and for water

$$\Delta\hat{H}=\int_{T_1}^{T_2}c_p dT+\frac{1}{\rho}(p_2-p_1) \tag{1.27}$$

In most building airflow and heat transfer applications, kinetic energy, potential energy and work effects are small, and many practical problems can be analysed by

$$q\Delta\hat{H}=Q \tag{1.28}$$

And it becomes $\rho c_p q\,(T_2-T_1)=Q$ if air is the fluid and c_p is a constant.

With regard to the integral mechanical energy balance, for constant density, we have another form of the Bernoulli's equation.

$$\frac{1}{2}\Delta v^2+g\Delta z+\frac{1}{\rho}\Delta p+W'+E'_v=0 \tag{1.29}$$

The Bernoulli equation is important in infiltration and natural ventilation calculations, duct sizing and design, and using the common flow measurement devices such as the Pitot tube and orifice meter.

Example 1.2.   Ventilation flow rate is conventionally measured as air change rate with a unit of ACH (air change per hour). For a room with air change rate of n, calculate the time required for changing the room air with outside air by 90 per cent, assuming that the room air is perfectly mixed.

Solution:
We call the room air at time $t=0$, 'old air' and the outside fresh air 'new air'. With this notation, at time $t=0$, the concentration c of 'old' air in the room is 100 per cent. As the room airflow is assumed perfectly mixed,

which means that the concentration c_2 of the extract air is the same as the concentration c of the room air. From Equation (1.23),

$$V\frac{dc}{dt} = -c v_{m2} A_2 \tag{1.30}$$

From the definition of air change rate,

$$n = \frac{v_{m2} A_2}{V}$$

Solution of Equation (1.30) with initial condition $t = 0$, $c = 1$, we have $c = e^{-nt}$. Thus, when $c = 0.1$, we have $t = -\frac{1}{n} \ln 0.1 \approx 2.3\frac{1}{n}$. It will take 2.3 times the nominal time $1/n$ to change the room air with outside air by 90 per cent. Thus, an air change rate of n ACH does not necessarily mean that the room air in a building can be changed with the fresh air by n times per hours. The concept of ACH is sometimes misleading. Etheridge and Sandberg (1996) suggested the concept of air change rate should be replaced by the specific airflow rate $(m^3 s^{-1} m^{-3})$.

Example 1.3. For a room of volume $V(m^3)$, the amount of pollutant generated in the room is $G(kgs^{-1})$. The outdoor air supply rate is $q(m^3 s^{-1})$ and the concentration of pollutant in the outdoor air is c_o. Calculate the indoor pollutant concentration at time t, assuming a perfect mixing flow in the room.

Solution:
Considering the source generation, the Equation (1.23) becomes

$$V\frac{dc}{dt} = G + q c_o - q c \tag{1.31}$$

The initial condition is $c = c_I$ at $t = 0$. We can obtain the general solution of Equation (1.31).

$$c = (c_G + c_o)(1 - e^{-nt}) + c_I e^{-nt} \tag{1.32}$$

where $n = q/V$ is the specific flow rate $(m^3 s^{-1} m^{-3})$ and $c_G = G/q$ the virtual generation concentration. Solution (1.32) is of very fundamental use in building ventilation, which can be simplified under some conditions. For example, when $c_o = 0$ and $G = 0$, the solution is $c = c_I e^{-nt}$, which is the simple decay equation (see Example 1.2), commonly used in measuring ventilation flow rates through a building envelop when a tracer gas technique is used (Etheridge and Sandberg, 1996).

Example 1.4. Air can move by a buoyant force, caused by density difference between indoor and outdoor air. This phenomenon is given by various names, such as stack effect, stack action and chimney effect. These names come from the comparison with the upward flow of gases in a smoke stack or chimney. In smoke control design of fire protection engineering, we often need to calculate the airflow rates because of the stack effect. Consider a simple building with two openings; derive the formula for calculating the airflow rate because of stack effect alone when the indoor and outdoor air temperatures are known.

Solution:
Because of its weight, a column of air produces a pressure at the bottom of the column. Imagine there are two columns of air, one in the room and another in outdoors. Assume that the room has two small openings. We take the middle of the lower opening as the datum level. The hydrostatic pressures at the datum level for both columns are p_o and p_i(Pa), respectively. Then at height, H (which equals $h_1 - h_2$), above the datum level, the pressure for both columns are respectively (see Fig. 1.7)

$$p_o(H) = p_o - \rho_o g H \tag{1.33}$$

$$p_i(H) = p_i - \rho_i g H \tag{1.34}$$

where

ρ_o – ambient air density (kg m^{-3})
ρ_i – room air density (kg m^{-3})
g – gravity acceleration (m s^{-2})
H – vertical distance between openings (m)

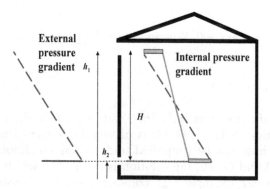

Figure 1.7 Pressure difference created by density difference between indoor and outdoor air.

The pressure differences across each opening are respectively:
bottom opening

$$\Delta p_b = p_o - p_i \qquad (1.35)$$

top opening

$$\Delta p_t = p_i(H) - p_o(H) = (p_i - \rho_i gH) - (p_o - \rho_o gH) = p_i - p_o + (\rho_o - \rho_i)gH$$
$$(1.36)$$

Thus, the total pressure difference between the two openings becomes

$$\Delta p_s = \Delta p_b + \Delta p_t = (\rho_o - \rho_i)\,gH \qquad (1.37)$$

Because of the simplified equation of state (the variations in pressure within a building is small compared with atmospheric pressure), we obtain

$$\rho_o T_o = \rho_i T_i \qquad (1.38)$$

Thus, we obtain the familiar equation for estimating stack pressure

$$\Delta p_s = \rho_o T_o gH \left(\frac{1}{T_o} - \frac{1}{T_i} \right) \qquad (1.39)$$

The following formula for buoyancy-driven flow rate in a building with two openings can be derived:

$$m_b = m_t = m \qquad (1.40)$$

$$m_b = \rho_o C_d A_b \sqrt{\frac{2\Delta p_b}{\rho_o}} = C_d A_b \sqrt{2\rho_o \Delta p_b} \qquad (1.41)$$

$$m_t = \rho_i C_d A_t \sqrt{\frac{2\Delta p_t}{\rho_i}} = C_d A_t \sqrt{2\rho_i \Delta p_t} \qquad (1.42)$$

We also have

$$\Delta p_b + \Delta p_t = \Delta p_s \qquad (1.43)$$

We derive

$$m = C_d A^* \sqrt{2\rho_o \Delta p_s} \qquad (1.44)$$

where

$$A^* = \left[\frac{1}{(A_t)^2} + \frac{1}{(A_b)^2} \frac{\rho_o}{\rho_i} \right]^{-\frac{1}{2}} \tag{1.45}$$

Example 1.5. A building has two openings with assisting wind, i.e. the wind pressure assists the stack-driven ventilation (see Figure 1.8). The heights of the two openings are relatively small compared with the building height. These openings are referred to as 'small openings'. We consider only steady-state conditions. The air temperature in the building is assumed to be uniform.

Use both the zonal pressure-based approach and the loop pressure equation approach to derive the following natural ventilation formula for combined buoyancy and assisting wind-driven flows.

$$q = C_d A^* \sqrt{2gH \frac{T_i - T_o}{T_o} + 2 \frac{\Delta P_w}{\rho_o}} \tag{1.46}$$

where

$$A^* = \left(\frac{1}{A_t^2} + \frac{1}{A_b^2} \right)^{-\frac{1}{2}} \tag{1.47}$$

Solution:
The continuity equation gives

$$q_t = q_b = q \tag{1.48}$$

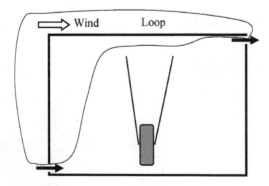

Figure 1.8 A two-opening building naturally ventilated by a stack force and an assisting wind.

Along the loop connecting the two openings in Figure 1.8, we can write the so-called loop pressure equation as follows (see also Chapter 4):

$$\Delta p_b + \Delta p_t = \frac{\rho_o (T_1 - T_o) gH}{T_o} + \Delta p_w \qquad (1.49)$$

We also have

$$\Delta p_b = \frac{\rho_o q^2}{2 (C_d A_b)^2} \qquad (1.50)$$

$$\Delta p_t = \frac{\rho_o q^2}{2 (C_d A_t)^2} \qquad (1.51)$$

Substitute Equations (1.50) and (1.51) into (1.49), we obtain

$$\frac{\rho_o q^2}{2 (C_d A_b)^2} + \frac{\rho_o q^2}{2 (C_d A_t)^2} = \frac{\rho_o (T_i - T_o) gH}{T_o} + \Delta p_w \qquad (1.52)$$

We then have

$$q = C_d A^* \sqrt{2gH \frac{T_i - T_o}{T_o} + 2 \frac{\Delta p_w}{\rho_o}} \qquad (1.53)$$

Thus, the derivation process for the loop equation method is simpler than that for the zonal pressure method.

The reader can derive the following formula for calculating ventilation flow rates for flows driven by combined buoyancy and opposing winds.

$$q = C_d A^* \sqrt{\left| 2gH \frac{T_i - T_o}{T_o} - 2 \frac{\Delta p_w}{\rho_o} \right|} \qquad (1.54)$$

The mass flow rate is adopted in Equation (1.40) to consider large air density variations in smoke flow, whereas the volumetric flow rate in Equation (1.48) can be used when the air density variation is small as in general ventilation applications.

Example 1.6. The basic physical principles of displacement ventilation are based on the properties of stratified flow. Ventilation air with lower temperature (usually around 19°C) than the mean room air temperature is introduced at floor level. Because of low supply velocity and large gravitational force, the supply airflow is of the gravity current type. Plumes are generated from the heat sources in the room, and a vertical temperature gradient is therefore generated. The ceiling is warmer than other surfaces, and this gives rise to radiation heat transfer from the ceiling, mainly to the floor. As a result, this makes the floor warmer than the air layer adjacent to the floor. The air temperature at floor level and the vertical temperature gradients in the room are important comfort parameters.

Assuming that the

- flow is divided into a cold gravity current zone and a stratified region (i.e. two control volumes);
- air temperature in the stratified region is linear;
- supply air spread over the floor without entrainment. This is a property of the cold gravity current in a stratified environment;
- surface temperature of the ceiling is equal to the near ceiling air temperature and the extract air temperature;
- radiation equations are linearized because of moderate temperature differences;
- room is perfectly insulated.

derive expressions for the exact air temperature, the floor surface temperature, the floor air temperature and vertical temperature gradient.

Solution:
The integral energy balance for the room as a whole is first written (see Fig. 1.9). The steady-state energy balance equation becomes (for negligible changes in kinetic and potential energy):

$$q\rho c_p \left(T_e - T_s\right) = E \tag{1.55}$$

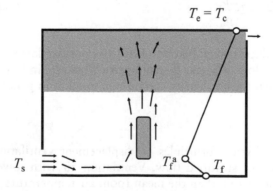

$$T_e = T_c$$

Figure 1.9 The three-node model of heat transfer process in a room ventilated by displacement ventilation.

where q is the supply airflow rate, and E is the total heat power in the room.

The integral energy balance for the floor surface and the cold gravity current are

$$h_r A (T_e - T_f) = h_f A (T_f - T_f^a) \tag{1.56}$$

$$q\rho c_p (T_f^a - T_s) = h_f A (T_f - T_f^a) \tag{1.57}$$

where the left-hand side of Equation (1.56) represents the radiative heat flux between the floor surface and the ceiling, and the radiative heat transfer coefficient is obtained with linearization, $h_r = 4T_o^3 \sigma$, T_o is calculated approximately by an assuming floor and ceiling temperature.

Form Equations (1.56) and (1.57)

$$\lambda = \frac{T_f^a - T_s}{T_e - T_s} = \left[\frac{q\rho c_p}{A} \left(\frac{1}{h_f} + \frac{1}{h_r} \right) + 1 \right]^{-1} \tag{1.58}$$

The mean vertical temperature gradient is then be calculated to be

$$s = (1 - \lambda) \frac{E}{\rho c_p q H} \tag{1.59}$$

The calculated values of the convective heat transfer number, λ, which are the relative increase in temperature of floor air, are found to be in rather good agreement with measurements as a function of ventilation airflow rates per unit floor area (q/A) (Mundt, 1995).

The temperatures of T_e, T_f^a and T_f can be calculated as

$$T_e = \frac{E}{\rho c_p q} + T_s \tag{1.60}$$

$$T_f^a = \lambda (T_e - T_s) + T_s \tag{1.61}$$

$$T_f = \frac{h_r T_e + h_f T_f^a}{h_r + h_f} \tag{1.62}$$

The above is the three-node model of displacement ventilation (Li et al., 1992). It should be noticed that the assumptions of equal ceiling surface and near ceiling air temperature may not be appropriate, and a near ceiling warm gravity zone can be added to the model. This is the four-node model. More complex models can also be found in Li et al. (1992).

Example 1.7. Thermal mass can regulate indoor air temperature and reduce cooling load of a building and it can be used together with night ventilation

(Allard and Santamouris, 1998). Consider a simple one-zone building model with a constant outdoor air ventilation flow rate. The basic assumptions include that

- the air temperature distribution in the building is uniform;
- the thermal mass has a uniform temperature distribution;
- the building envelope is perfectly insulated;
- the thermal mass materials are not in equilibrium with the indoor air;
- all heat gain and heat generation in the building can be lumped into one heat source term, E.

What are the parameters that affect the phase shift and attenuation of indoor air temperature amplitude in this building?

Solution:
There are two basic heat balance equations, one for the room air and one for the thermal mass. Considering Figure 1.10,

$$\rho c_p q(T_o - T_i) + h_M A_M (T_M - T_i) + E = 0 \tag{1.63}$$

$$M c_M \frac{dT_M}{dt} + h_M A_M (T_M - T_i) = 0 \tag{1.64}$$

where $T_o = \tilde{T}_o + \Delta \tilde{T}_o \sin(\omega t)$. $\Delta \tilde{T}_o$ and $\tilde{T}_o$ are independent of time and $\Delta \tilde{T}_o \geq 0$. The outdoor temperature changes sinusoidally with a period of 24 h. The subscript M denotes variables related to thermal mass.

From Equation (1.63), we obtain

$$T_M = \left(1 + \frac{\rho c_p q}{h_M A_M}\right) T_i - \frac{\rho c_p q}{h_M A_M} T_o - \frac{E}{h_M A_M} \tag{1.65}$$

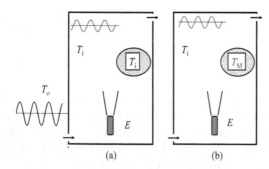

(a)   (b)

Figure 1.10 A simple two-opening one-zone building model with periodic outdoor air temperature variation. The shaded area represents the thermal mass. (a) The thermal mass is in equilibrium with the room air. (b) The thermal mass is not in equilibrium with the room air.

Substituting Equation (1.65) into (1.64), and after some manipulation, we obtain

$$\omega\tau\frac{dT_i}{d(\omega t)}+\frac{\lambda}{1+\lambda}T_i=\frac{\lambda}{1+\lambda}\left(\tilde{T}_o+T_E\right)+\frac{\lambda}{1+\lambda}\Delta\tilde{T}_o$$

$$\left[\sin(\omega t)+\left(\frac{\omega\tau}{\lambda}\right)\cos(\omega t)\right] \tag{1.66}$$

where $\lambda=\dfrac{h_M A_M}{\rho c_p q},\ \tau=\dfrac{MC_M}{\rho c_p q}$ and $T_E=\dfrac{E}{\rho c_p q}.$ $\tag{1.67}$

The time constant τ represents the relative amount of thermal storage capacity. The convective heat transfer number λ measures the relative strength of convective heat transfer at the thermal mass surface. A large value of the convective heat transfer number means that the convective heat transfer is very effective compared with the flow mixing in the room. In this case, the thermal mass can be considered to be in equilibrium with the room air temperature. The role of the convective heat transfer number is analogous to the Biot number in heat transfer. Recall that a small value of the Biot number means that the external resistance (convective heat transfer) is large compared with the internal resistance (heat conduction), and in this case, the internal temperature distribution can be assumed to be uniform. Similarly, a large value of the convective heat transfer number means that the convective heat transfer is very effective compared with the flow mixing in the room. In this case, the thermal mass can be considered to be in equilibrium with the room air temperature.

The general solution for Equation (1.66) is

$$T_i(\omega t)=\tilde{T}_o+T_E+\sqrt{\frac{\lambda^2+\omega^2\tau^2}{\lambda^2+\omega^2\tau^2(1+\lambda)^2}}\Delta\tilde{T}_o\sin(\omega t-\beta)+Ce^{-[\lambda/\omega\tau(1+\lambda)]\omega t}$$

$$\tag{1.68}$$

where C is an integrating constant and $\beta=\tan^{-1}\left\{\frac{\lambda^2\omega\tau}{[\lambda^2+\omega^2\tau^2(1+\lambda)]}\right\}$. As the convective heat transfer number λ becomes infinity, $\beta=\tan^{-1}(\omega\tau)$.

After sufficient long time, the solution approaches to a periodic one as

$$T_i(\omega t)=\tilde{T}_o+T_E+\sqrt{\frac{\lambda^2+\omega^2\tau^2}{\lambda^2+\omega^2\tau^2(1+\lambda)^2}}\Delta\tilde{T}_o\sin(\omega t-\beta) \tag{1.69}$$

The first term on the right side of Equation (1.69) is the mean outdoor temperature and the second term is the steady-state air temperature rise because of steady heat source. The mean indoor air temperature $(\tilde{T}_o + T_E)$ is not a function of the convective heat transfer number and the time constant of the system. The third term is the periodic fluctuating component with its amplitude depending on the outdoor temperature fluctuation $\Delta \tilde{T}_o$, the time constant τ and the convective heat transfer number λ. β is the phase lag of the indoor air temperature with respect to the outdoor temperature.

Analytical solution represented by Equation (1.68) is plotted in Figures 1.11 and 1.12 for the phase shift and the fluctuation amplitude of the indoor air temperature, respectively.

It is not difficult to understand that when the convective heat transfer number is small (between 0.1 and 10, which are typical practical values), the phase shifts are much smaller than those with very large convective heat transfer numbers. However, it is not obvious that for a fixed value of the convective heat transfer number, the phase shift first increases exponentially as the time constant increases, then drops as the time constant further increases, approaching zero as the time constant approaches infinity. In Figure 1.12, it can be seen that for small convective heat transfer numbers (0.1–1), the fluctuation amplitude of the indoor air temperature, normalized by the outdoor fluctuation amplitude, becomes constant as the time constant becomes very large. This suggests that the convective heat transfer between the mass and nearby air is an important aspect in thermal mass design, which is known well by engineers.

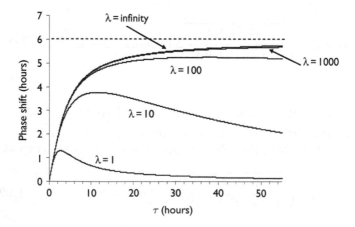

Figure 1.11 The phase shift of the indoor air temperature as a function of the time constant τ and the convective heat transfer number.

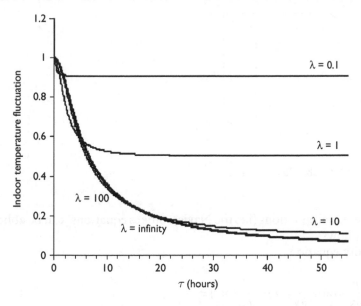

Figure 1.12 The non-dimensional indoor air temperature fluctuation $\Delta \tilde{T}_{in}$ (normalized by the outdoor air temperature fluctuation $\Delta \tilde{T}_o$) as a function of the time constant τ and the convective heat transfer number.

1.3.2 Differential equations and boundary conditions

The integral equations of fluid flow use only averaged or lumped quantities; hence, they cannot generally be applied to the problems when detail of the flow structure is needed. Here we shall introduce the differential equations of fluid flow, in which the conservation principles are applied over a small control volume (infinitesimal fluid element).

For simplicity, a small cube fluid element is considered as a control volume, which means the resulting equations will all be expressed in terms of a rectangular Cartesian coordinate system (Figure 1.13). Indeed, a general shape of the control volume can also be used, and the differential equations of fluid flow in curvilinear coordinates can also be derived, as for application in buildings with complex geometries, such as large enclosures such as theatres, stadiums and atria.

No general analytical solutions are available for the differential airflow equations, and numerical methods such as CFD can therefore be applied to solve these. For high Reynolds numbers, turbulence modeling is generally required for complex flows (Wilcox, 1993). Textbooks on CFD include Roache (1972), Fletcher and Srinivas (1992) and Ferziger and Peric (1996).

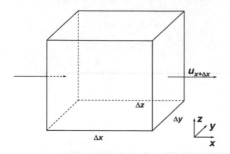

Figure 1.13 A cubic fluid element.

The differential equations (i.e. the Navier–Stokes equations) can be abbreviated as follows:

Continuity equation

$$\frac{\partial \rho}{\partial t} + \frac{\partial}{\partial x}(\rho u) + \frac{\partial}{\partial y}(\rho v) + \frac{\partial}{\partial z}(\rho w) = 0 \tag{1.70}$$

Momentum equations

$$\rho \frac{Du}{Dt} = -\frac{\partial p}{\partial x} + \frac{\partial}{\partial x}\left[2\mu\frac{\partial u}{\partial x} - \frac{2}{3}\mu\nabla\cdot v\right] + \frac{\partial}{\partial y}\left[\mu\left(\frac{\partial u}{\partial y} + \frac{\partial v}{\partial x}\right)\right]$$
$$+ \frac{\partial}{\partial z}\left[\mu\left(\frac{\partial w}{\partial x} + \frac{\partial u}{\partial z}\right)\right] + \rho g_x$$

$$\rho \frac{Dv}{Dt} = -\frac{\partial p}{\partial y} + \frac{\partial}{\partial y}\left[2\mu\frac{\partial v}{\partial y} - \frac{2}{3}\mu\nabla\cdot v\right] + \frac{\partial}{\partial x}\left[\mu\left(\frac{\partial v}{\partial x} + \frac{\partial u}{\partial y}\right)\right]$$
$$+ \frac{\partial}{\partial z}\left[\mu\left(\frac{\partial w}{\partial y} + \frac{\partial v}{\partial z}\right)\right] + \rho g_y \tag{1.71}$$

$$\rho \frac{Dw}{Dt} = -\frac{\partial p}{\partial z} + \frac{\partial}{\partial z}\left[2\mu\frac{\partial w}{\partial z} - \frac{2}{3}\mu\nabla\cdot v\right] + \frac{\partial}{\partial x}\left[\mu\left(\frac{\partial w}{\partial x} + \frac{\partial u}{\partial z}\right)\right]$$
$$+ \frac{\partial}{\partial y}\left[\mu\left(\frac{\partial w}{\partial y} + \frac{\partial v}{\partial z}\right)\right] + \rho g_z$$

where

$$\frac{D\phi}{Dt} = \frac{\partial \phi}{\partial t} + v\cdot\nabla\phi \tag{1.72}$$

ϕ is a general variable. In a physical sense, $\frac{D\phi}{Dt}$, is the time derivation of a quantity evaluated on a path following the fluid motion.

The energy equation is

$$\rho c_{\mathrm{p}} \frac{DT}{Dt} = \frac{\partial}{\partial x}\left(k\frac{\partial T}{\partial x}\right) + \frac{\partial}{\partial y}\left(k\frac{\partial T}{\partial y}\right) + \frac{\partial}{\partial z}\left(k\frac{\partial T}{\partial z}\right) + \dot{Q} \qquad (1.73)$$

where $\dot{Q}$ ($\mathrm{W\,m^{-3}}$) is the heat being generated within the fluid.

The concept of mass conservation is very important in airflow analysis. All the fluid flow analysis should satisfy Equation (1.70). Continuity equation of mass is an inherent part of the statement of any fluid flow problems.

For incompressible flow, the continuity, momentum and energy equations give five differential equations for the pressure, the temperature and the three components of the velocity. These governing equations are generally second-order partial differential equations. The boundary conditions must be provided to specify any practical fluid flow problem. This may be understood by the fact that integration of these equations will lead to constants of integration, and these constants can be calculated by using the fluid flow information at boundaries, i.e. 'boundary conditions'.

The most common boundaries are solid walls, and inflow and outflow boundaries and symmetric planes. It should be mentioned that outflow boundary conditions are generally difficult to specify, and there is no universal outflow boundary conditions. In describing the boundary conditions, care has to be exercised for each specific problem.

When a flow is symmetrical, only half of the domain needs to be solved. At symmetry planes and lines, the normal gradient for all quantities is zero. In addition, velocity components normal to symmetry planes or lines and scalar fluxes are zero. In many situations, even when both the geometry and the boundary conditions are symmetrical, the flow may not always be symmetrical (Chen and Jiang, 1992).

The concept of stream function is useful to represent and interpret fluid flow solutions. The stream function exists only for some specific types of flows. Flow visualization can show the streamlines, hence produce flow patterns. There are three common types of flow visualization in which passive dye or smoke is introduced at a point in the fluid.

- if the dye is introduced once and observed (or photographed) continuously, we see the trajectory of an individual element of the fluid (or dye). This trajectory is called the particle path.
- if dye is introduced continuously and observed (or photographed) at one point in time, we see the location of all fluid elements that have previously passed the point where the dye is introduced. This is called the streakline.
- if the flow field is filled with visible particles, during a short time interval, we may see the change in position of many particles, from

which we can synthesize a continuous line whose tangent at any point is along the local velocity direction. This is the streamline.

If the flow is unsteady, the streamlines, particle paths and streaklines are all different. In contrast, for a steady flow, these are all identical. The significance of these 'lines' in a fluid flow is that they are often more easily visualized than velocity components, and there is a direct relation between these lines and the velocity components.

1.4   Turbulence and its modeling

Most airflows encountered in buildings are turbulent, and they differ considerably from laminar flows. Steady laminar flows will become unstable when the Reynolds number or the Rayleigh number exceeds a certain numerical value (Tennekes and Lumley, 1972). This flow instability could lead the flow through transition to turbulence, with various scales of eddy motion. In theory, the Navier–Stokes equation of motion should be valid for turbulent flows, because even the smallest eddy size in the flow is generally much greater than the mean free path of the molecules in the fluids. Thus, turbulent flow is not at molecular level.

Simply speaking, turbulent motion is irregular and chaotic. Experiments show that at any fixed point in a fully developed turbulent flow, the instantaneous velocity, pressure and/or temperature fluctuate about a mean value, over a sufficiently long time period. Statistical methods are usually used for describing such flows.

When the interest is not in the detail of turbulence structure but in the transport of momentum, heat and mass by turbulence, and the only wish is to predict the mean velocity field in a given flow situation, a statistical approach is rather attractive. However, statistical averages may result in losing much relevant information of the turbulent flow, in particular if the flow is not completely disorganised. Strictly, time-average flow does not exist but is created by scientists for convenience as an effort to predict the effects of turbulence. Here, we shall introduce some basic concepts in turbulence theory and its modeling. The turbulence models to be discussed provide one of two important elements in CFD, i.e. the equations to solve.

1.4.1   Transition to turbulence

In any real flow situation, there is always the possibility of imperfections or disturbances, for example irregularities in the inflow. At a low Reynolds number, the viscosity may be able to damp out the disturbances. At a sufficiently high Reynolds number, all the shear flows in the flow field are unstable and will undergo transition towards turbulent motion.

Initiation of any instability in some flows may be analysed by a linear stability theory. For a given steady flow v, which satisfies the fluid flow equations, a perturbation of some type is added. The flow then becomes

$$v = \bar{v} + v'(x, t) \tag{1.74}$$

$$|v'| << |v|, \tag{1.75}$$

In linear stability theory, we assume and on substituting Equation (1.74) into the fluid flow Equations (1.70–1.73), the equations for determining v' can be obtained for a known basic flow v. Because of the assumption of a small amplitude, the product of terms containing v' is cancelled, and this results in a linear system of equations for v'. This theory can be expected to provide information on whether the disturbances will grow, what types of disturbances will grow and the critical Reynolds number at which this occurs. However, this linear theory cannot indicate any growth rate of the disturbances as the linearity assumption will not be valid then. Most of the knowledge about the critical Reynolds number comes from experimental investigations, in which controlled disturbances are introduced into the flow (Bejan, 2004).

There are two types of turbulent flows that are fundamental in engineering applications. These are the free shear flows (Figure 1.14) and wall shear flows (Figure 1.15). For the free shear flows, the extent of the turbulent region always grows downstream. It is generally accepted that at distances far from the origin, these flows develop some universal characteristics, so-called self-preserving or self-similar. In a similar state, the flow depends only on local quantities, for example mean velocity. For the wall shear flows, the presence of a wall has a dominant effect. Both flows are very important in building airflows.

1.4.2 Statistical description of turbulence and time-averaged equations

The statistical description of turbulence (Landahl and Mollo-Christensen, 1992) considers that the flow variables ϕ, u, p and t consist of a mean part and a fluctuating part. For example, $\phi = \bar{\phi} + \phi'$, where an over bar denotes averaging. In theory, we consider the ensemble average that takes the average over many identical experiments. In practice, the average is usually a time average. We define the average of u and the r-th moment of u as

$$\bar{u} = \lim_{N \to \infty} \frac{1}{N} \sum_{j=1}^{N} u_j \tag{1.76}$$

$$\overline{u^r} = \lim_{N \to \infty} \frac{1}{N} \sum_{j=1}^{N} u_j^r \tag{1.77}$$

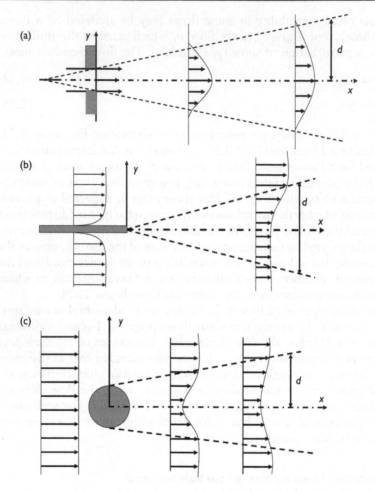

Figure 1.14 Examples of free shear flows: (a) a jet flow; (b) a mixing layer flow; (c) a wake flow.

One can also present these definitions by the concept of probability distribution function $P(u)$. $P(u)$ is defined for a velocity component u at one point, so that the probability that the fluctuation velocity is between u and $u+du$ is $P(u)du$. Thus

$$\int_{-\infty}^{\infty} P(u)du = 1 \tag{1.78}$$

Equations (1.76) and (1.78) become

$$u_{\mathrm{m}} = \overline{u} = \int_{-\infty}^{\infty} uP(u)du \tag{1.79}$$

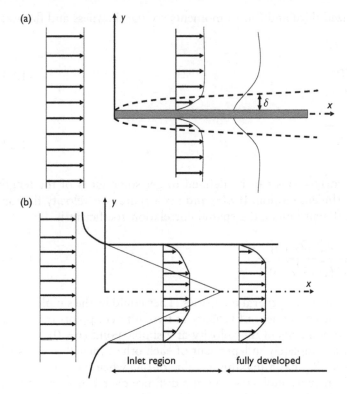

Figure 1.15 Examples of wall shear flows: (a) flat boundary layer; (b) channel or pipe flows.

$$\overline{(u - u_{\mathrm{m}})^r} = \int_{-\infty}^{\infty} (u - u_{\mathrm{m}})^r P(u) \, du \tag{1.80}$$

The second moment is generally considered as the fluctuation intensity u_{rms},

$$u_{\mathrm{rms}} = \left[\overline{(u - u_{\mathrm{m}})^2} \right]^{\frac{1}{2}} \tag{1.81}$$

The turbulence intensity is generally defined as

$$TI = \frac{(u_{\mathrm{rms}}^2 + v_{\mathrm{rms}}^2 + w_{\mathrm{rms}}^2)^{\frac{1}{2}}}{(u_{\mathrm{m}}^2 + v_{\mathrm{m}}^2 + w_{\mathrm{m}}^2)^{\frac{1}{2}}} \tag{1.82}$$

Turbulence intensity is closely related to the turbulent kinetic energy $k = 0.5\overline{q^2}$ where

$$\overline{q^2} = u_{\mathrm{rms}}^2 + v_{\mathrm{rms}}^2 + w_{\mathrm{rms}}^2.$$

The normalized third and fourth moments are the skewness and flatness, respectively.

$$S = \frac{\overline{(u - u_{\mathrm{m}})^3}}{u_{\mathrm{rms}}^3} \tag{1.83}$$

$$F = \frac{\overline{(u - u_{\mathrm{m}})^4}}{u_{\mathrm{rms}}^4} \tag{1.84}$$

The spatial correlations can be defined to get some ideas of the length scales of the turbulent motion. If $u(x)$ and $v(x+r)$ are two velocity fluctuations at two different points, the spatial correlation coefficient is

$$R_{uv}(r) = \frac{\overline{u(x).v(x+r)}}{[\overline{u^2(x).v^2(x+r)}]^{\frac{1}{2}}} \tag{1.85}$$

Here u and v are quite general quantities. They could be the simultaneous values of the same component of velocity at two different points $(r \neq 0)$, or two different components of the velocity at a simple point $(r = 0)$.

If u and v are completely independent of each other, then $R_{uv} = 0$, such as when $|r| \to \infty$. R_{uv} has a maximum value of one when $|r| = 0$.

The correlation curve indicates how the distance over which the motion at one point affects that at another. Roughly speaking, the correlation coefficients $R_{uv}(r)$ measures the strength of eddies whose length in the direction of r is greater than $|r|$, and any eddies that are smaller than r will not be included. This is a very rough (imprecise) concept as R_{uu}, R_{vv} and R_{ww} will be different for the same r.

Similar to the spatial correlation, the correlation of the same fluctuating variable measured at two different times at the same point can also be defined, which is known as autocorrelation.

$$R_{uu}(\tau) = \frac{\overline{u(t)u(t+\tau)}}{[\overline{u^2(t)u^2(t+\tau)}]^{\frac{1}{2}}} \tag{1.86}$$

In fact, a general definition of spatial-time correlation $R_{uv}(r, \tau)$ can be defined as

$$R_{uv}(r, \tau) = \frac{\overline{u(x, t)v(x+r, t+\tau)}}{[\overline{u^2(x, t)v^2(x+r, t+\tau)}]^{\frac{1}{2}}} \tag{1.87}$$

$R_{uu}(\tau)$ can be used to define some time scales of turbulence. $R_{uu}(\tau)$ is usually easier to measure than $R_{uv}(r)$. Taylor's hypothesis can be used to relate $R_{uu}(\tau)$ and $R_{uv}(r)$. It assumes that the fluctuation intensity u_{rms} is small compared with the mean velocity u_m, which means that the turbulent eddy is advected more rapidly by u_m past the measuring point than its shape being changed. Thus

$$R_{uv}(r) = R_{uu}(\tau) \tag{1.88}$$

where $r = -u_m \tau$.

A different way to study the length and time scales of turbulence is to use the Fourier transforms of the auto-correlation, $R_{uu}(\tau)$, and the spatial correlation $R_{uv}(r)$, i.e. frequency spectrum, $\phi(\omega)$ and $E(\kappa)$,

$$\overline{u^2} = \int_0^\infty \phi(\omega)d\omega \tag{1.89}$$

$$k = \frac{1}{2}q^2 = \int_0^\infty E(\kappa)d\kappa \tag{1.90}$$

Eddies in the dissipation range are generally isotropic, i.e. its statistical properties do not vary with direction. In contrast, the energy-containing ranges can be significantly anisotropic.

In the dissipation range, the kinetic energy is dissipated into heat by the smallest eddies. Their typical scales are the Kolmogorov scales.

$$l_k = \left(\frac{\nu^3}{\varepsilon}\right)^{\frac{1}{4}}, \quad t_k = \left(\frac{\nu}{\varepsilon}\right)^{\frac{1}{2}}, \quad \text{and } v_k = (\nu\varepsilon)^{\frac{1}{4}}, \tag{1.91}$$

In the inertial sub-range, Kolmogorov hypothesized (Landahl and Mollo-Christensen, 1992) that the turbulence is in a statistical equilibrium, which is determined by the wave number and the dissipation rate, ε. Dimensional analysis shows

$$E(\kappa) = \alpha\varepsilon^{\frac{2}{3}}\kappa^{-\frac{5}{3}} \tag{1.92}$$

This is the well-known Kolmogorov's $\kappa^{-\frac{5}{3}}$ law.

There are two useful turbulent scales that can be defined from the correlation. The first is the Taylor integral scale as a typical length scale of the energy-containing eddies

$$\Lambda = \int_0^\infty R_{uv}(r)dr \tag{1.93}$$

The second is the Taylor microscale λ, and it is far larger than the Kolmogorov length scale l_k,

$$\lambda = \frac{-2}{\dfrac{\partial^2 R_{uu}}{\partial r^2}} = \frac{2\overline{u^2}}{\left(\dfrac{\partial u}{\partial r}\right)^2} \tag{1.94}$$

Using Taylor's hypothesis, Λ and λ can also be determined through the autocorrelation, which defines the corresponding Taylor's time scales.

$$\tau_0 = \int_0^\infty R_{uu}(\tau)d\tau \tag{1.95}$$

$$t_0 = \frac{2\overline{u^2}}{\left(\dfrac{\partial u}{\partial t}\right)^2} \tag{1.96}$$

t_0 can also be considered as the intersection of r-axis and the parabola of R_{uu} at $\tau = 0$, i.e.

$$R_{uu} = 1 - \frac{r^2}{\lambda^2} \tag{1.97}$$

Examples of the measured skewness, flatness, turbulence energy spectrum, Taylor integral scales and Taylor microscales in room airflows can be found in Li et al. (1992), Chao and Wan (2006), Etheridge and Sandberg (1996).

1.4.3 Reynolds-averaged approach for turbulence modeling

The governing equations of incompressible fluid flow can be summarized using the Cartesian tensor form such as

$$\frac{\partial u_i}{\partial x_i} = 0 \tag{1.98}$$

$$\frac{\partial u_i}{\partial t} + \frac{\partial(u_i u_j)}{\partial x_j} = -\frac{1}{\rho}\frac{\partial p}{\partial x_i} + \nu\frac{\partial^2 u_i}{\partial x_j \partial x_j} - \beta g_i \Delta T \tag{1.99}$$

$$\frac{\partial T}{\partial t} + \frac{\partial(u_j T)}{\partial x_j} = \alpha\frac{\partial^2 T}{\partial x_j \partial x_j} \tag{1.100}$$

By time-averaging Equations (1.98–1.100), we obtain the Reynolds-averaged equations of fluid flow.

$$\frac{\partial \bar{u}_i}{\partial x_i} = 0 \tag{1.101}$$

$$\frac{\partial \bar{u}_i}{\partial t} + \frac{\partial(\bar{u}_i \bar{u}_j)}{\partial x_j} = -\frac{1}{\rho}\frac{\partial \bar{p}}{\partial x_i} + \nu\frac{\partial^2 \bar{u}_i}{\partial x_j \partial x_j} - \frac{\partial(\overline{u_i' u_j'})}{\partial x_j} - \beta\Delta\bar{\theta}g_i \tag{1.102}$$

$$\frac{\partial \bar{\theta}}{\partial t} + \frac{\partial(\bar{u}_j \bar{\theta})}{\partial x_j} = \alpha\frac{\partial^2 \bar{\theta}}{\partial x_j \partial x_j} - \frac{\partial(\overline{u_j' \theta'})}{\partial x_j} \tag{1.103}$$

Comparing Equations (1.98–1.100) with (1.101–1.103), aside from replacement of instantaneous quantities by their mean, the only difference is the appearance of the correlations $\overline{u'_i u'_j}$ and $\overline{u'_i \theta'}$ in Equations (1.102) and (1.103), respectively.

The quantity $-\rho \overline{u'_i u'_j}$ is known as the Reynolds-stress tensor, $\tau_{ij} = -\rho \overline{u'_i u'_j}$, which is a symmetric tensor with six independent components. The quantity $-\rho \overline{u'_i \theta'}$ is the eddy heat flux that has three independent components. These are nine additional unknowns for non-isothermal flows and six for isothermal flows, but at this stage no additional equations exist.

If we multiply the momentum equation by the fluctuating components u'_i and take the time average of the product, we can derive a partial differential equation for the Reynolds-stress tensor as follows:

$$\frac{\partial \tau_{ij}}{\partial t} + \frac{\partial (\overline{u_k}\tau_{ij})}{\partial x_k} = -\frac{\tau_{ik}}{\rho}\frac{\partial \overline{u_j}}{\partial x_k} - \frac{\tau_{jk}}{\rho}\frac{\partial \overline{u_i}}{\partial x_k} + G_{ij} + \varepsilon_{ij} - \Pi_{ij} + \frac{\partial}{\partial x_k}\left(v\frac{\partial \tau_{ij}}{\partial x_k} + C_{ijk}\right)$$

$$(1.104)$$

where

$$\Pi_{ij} = \overline{p'\left(\frac{\partial u'_i}{\partial x_j} + \frac{\partial u'_j}{\partial x_i}\right)} \tag{1.105}$$

$$G_{ij} = -\beta(g_i\overline{u'_j\theta'} + g_j\overline{u'_i\theta'}) \tag{1.106}$$

$$\varepsilon_{ij} = \frac{2\mu}{\rho}\overline{\frac{\partial u'_i}{\partial x_k}\frac{\partial u'_j}{\partial x_k}} \tag{1.107}$$

$$C_{ijk} = \overline{\rho u'_i u'_j u'_k} + \overline{p'u'_i}\delta_{jk} + \overline{p'u'_j}\delta_{ik} \tag{1.108}$$

These are six new equations for each component of the Reynolds-stress tensor. However, 22 new unknowns are generated. In fact, equations can again be derived for these new unknowns, but further more new unknowns will be produced. In the statistical theory of turbulence, we try to approximate the unknown correlations in terms of mean or lower-order correlations that are known. This is known as the closure problem of turbulence modeling.

Let us consider modeling the Reynolds stress $-\rho \overline{u'v'}$ as an example. It is helpful to understand first the physical meaning of $-\rho \overline{u'v'}$. The term $-\rho \overline{u'v'}$ was originated from the convection term (ρuv) during the averaging process, which is the rate of momentum transported by convection.$-\rho \overline{u'v'}$ may be simply understood as the rate of momentum transported by turbulence. Similar to the viscous shear stress, the viscous momentum flux can be represented by $-\rho \overline{u'v'}$, which may also be interpreted as eddy shear stress.

As a first turbulence model, Boussinesq introduced the concept of eddy viscosity ν_t

$$-\rho\overline{u'v'} = \rho\nu_t\frac{\partial \overline{u}}{\partial y} \tag{1.109}$$

The eddy viscosity ν_t is a flow property, not a fluid property. It depends on the turbulence in the flow and hence is a function of position for all practical flows. Thus, the constant eddy viscosity model is not a practical tool.

The first practical turbulence model is the Prandtl's mixing length model, which relates ν_t to a mixing length l by the equation

$$\nu_t = l^2\left|\frac{du}{dy}\right| \tag{1.110}$$

The equation is derived by analogy with the gas kinetic theory model for molecular viscosity. The mixing length is much like the mean free path of molecules in a gas. At least, two assumptions have been made in deriving Equation (1.110). One is that Boussinesq approximation is valid, and another is that eddies in turbulence are not changed by mean shear. Both assumptions are not fully satisfied in practical flows. However, the model works well for free shear flows such as jets, wakes and mixing layers. But the values of the mixing length are all different and have to be tuned individually. Generally, the mixing length is also assumed to be a constant across the layer and proportional to the width of the layer.

In the following, we will derive the velocity distribution in a turbulent boundary layer with a zero pressure gradient, i.e. the well-known law of the wall.

Unlike the free shear flows, Prandtl originally suggested for the flow in turbulent boundary layer, the mixing length is proportional to the distance from the wall. Considering the inner region very close to the wall, the convection terms can be neglected and Equation (1.109) is simplified into Equation (1.111).

$$\frac{\partial}{\partial y}\left[(\mu+\mu_t)\frac{\partial \overline{u}}{\partial y}\right] = 0 \tag{1.111}$$

Integrating Equation (1.111) over a small control volume including the wall surface (at wall $\mu_t = 0$) gives:

$$(\mu+\mu_t)\frac{\partial \overline{u}}{\partial y} = \mu\left(\frac{\partial \overline{u}}{\partial y}\right) = \tau_w = \rho u_\tau^2 \tag{1.112}$$

where $u_\tau = \sqrt{\tau_w/\rho}$ is known as the friction velocity. If we introduce the dimensionless velocity and normal distance

$$\overline{u}^+ = \frac{\overline{u}}{u_\tau}, y^+ = \frac{\rho u_\tau y}{\mu} \tag{1.113}$$

We obtain for the viscous sublayer, where $\mu_t << \mu$,

$$\overline{u}^+ = y^+ \tag{1.114}$$

and for the fully turbulent sublayer, where $\mu_t >> \mu$,

$$\overline{u}^+ = \frac{1}{\kappa}\ln y^+ + B \tag{1.115}$$

Equation (1.115) is the classical law of the wall. The coefficient κ is known as the Karman constant, and B is a dimensionless constant. It is found experimentally that

$$\kappa \approx 0.41 \text{ and } B = 5.0 \tag{1.116}$$

It will be shown later that the law of wall has been used widely as boundary conditions for some more complicated turbulence models, although the approach does not produce satisfactory results for many flows especially for separated flows. Prandtl's mixing-length hypothesis belongs to a class of turbulence models known as the algebraic models.

1.4.4 Turbulent energy equation models

The Boussinesq eddy-viscosity concept introduced in Section 1.4.3 can be generalized as

$$\tau_{ij} = -\rho\overline{u'v'} = \mu_t S_{ij} - \frac{2}{3}\rho k \delta_{ij} \tag{1.117}$$

where S_{ij} is the mean strain-rate tensor,

$$S_{ij} = \frac{\partial \overline{u}_i}{\partial x_j} + \frac{\partial \overline{u}_j}{\partial x_i} \tag{1.118}$$

k is the kinetic energy of turbulence that is related to the trace of the Reynolds stress tensor τ_{ij}.

$$k = -\frac{1}{2}\frac{\tau_{ii}}{\rho} = \frac{1}{2}\overline{u'_i u'_i} = \frac{1}{2}\left(\overline{u'^2} + \overline{v'^2} + \overline{w'^2}\right) \tag{1.119}$$

Kolmogorov in 1942 and Prandtl in 1945 proposed that the eddy viscosity is a function of the characteristic velocity $\sqrt{k}$ and length scales of the turbulence l. By dimensional arguments,

$$\mu_t \propto \rho k^{\frac{1}{2}} l \qquad (1.120)$$

This gives a whole new class of turbulence models. The basic idea is to model the turbulence kinetic energy and the length scale or its equivalent. Differential equations can be derived for these quantities and proper models may then be developed and generally used. Such models are known as turbulence energy equation models.

Using an analogy to the Boussinesq eddy-viscosity concept, we introduce

$$-\overline{u_j'\theta'} = \frac{\nu_t}{\sigma_\theta}\frac{\partial\overline{\theta}}{\partial x_j} \qquad (1.121)$$

$$-\frac{1}{2}\overline{u_i'u_i'u_k'} - \frac{1}{\rho}\overline{p'u_j'} = \frac{\nu_t}{\sigma_k}\frac{\partial k}{\partial x_j} \qquad (1.122)$$

where σ_θ is the turbulent Prandtl number and σ_k is another closure coefficient.

Thus, the modeled turbulence kinetic energy equation after using the above relationships becomes

$$\frac{\partial k}{\partial t} + \frac{\partial(u_j k)}{x_j} = \frac{\partial}{\partial x_j}\left[\left(\nu + \frac{\nu_t}{\sigma_k}\right)\frac{\partial k}{\partial x_j}\right] - \overline{u_i'u_j'}\frac{\partial\overline{u_i}}{\partial x_j} + \beta g_j \frac{\nu_t}{\sigma_t}\frac{\partial\overline{\theta}}{\partial x_j} - \varepsilon \qquad (1.123)$$

where on the right-hand side

- the first term represents the diffusion of turbulence energy by molecular diffusion, turbulent diffusion and turbulent pressure diffusion.
- the second term represents the production of turbulence kinetic energy by shear, i.e. from the mean flow to the turbulence.
- the third term represents the production of turbulence kinetic energy by buoyancy, i.e. from the potential energy to the turbulence.
- and the last term ε represents the dissipation rate of the turbulence (per unit mass), i.e. from the turbulence kinetic energy to thermal internal energy.

This is the basic concept of all the turbulence energy equation models. If the length scale l is related to some flow dimensions, we have the so-called one-equation models. One such model is for thin shear flows in which case the length scale l is taken as proportional to the mixing length. By dimensional arguments, the dissipation rate in Equation (1.123) is modeled as

$$\varepsilon = C_D \frac{k^{\frac{3}{2}}}{l} \tag{1.124}$$

where C_D is a model constant.

Advantages of the one-equation model over a mixing length model are very limited. To develop more general turbulence models, the transport of the length scale or its equivalent must be modeled. This is the two-equation models of turbulence. There are many forms of the two-equation models, mainly depending on the transport quantities in the second transport equation. The two most popular models are the $k - \varepsilon$ model of Launder and Spalding (1974) (so-called the standard $k - \varepsilon$ model) and the $k - \omega$ model of Wilcox (1993). Historically, these two equations were first proposed by Kolmogorov in 1942 and Chou in 1945, respectively.

In the $k - \varepsilon$ model,

$$\nu_t \propto \frac{k^2}{\varepsilon}, \quad l \propto \frac{k^{\frac{3}{2}}}{\varepsilon} \tag{1.125}$$

In the $k - \omega$ model,

$$\nu_t \propto \frac{k}{\omega}, \quad l \propto \frac{k^{\frac{1}{2}}}{\omega} \tag{1.126}$$

The standard $k - \varepsilon$ (Launder and Spalding, 1974) is as follows:
Eddy viscosity

$$\nu_t = C_\mu \frac{k^2}{\varepsilon} \tag{1.127}$$

Turbulence kinetic energy

$$\frac{\partial k}{\partial t} + \frac{\partial(u_j k)}{x_j} = \frac{\partial}{\partial x_j}\left[\left(\nu + \frac{\nu_t}{\sigma_k}\right)\frac{\partial k}{\partial x_j}\right] - P_k - \varepsilon \text{ and } P_k = \overline{u'_i u'_j}\frac{\partial \overline{u_i}}{\partial x_j} \tag{1.128}$$

Dissipation rate of turbulence

$$\frac{\partial \varepsilon}{\partial t} + \frac{\partial(u_j \varepsilon)}{x_j} = \frac{\partial}{\partial x_j}\left[\left(\nu + \frac{\nu_t}{\sigma_\varepsilon}\right)\frac{\partial \varepsilon}{\partial x_j}\right] - C_{1\varepsilon}\frac{\varepsilon}{k}P_k - C_{2\varepsilon}\frac{\varepsilon^2}{k} \tag{1.129}$$

Closure coefficients

$$C_{1\varepsilon} = 1.44, \quad C_{2\varepsilon} = 1.92, \ C_\mu = 0.09, \ k = 1.0, \ \sigma_\varepsilon = 1.3 \tag{1.130}$$

In the above equations, the closure coefficients are generally obtained by applying the model to some simple flows, for example grid turbulence or boundary layers. This is one of the reasons why the two-equation models are still not very general turbulence models although these have produced satisfactory results for many practical flows.

In the wall function method, the solutions are generally sensitive to the distance between matching points, where the wall-function being applied close to the wall surface. It is also known that the law of wall is not valid for separated flows and many 3D flows. To avoid this difficulty, a number of so-called low-Reynolds-number two-equation models have been developed, by the addition of viscous damping functions. The low-Reynolds-number $k - \varepsilon$ models have the same form as the standard $k - \varepsilon$ model except the correction damping functions (Patel *et al.*, 1985).

The two-equation models described above, especially the $k - \varepsilon$ models, have become the cornerstone of turbulent flow simulation since 1960s. These models are available in most of the commercial CFD codes in one form or another. However, the Boussinesq eddy-viscosity approximation in the turbulence energy models assumes that the eddy-viscosity is the same for all Reynolds stresses (isotropic assumption). These models can still fail in many applications.

1.4.5   Reynolds stress turbulence models

Experiments show that the simple linear relationship between the Reynolds stress tensor and the mean strain-rate tensor is not valid for some flows, for example flow over curved surfaces and flow in ducts with secondary motions. A natural way to develop more elaborate turbulence models is to model the individual Reynolds stresses. This will not only, hopefully, give better prediction of the mean quantities but also make it possible to get more detailed information about the distribution and development of the turbulence quantities.

Development of the Reynolds stress models begins with the exact transport equations of the Reynolds stress tensor. These are

$$\frac{\partial \overline{u_i' u_j'}}{\partial t} + \frac{\partial (\overline{u_k}\,\overline{u_i' u_j'})}{\partial x_k} = \frac{\partial}{\partial x_k} \left[\nu \frac{\partial \overline{u_i' u_j'}}{\partial x_k} + C_{ijk} \right] + P_{ij} + G_{ij} + \varepsilon_{ij} + \Pi_{ij} \quad (1.131)$$

where

$$C_{ijk} = -\overline{u_i' u_j' u_k'} - \frac{1}{\rho}(\overline{p' u_i'}\delta_{jk} + \overline{p' u_j'}\delta_{ik}), \text{ diffusion}$$

$$P_{ij} = -\overline{u_i' u_k'}\frac{\partial \overline{u_j}}{\partial x_k} - \overline{u_j' u_k'}\frac{\partial \overline{u_i}}{\partial x_k}, \text{ shear production}$$

$$G_{ij} = -\alpha(g_i\overline{u_j'\theta'} + g_j\overline{u_i'\theta'}), \quad \text{buoyancy-force production}$$

$$\Pi_{ij} = \frac{1}{\rho}\overline{p'(\frac{\partial u_i'}{\partial x_j} + \frac{\partial u_j'}{\partial x_i})}, \quad \text{pressure-strain}$$

$$\varepsilon_{ij} = -2\nu\overline{\frac{\partial u_i'}{\partial x_k}\frac{\partial u_j'}{\partial x_k}}, \quad \text{dissipation}$$

In this equation, the production terms P_{ij} and G_{ij} do not need to be modeled. In the Reynolds stress models, the turbulent heat flux $\overline{u_i'\theta'}$ is also governed by an equation similar to that of the Reynolds stress. For simplicity, this equation is not discussed here. The turbulent diffusion term can be modeled as

$$C_{ijk} = C_s\overline{u_k'u_l'}\frac{k}{\varepsilon}\frac{\partial \overline{u_i'u_j'}}{\partial x_l} \tag{1.132}$$

As dissipation occurs at the smallest scales, the Kolmogorov hypothesis of local isotropy is used.

$$\varepsilon_{ij} = \frac{2}{3}\rho\varepsilon\delta_{ij} \tag{1.133}$$

where

$$\varepsilon = \nu\overline{\frac{\partial u_i'}{\partial x_k}\frac{\partial u_j'}{\partial x_k}} \tag{1.134}$$

The same ε -equation in the $k-\varepsilon$ models can be used to determine the dissipation rate of turbulence kinetic energy.

$$\frac{\partial \varepsilon}{\partial t} + \frac{\partial(\overline{u_j}\varepsilon)}{x_j} = C_\varepsilon\frac{\partial}{\partial x_j}\left(\overline{u_k'u_j'}\frac{k}{\varepsilon}\frac{\partial \varepsilon}{\partial x_j}\right) + \frac{1}{2}C_{1\varepsilon}\frac{\varepsilon}{k}(P_{kk}+G_{kk}) - C_{2\varepsilon}\frac{\varepsilon^2}{k} \tag{1.135}$$

The tensor Π_{ij} is called the pressure-strain redistribution term, and it is the most difficult term to model. For simplicity, here we present one possible and simple model

$$\Pi_{ij} = \Pi_{ij1} + \Pi_{ij2} + \Pi_{ij3} + \Pi_{ijw} \tag{1.136}$$

where

$$\Pi_{ij1} = -C_1\varepsilon a_{ij}$$

$$\Pi_{ij2} = -C_2\left[\left(P_{ij} - \frac{1}{3}\delta_{ij}P_{kk}\right) - \left(C_{ij} - \frac{1}{3}\delta_{ij}C_{kk}\right)\right]$$

$$\Pi_{ij3} = -C_3 \left(F_{ij} - \frac{1}{3} \delta_{ij} F_{kk} \right) \tag{1.137}$$

$$\Pi_{ijw} = C_1' \frac{\varepsilon}{k} \left(\overline{u_k' u_m'} n_k n_m \delta_{ij} - \frac{3}{2} \overline{u_k' u_i'} n_k n_j - \frac{3}{2} \overline{u_k' u_j'} n_k n_i \right)$$

$$+ C_2' \left(\Pi_{km2} n_k n_m \delta_{ij} - \frac{3}{2} \Pi_{ik2} n_k n_j - \frac{3}{2} \Pi_{kj2} n_k n_i \right)$$

$$+ C_2' \left(\Pi_{km3} n_k n_m \delta_{ij} - \frac{3}{2} \Pi_{ik3} n_k n_i \right) \frac{k^{\frac{3}{2}}}{C_l \varepsilon x_n} \tag{1.138}$$

The above model is only one of few other Reynolds stress models developed in the literature. Launder *et al.* (1975) referred to the above model as the Basic Reynolds Stress Model.

The closure coefficients

$$C_s = 0.22;\ C_1 = 1.8;\ C_2 = 0.6;\ C_3 = 0.6;\ C_1' = 0.5;\ C_2' = 0.3;$$
$$C_3' = 0.0;\ C_l = 0.15;\ C_\varepsilon = 0.15;\ C_{1\varepsilon} = 1.44;\ C_{2\varepsilon} = 1.92 \tag{1.139}$$

Examples of application and evaluation of the Reynolds stress models in building ventilation can be found in Chen (1996).

1.4.6   Large eddy simulation

If we try to solve the complete time-dependent solution of the Navier–Stokes and continuity equations, we have to deal with the so-called direct numerical simulation (DNS). In principle, the computational domain must be sufficiently large to accommodate the largest turbulence scales, and the grid must also be sufficiently fine to resolve the smallest eddies whose size is of the order of the Kolmogorov length scale η, and similarly, the time step should be at least of the same order as the Kolmogorov time scale τ.

If we compute only the large eddies and those small eddies not resolved or modeled, we have the so-called large eddy simulation (LES). It is believed that the large eddies contain most of the turbulence energy, and they are directly affected by the boundary conditions. The small-scale turbulence has nearly universal characteristics and is more isotropic. Thus, the small eddies are easier to be modeled than large eddies.

It should be noted that both methods require significant computer resources. Wilcox (1993) summarized the required number of grid points for a turbulent channel flow with DNS and LES, when a stretched grid is used.

$$N_{\text{DNS}} \approx (3Re_\tau)^{\frac{9}{4}} \tag{1.140}$$

$$N_{\text{LES}} \approx \left(\frac{0.4}{Re_\tau^{1/4}} \right) N_{\text{DNS}} \tag{1.141}$$

where $Re_\tau = u_\tau H/2\nu$, H is the height of the channel. About an order of magnitude less number of grids is needed for LES than DNS. In fact, as the grid size is larger in LES than in DNS, the time step in LES can also be larger.

The Reynolds averaged approach of turbulence models can be considered as a single-scale approach. Roughly speaking, we are looking for some kind of laminar representation of the turbulent flow. In the Reynolds averaged approach, all the scales of turbulence have to be modeled, whereas in the sub-grid scale (SGS) stress models, only those small scales are to be modeled. Another difference is that the length scale is readily available in the SGS approach, i.e. the filter width. This may be the main reason that the two-equation models of turbulence discussed earlier have not been widely used as SGS models.

The simplest SGS model is the Smagorinsky model. Similar to the Boussinesq approximation, the model assumes

$$\tau_{ij} = 2\mu_T S_{ij}; \quad S_{ij} = \frac{1}{2}\left(\frac{\partial \overline{u_i}}{\partial x_j} + \frac{\partial \overline{u_j}}{\partial x_i}\right) \tag{1.142}$$

where the Smagorinsky eddy viscosity is

$$\mu_T = \rho(C_s\Delta)^2\sqrt{S_{ij}S_{ij}} \tag{1.143}$$

Similar to the mixing-length model, the value of the Smagorinsky coefficient C_s is not universal. Its value varies from flow to flow. Generally, C_s is between 0.10 and 0.24. Some other SGS models are the one-equation model of Lilly(1966), where an equation for the SGS kinetic energy is solved, the analogy of second-order closure model by Dearoff (1973) and the dynamics SGS model of Germano et al. (1990). Examples of using LES in building ventilation can be found in Kato et al. (1992).

1.5 Jets, plumes and gravity currents

A good starting point of analysing the building airflow is to understand basic flow elements such as wall boundary layers, thermal plumes, supply air jets and so on.

The conventional design of air distribution in a room is often based on the data obtained from the physical model tests or on the study of jets, plumes and boundary layer flows that defines the airflow in the room. The primary air streams are in general assumed to be in an infinite or semi-infinite space. When a jet or a plume is discharged into an infinite space, the continuity is satisfied by entrainment of air from infinity, and this air is returned back to infinity. Therefore, the pressure in the ambient is assumed to be constant (Etheridge and Sandberg, 1996). However, in rooms considered as confined spaces, and the room air is partially or fully encircled by these primary streams, the room airflow cannot be readily determined from the primary streams but depends on them, while they in turn are influenced by the room air. These interactions cause the primary air streams themselves to change.

Most of the basic flow elements of interest can be analysed by the boundary layer approximations, for example jets and plumes. Both the wall shear flows (Figure 1.15) and free shear flows (Figure 1.14) have been investigated extensively in fluid mechanics. One useful analytical treatment of these flows is the combination of similarity analysis, integral methods and dimensional analysis. This treatment presents a good example of combining the integral methods and the differential methods discussed earlier. Gradually, the reader will find that the distinction between the integral and the differential methods described earlier is only useful for presentation purposes. A common idea shared in these methods is to first identify the significant physical transport mechanism and to then apply the conservation principles upon an approximate length or time scales for a particular engineering problem. The fundamental ideas in these 'classical' methods are very helpful for a good understanding of the flow features and possibly 'modern' methods, such as CFD.

Let us first consider a jet flow into a stagnant room, and we consider two directions of air supply, the vertical and the horizontal, when the supply air temperature differs from that of the room air. Here, the buoyancy force will influence the flow. The Archimedes number is introduced in Section 1.2.2 to define the ratio between the force because of thermal convection and that because of advection,

$$Ar = \frac{g\beta L(T_o - T_a)}{U^2} = \frac{Gr}{Re^2} \tag{1.144}$$

where T_a is the room air temperature, T_o is the supply air temperature and U is the supply air velocity.

For vertical flow streams, there are at least four basic types of flows (see Figure 1.16):

- pure jet (Figure 1.16a): the effect of buoyancy is negligible and $Ar = 0$. It can also be termed as non-buoyant jet or simply jet.
- warm buoyant jet (Figure 1.16b): the buoyancy force acts in the direction to the jet. It is also called the forced plume or positive buoyant jet.
- cold buoyant jet (Figure 1.16c): the buoyancy force acts in the opposite direction to the jet. It is also called negative buoyant jet.
- pure plume (Figure 1.16d): the effect of initial momentum is negligible and the buoyancy effect is dominant, $Ar >> 0$.

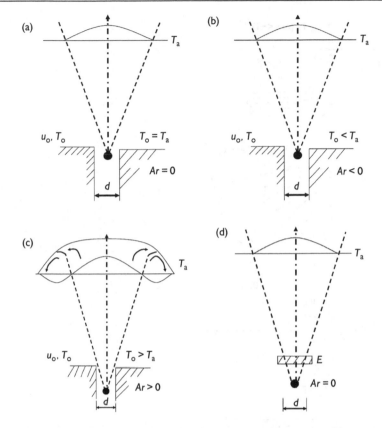

Figure 1.16 Basic vertically supplied flow streams: (a) pure jet, (b) positive buoyant jet (forced plume), (c) negative buoyant jet, (d) pure plume.

For horizontal flows, the situation is different because the buoyancy force acts perpendicular to the flow direction. There are at least five types of flows (see Figure 1.17).

- pure jet (Figure 1.17c): the effect of buoyancy is negligible and $Ar = 0$. There is certainly no physical difference between a horizontal pure jet and a vertical pure jet.
- warm buoyant jet (Figure 1.17e): the supply air is warmer and the buoyancy force acts in the upward direction $(0 \leq Ar \leq 1)$.
- cold buoyant jet (Figure 1.17b): the supply air is colder and the buoyancy force acts in the downward direction $(-1 \leq Ar \leq 0)$.
- warm gravity current (Figure 1.17d): the supply air is warmer and the buoyancy force dominant in the upward direction $(Ar >> 1)$.
- cold gravity current (Figure 1.17a): the supply air is colder and the buoyancy force dominant in the downward direction $(Ar << -1)$.

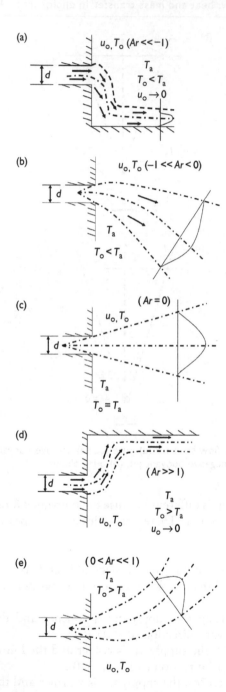

(a) $u_o, T_o \ (Ar \ll -1)$

T_a
$T_o < T_a$
$u_o \to 0$

(b) $u_o, T_o \ (-1 \ll Ar < 0)$

T_a
$T_o < T_a$

(c) $(Ar = 0)$

u_o, T_o

T_a
$T_o = T_a$

(d) $(Ar \gg 1)$

T_a
$T_o > T_a$
$u_o \to 0$

u_o, T_o

(e) $(0 < Ar \ll 1)$

T_a
$T_o > T_a$

u_o, T_o

Figure 1.17 Basic horizontally supplied flow streams: (a) cold gravity current, (b) cold buoyant jet, (c) (pure) jet, (d) warm gravity current, (e) warm buoyant jet.

Except in gravity current, all other flows are free flows which are not confined by solid walls. In free turbulent flows, the molecular diffusion may be wholly neglected. In gravity currents or wall jets to be discussed later, like the wall turbulent boundary layers, the molecular diffusion must be taken into account close to walls.

From an engineering viewpoint, the jet width, the decay in the centreline velocity and temperature, and the distribution of velocity and temperature across the jet are of practical interests. It is beyond the scope of this chapter to provide a full analysis of jets, plumes and gravity currents. Interested readers can refer to Abramovich(1963), Turner (1973), Chen and Rodi (1980) and List (1982).

Vertical buoyant jets or forced plumes are generated by a steady release of mass, momentum and positive buoyancy from a source situated in a stagnant uniform environment. As discussed, there are two limiting cases, namely pure jet from a source of momentum only and the pure plume from a source of buoyancy only. As expected, flows in vertical buoyant jets are governed by the relative importance of the buoyancy and the momentum fluxes at the source. Under certain conditions, the flow may behave first like a pure jet, then the transition region, and finally the pure plume region. Both the jet and plume regions are expected to follow their respective similarity characteristics, but no self-similarity exists in the transition region. A universal scaling law for all three regions of plane and axis-symmetrical vertical buoyant jets has been proposed by Chen and Rodi (1980).

When air supplied from an opening that is bounded by a flat wall surface on one side, and the velocity is directed parallelly to the wall surface, a wall jet occurs. The wall jet is referred to as the semi-contained jet in Abramovich (1963). When there is a distance between the nozzle and the wall surface, the entrainment of air on the wall side causes a pressure differences across the jet. The jet is curved towards the wall. This is known as the Coanda effect.

It is known that a wall jet has a longer throw than a free jet. A wall jet along the ceiling of a room can be applied to avoid high velocity in the occupied region, i.e. draught. Two simple wall jets are the plane wall jet and the round wall jet. Integral analysis with self-similarity observations in the fully developed region can be carried out to study the characteristic velocity decay of wall jet and its growth (see Awbi, 2003).

1.6 Solution multiplicity of building airflows

Building airflows are often non-linear. Multiple states could exist as shown in simple buildings. The existence of multiple solutions presents a challenge to the conventional multi-zone or CFD modeling methodologies, which can only give one of the solutions for the same set of physical and numerical parameters depending on initial conditions. A few examples of solution multiplicity are available on the existence of multiple solutions in simple buildings (Nielsen, et al., 1979; Nitta, 1996; Li and Delsante, 1998; Hunt and Linden, 2000; Gladstone and Woods, 2001).

Let us demonstrate the existence of solution multiplicity by considering a simple model of natural ventilation in a single-zone building with two openings (Li and Delsante, 2001). We consider the opposing wind situation. In Example 1.5, we showed how to derive the combined natural ventilation flow rate because of combined buoyancy and opposing wind. The ventilation flow rate is calculated as

$$q = C_d A^* \sqrt{\left| 2gH\frac{T_i - T_o}{T_o} - 2\frac{\Delta p_w}{\rho_o} \right|} \tag{1.145}$$

The solution shows that if the wind force is stronger, the flow rate is calculated as

$$q = C_d A^* \sqrt{2\frac{\Delta p_w}{\rho_o} - 2gH\frac{T_i - T_o}{T_o}} \tag{1.146}$$

If the thermal buoyancy is stronger, the flow rate is calculated as

$$q = C_d A^* \sqrt{2gH\frac{T_i - T_o}{T_o} - 2\frac{\Delta p_w}{\rho_o}} \tag{1.147}$$

This solution behaviour corresponds to two flow patterns, one is a wind-dominated flow and the other is buoyancy-dominated (see Figure 1.18). For both flow patterns, the airflow is assumed to be fully mixed.

We now ask the question what happens if the air temperature is not given and what would be the switching criteria from one flow pattern to another.

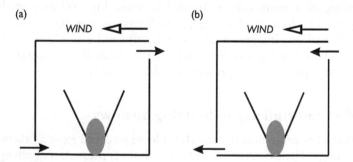

Figure 1.18 A one-zone building with two openings with possible flow states: (a) upward flow and (b) downward flow. The room is assumed to be fully mixed.

Again, we use the following two air change parameters to characterize, respectively, the effects of thermal buoyancy and wind force:

$$\alpha = (C_d A^*)^{\frac{2}{3}} (BH)^{\frac{1}{3}} \tag{1.148}$$

$$\gamma = \frac{1}{\sqrt{3}} (C_d A^*) \sqrt{2\Delta p_w} \tag{1.149}$$

Both air change parameters have the same dimension as the ventilation flow rate. In Equations (1.148) and (1.149), B is the buoyancy flux $(Eg/\rho c_p T_o)$, $A^* = A_t A_b / \sqrt{A_t^2 + A_b^2}$ is the effective area, and ΔP_w is the wind pressure difference between the two openings, which is always taken to be non-negative:

$$\Delta p_w = \frac{1}{2} C_{pt} v_t^2 - \frac{1}{2} C_{pb} v_b^2 \tag{1.150}$$

The governing equation for the heat balance is as follows

$$c_M M \frac{dT_i}{dt} + \rho c_p |q| (T_i - T_o) = E \tag{1.151}$$

The ventilation flow because of the combined wind and thermal buoyancy force can be written as follows, considering the signs of the flow rate. A positive flow rate represents an upward flow and a negative one represents a downward flow.

$$q|q| = (C_d A^*)^2 \left[2gH \frac{T_i - T_o}{T_o} - 2\Delta p_w \right] \tag{1.152}$$

The general equation for the ventilation flow rate q can be obtained with the earlier definitions of the two air change parameters, after some manipulation of the Equations (1.151) and (1.152):

$$2\omega |q| \frac{dq}{dt} = -q^3 - 3\gamma^2 |q| + 2\alpha^3 \tag{1.153}$$

where α and γ are defined in Equations (1.148 and 1.149) and ω is a thermal mass parameter, defined as:

$$\omega = \frac{c_M M}{\rho c_p} \tag{1.154}$$

The solution for opposing winds is complex. The governing equations can be written for upward flows and downward flows, respectively.

For upward flows, $q > 0$,

$$2\omega q \frac{dq}{dt} = -q^3 - 3\gamma^2 q + 2\alpha^3 \qquad (1.155)$$

For downward flows, $q < 0$,

$$-2\omega q \frac{dq}{dt} = -q^3 + 3\gamma^2 q + 2\alpha^3, \text{ i.e. } 2\omega q \frac{dq}{dt} = q^3 - 3\gamma^2 q - 2\alpha^3 \quad (1.156)$$

For some range of parameters, there could be three real fixed points for each of the equations, but not all of them are physical.

For upward flows, the fixed points can be determined as follows

$$0 = -q^3 - 3\gamma^2 q + 2\alpha^3 \text{ or } 0 = q^3 + 3\gamma^2 q - 2\alpha^3$$

Thus,

$$\left(\frac{q}{\gamma}\right)^3 + 3\frac{q}{\gamma} - 2\left(\frac{\alpha}{\gamma}\right)^3 = 0 \qquad (1.157)$$

For all values of α/γ, there is one real root and two imaginary roots. For downward flows,

$$q^3 - 3\gamma^2 q - 2\alpha^3 = 0 \text{ or } \left(\frac{q}{\gamma}\right)^3 - 3\frac{q}{\gamma} - 2\left(\frac{\alpha}{\gamma}\right)^3 = 0 \qquad (1.158)$$

When $\alpha/\gamma < 1$, there are three real and unequal roots, and when $\alpha/\gamma = 1$, there are also three real roots, but at least two of them are equal, and finally when $\alpha/\gamma > 1$, there is only one real root, plus two conjugate imaginary roots.

Only the real roots are of our interests here, and they are shown in Figure 1.19.

The behaviour of the flow rate as a function of α and γ reveals that for $\beta \neq 0$ the general form of the solutions is shown in Figure 1.20. The flow exhibits hysteresis: for values of α between α_A and α_B, there are three possible solutions for the flow, one upward and two downward. The flow rate adopted by the system for a given value of the buoyancy flux depends on the path taken to arrive at that value (either decreasing it from an initial high value, along the curve E–A, or increasing it from an initial low value, along the curve C–B).

The autonomous dynamical system shown by Equation (1.153) can be written in a more general form, that is,

$$\frac{dq}{dt} = f(q, \alpha, \gamma) \qquad (1.159)$$

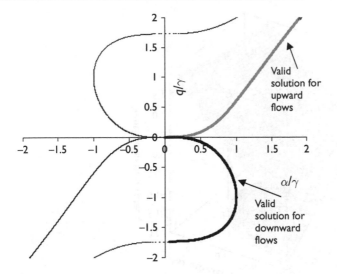

Figure 1.19 Real solutions for the governing equations of both upward flows and downward flows at steady state.

where α and γ are two control parameters.

$$q > 0, \quad f(q) = -q^2 - 3\gamma^2 + \frac{2\alpha^3}{q} \tag{1.160}$$

$$q < 0, \quad f(q) = q^2 - 3\gamma^2 - \frac{2\alpha^3}{q} \tag{1.161}$$

As shown earlier, there are three equilibria for an α/γ ratio between 0 and 1. The stability of the equilibria can be shown by examining whether a small perturbation grows or decays. A small perturbation grows exponentially if the slope $f'(q^*) > 0$ and decays if $f'(q^*) < 0$, where q^* is the fixed point of the system, i.e. the solution of $f(q) = 0$:

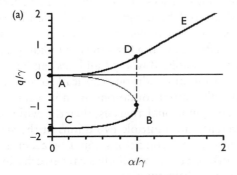

Figure 1.20 Equilibrium ventilation flow rate as a function of the thermal buoyancy and wind air change parameters: (a) normalized 2D view and (b) 3D view.

(b)

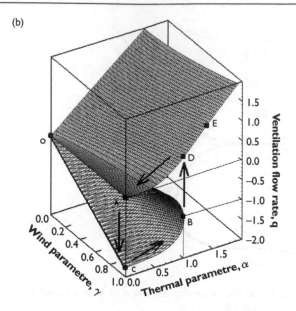

Figure 1.20 (Continued).

$$q > 0, \quad f'(q) = -2q - \frac{2\alpha^3}{q^2} \tag{1.162}$$

$$q < 0, \quad f'(q) = 2q + \frac{2\alpha^3}{q^2} \tag{1.163}$$

We find that $f'(q^*) > 0$ for the solution curve AB (excluding point A) in Figure 1.20, and thus, it is unstable. Also we find that $f'(q^*) < 0$ for the solution curves CB and ADE (also excluding point A), and thus, they are stable. Therefore, alternate stable states exist for an α/γ ratio between 0 and 1.

The existence of the multiple solution behaviour is an inherent feature of ventilation flows, as the ventilation process is non-linear. In general, it is known that airflows in and around buildings are highly non-linear. Understanding, prediction and control of airflows are essential for thermal comfort, indoor air quality, energy efficiency, and fire safety in buildings. As non-linear dynamical systems, air and smoke flows in buildings can be very sensitive to perturbations. Different initial conditions can lead to different stable solutions. This means that if the initial conditions are not accurately specified in a simulation tool for computational analysis, then the evolution of the state can also be very different, and multiple evolutions of the state of flow must be considered. Even if initial conditions could be accurately specified (which, in general, cannot), relatively small physical perturbations can also lead the airflows to switch to different states.

The example that we dealt with here is a very simple building model. We also neglected the effect of non-uniformity of the indoor air temperature distribution. We assumed that the two openings had a relatively small height, so that the airflows through both openings were unidirectional. Our chosen mathematical model also implied that the kinetic energy term through the indoor space was negligible (Axley et al., 2002). All these assumptions were shown to be non-critical in affecting the solution multiplicity for the simple two-opening building. We assumed that the Bousinnesq approximation was valid, so that our analysis was not directly applicable to smoke-control problems. There are also other possible approximations that could limit the applicability of the present analysis to realistic problems. However, the simple fact is that most existing multi-zone analysis models make similar (if not the same) assumptions. This suggests the importance and implications of the present results to the application of these multi-zone analysis models. If these models are used for a building design, and the designers are not aware of the possible existence of multiple solutions, the designers may obtain only one of the possible solutions if multiple solutions exist. There is at least one question to be asked – is the design reliable?

Non-linear dynamics have been applied into many disciplines such as finance, biology, fluid mechanics, chemistry, population dynamics, etc. The application of non-linear dynamics to building airflows is relatively new. There are many remaining questions, some are quite fundamental. For example, most of the existing studies are only for very simple situations and cases involving ideal buildings. Thus, if dynamic flow phenomena in buildings exist, it is reasonable to ask:

- how significant are these to the airflow design?
- how can these be controlled?
- what are the main methods that are available or can be developed for effectively analysing these non-linear dynamic behaviours of airflow systems in building?

To answer these and other related questions, more research is clearly needed.

1.7 Experimental methods

1.7.1 Characteristic scales and non-dimensional parameters revisited

The characteristic scales and non-dimensional parameters presented in Section 1.22 are very fundamental in building ventilation. Some of these characteristic scales and non-dimensional parameters are derived here in a different manner with a view that such a derivation may assist us to understand building airflows. Additionally, scaling parameters at thermal boundaries, including the radiative effects, are also crucial for similarity requirement in experimental studies.

The distribution of mean air velocity, pressure and temperature in rooms is governed by the continuity, momentum and energy equations (Equations 1.101–1.103) with corresponding boundary conditions. The Boussinesq approximation is adopted here for the buoyancy effect. The above equations can be non-dimensionalized by dividing the variables by the corresponding reference values, i.e. x_0, u_0, ΔT_0, p_0 (which may be written as $\rho_0 u_0^2$), t_0 (which may be written as x_0/u_0) and Q_0 (which may be written as $\rho c_p u_0 \Delta T_0/x_0$). Equations (1.101–1.103) become

$$\frac{\partial u_i^*}{\partial x_i^*} = 0 \tag{1.167}$$

$$\frac{\partial u_i^*}{\partial t^*} + \frac{\partial \left(u_i^* u_j^*\right)}{\partial x_j^*} = -\frac{\partial p^*}{\partial x_i^*} + \frac{1}{Re}\frac{\partial^2 u_i^*}{\partial x_j^* \partial x_j^*} + \frac{\partial}{\partial x_j^*}\left(-\overline{u_i'^* u_j'^*}\right) - Ar\Delta T^* \tag{1.168}$$

$$\frac{\partial T^*}{\partial t^*} + \frac{\partial \left(T^* u_j^*\right)}{\partial x_j^*} = \frac{1}{Pr\,Re}\frac{\partial^2 T^*}{\partial x_j^* \partial x_j^*} + \frac{\partial}{\partial x_j^*}\left(-\overline{u_j'^* T^*}\right) + Q^* \tag{1.169}$$

A heat source term is added in Equation (1.169). If the dimension of the diffuser, L, is chosen as x_0, then u_0 is the supply velocity U_s and ΔT_0 is the temperature difference between the return and supply air. The three dimensionless parameters in the above equations become

$$Pr = \frac{\mu c_p}{k} \qquad Re = \frac{\rho U_s L}{\mu} \qquad Ar = \frac{g\beta L \Delta T_0}{U_s^2} \tag{1.170}$$

We will also derive scale parameters presenting radiative effects on the thermal boundary, which are particularly meaningful in buoyancy-controlled ventilation. Multiple thermal transport modes exist on the boundary, i.e. radiation, convection and conduction. It is therefore important to evaluate the dimensionless parameters for prototype conditions and to indicate which of these modes are significant.

The temperature and heat flux at the wall–air interface are usually unknown a priori. What we do know is the outdoor environment, which gives rise to a conjugate heat-transfer problem that includes the indoor air domain and the room walls. The distribution of temperature at the wall–air interface generally depends on the heat conduction q_i through the wall, the thermal convection near the wall surface and the radiation between different surfaces. For a general surface i in a room of N

small isothermal grey surfaces with equal areas, the heat balance equation becomes

$$q_i = \lambda \frac{\partial T}{\partial n} + q_{r,w} \tag{1.171}$$

$$q_i = \lambda \frac{\partial T}{\partial n} + \sigma \varepsilon T_{wi}^4 - \sigma \varepsilon \sum_{j=1}^{N} G_{ij} T_{wj}^4 \tag{1.172}$$

Equation (1.171) is generally valid, whereas Equation (1.172) is valid only in the case of surface radiation. The reflectivity of the room surfaces can be ignored, because ordinary room surfaces have an approximate emissivity of 0.9. Equation (1.172) can be written as

$$q_i = \lambda \frac{\partial T}{\partial n} + \sigma \varepsilon T_{wi}^4 - \sigma \varepsilon \sum_{j=1}^{N} F_{ij} T_{wj}^4 \tag{1.173}$$

As the temperature differences are moderate, we linearize the term $T_{wi}^4 - T_{wj}^4$ as $4T_0^3 \left(T_{wi} - T_{wj} \right)$, where T_0 is the average temperature. The equation is non-dimensionalized by defining the reference values ΔT_0, L, $F_0 (= 1)$ and a reference heat flux $q_0 = \rho c_p U_s \Delta T_0$ in the case of forced convection and $q_0 = h \Delta T_0$ in the case of natural convection. We then have
Forced convection:

$$q_i^* = \frac{1}{Pr\,Re} \frac{\partial T^*}{\partial n^*} + \frac{1}{Pl\,Pr\,Re} \sum_{j=1}^{N} F_{ij}^* \left(T_{wi}^* - T_{wj}^* \right) \tag{1.174}$$

Natural convection:

$$q_i^* = \frac{1}{Nu} \frac{\partial T^*}{\partial n^*} + \frac{1}{PlNu} \sum_{j=1}^{N} F_{ij}^* \left(T_{wi}^* - T_{wj}^* \right) \tag{1.175}$$

where

$$Nu = \frac{hL}{\lambda} \qquad Pl = \frac{\lambda}{4T_0^3 L \varepsilon \sigma} \tag{1.176}$$

Another probable choice of reference heat flux is the conductive heat flux $q_0 = \lambda \Delta T_0 / L$; the equation then becomes

$$q_i^* = \frac{\partial T^*}{\partial n^*} + \frac{1}{Pl} \sum_{j=1}^{N} F_{ij}^* \left(T_{wi}^* - T_{wj}^* \right) \tag{1.177}$$

Equation (1.177) has been derived by Nielsen (1974).

Table 1.1 Estimated ranges of thermal and flow parameters in rooms ventilated by a
mixing or displacement system

Type of system	Parameters	Ranges
Mixing system	Supply air temperature	9–40°C
	Supply air speed	2.0–8.0 m s^{-1}
	Reynolds number (based on supply data)	2000–10000
	Archimedes number	< 0.01
Displacement system	Supply air temperature	15–20°C
	Supply air speed	0.1–0.4 m s^{-1}
	Archimedes number	> 0.5
Both systems	Supply airflow rate	< 20 ACH
	Air speed in the occupied zone	0.15–0.30 m s^{-1}
	Air temperatures in the occupied zone	16–26°C
	Vertical temperature gradient in the occupied zone	< 3K m^{-1}
	Surface temperature of thermal radiators	< 50°C
	Ratio of radiative to convective heat transfer from a surface	0.5–4
	Convective heat transfer coefficient of room surfaces	0.5–10 W m^{-2} K^{-1}

An elementary analysis of the characteristics of indoor airflow is shown in
Table 1.1. This would constitute necessary information for anyone wishing
to simulate or study indoor airflow.

1.7.2 Small-scale and full-scale experimental methods

Experimental methods commonly used for studying room air distribution
include full-scale measurement and reduced-scale measurement. The data
obtained from model studies of buildings can be used in designing air
distribution systems. Detailed data are also essential for evaluating and
improving the numerical simulation of indoor airflow. This section gives
a brief account of the current understanding and application of indoor
airflow measurement as well as low air velocity measurement techniques,
but for fuller treatment, the reader should refer to other texts, for example
Etheridge and Sandberg (1996) and Awbi (2003).

Small-scale model

A broad range of ventilation conditions may be more conveniently and
economically investigated by means of small-scale studies. The measured
results can be extrapolated to rooms of different sizes if a proper scaling
method is employed.

The airflow field in two geometrically similar rooms (e.g. the reduced-
scale model and its full-scale equivalent) are completely similar if they have

the same distributions of dimensionless velocity (U^*), temperature (T^*), pressure (P^*) and turbulent stress. This would require not only the equality of the dimensionless numbers Re, Pr and Ar (which is necessary for kinematic and thermal similarity) but also the similarity of the boundary conditions. As the model and prototype are geometrically similar (the requirement of geometrically similar boundaries is automatically satisfied), the size of the supply and exhaust air terminal devices must be carefully scaled down. As the surface roughness could have a significant effect in the development of the boundary layers in the internal room surfaces and hence the heat convection to these surfaces, care should be taken to apply the same scaling factors to the roughness of the internal surface. Hydraulic similarity of the boundaries can be achieved by accurate scaling of the flow boundaries (e.g. air supply and return) in order to ensure a similar flow pattern and turbulence level at the boundaries of the model and prototype.

The complexity of the internal boundary similarity depends on the type of the thermal boundary conditions. For similarity of the dimensionless temperature gradient at the boundary of model and prototype, $Pe = PrRe$ must be equal for the two enclosures, and this condition plus the same distribution of dimensionless heat flux q_i^* would be sufficient in cases where radiation is negligible. In a case where radiation is significant, the surface Plank number Pl_s must be identical. The emissivity for long-wave radiation is about 0.9 for common surfaces in a room. It is therefore very difficult to appreciably reduce this coefficient in the model (Nielsen, 1974).

An ideal reduced-scale study would satisfy the complete similarity conditions as above. Basically, the requirement of an equal Prandtl number for two different fluids that would also satisfy other modeling parameters cannot be achieved (Awbi, 2003), and, in addition, the requirement of the equality of Re and Ar leads to contradictory scaling factors. Therefore, reduced-model studies are usually conducted, so that the model can predict the overall room airflow pattern and the distribution of air velocities and temperature within the regions in which one is most interested (e.g. the occupied regions).

Water is sometimes used for obtaining high Re flow in reduced model studies under isothermal conditions, but it is probably not suitable for non-isothermal flows, because the value of Pr for water decreases sharply with increasing temperature, and the water is also opaque to long-wave radiation. Air is still used for modeling non-isothermal flows. The Reynolds number is in general ignored; assuming that the Re value is higher than a threshold, the jet flow becomes fully turbulent and no longer depends on the Reynolds number. When the radiation heat transfer at the boundaries is to be considered, the requirements of equal Ar and equal $1/(Re.Pl_s)$ can be combined to find a proper scale factor and model parameter. However, the entire room air flow under realistic ventilation conditions is generally not fully turbulent. In this case, the air movement within the room would still be dependent on the diffuser Reynolds number because the viscous effect cannot be neglected. Thus in practice, care must be taken to choose a proper scaling method.

Full-scale model

Currently, measurement in full-scale rooms is the only reliable method of evaluating room airflow. This can be seen from the difficulty of scaling different variables in small-scale studies and the need for evaluation of numerical simulations. Indeed, advances in numerical simulation are limited by the lack of compatible, adequately detailed experimental data. This may be overcome by building a full-scale physical model to evaluate the air temperatures, radiant temperatures and air velocities in the entire room, especially in the occupied zone. Full-scale studies can be done both in the laboratory and in field investigations. Several well-known full-scale studies have been summarized in Awbi (2003) and Croome and Roberts (1980). These data are valuable and have resulted in empirical relations that form the basis of the design procedures, see Awbi (2003) and Croome and Roberts (1980).

References

Abramovich, G. N. (1963), *The Theory of Turbulent Jets*. MIT Press.

Allard, F. and Santamouris, M. (1998), *Natural Ventilation in Buildings*. James and James, London.

Awbi, H. B. (2003), *Ventilation of Buildings*. Taylor & Francis, New York.

Axley, J., Wurtz, E. and Mora, L. (2002), Macroscopic airflow analysis and the conservation of kinetic energy. In the Proceedings of Roomvent 2002, The 7th International Conference on Air Distribution in Rooms, Copenhagen, 8–11 September 2002.

Bejan, A. (2004), *Convection Heat Transfer*. Wiley, Hoboken, 3rd ed.

Bird, R. B., Stewart, W. E. and Lightfoot, E. N. (2002), *Transport Phenomena*. Wiley, New York, 2nd ed.

Chao, C. Y. H. and Wan, M. P. (2006), A study of the dispersion of expiratory aerosols in unidirectional downward and ceiling-return type airflows using a multiphase approach. *Indoor Air*, Online early issue, June, 2006.

Chen, C. J. and Rodi, W. (1980), *Vertical Turbulent Buoyant Jets: A Review of Experimental Data*. Pergamon Press, Oxford.

Chen, Q. and Jiang, Z. (1992), Significant questions in predicting room air motion. *ASHRAE Transactions*, 98 (Part 1), 929–939.

Chen, Q. (1996), Prediction of room air motion by Reynolds-stress models. *Building and Environment*, 31, 233–244.

Croome, D. J. and Roberts, B. M. (1980), *Air Conditioning and Ventilation of Buildings*. Pergamon, Oxford.

Danckwerts, P. V. (1952), Continuous flow systems: distribution of residence time. *Chemical Engineering Science*, 2, 1–13.

Deardorff, J. W. (1973), The use of subgrid transport equations in a three-dimensional model of atmospheric turbulence. *ASME Journal of Fluids Engineering*, 95, 429–438.

Etheridge, D. and Sandberg, M. (1996), *Building Ventilation: Theory and Measurement*. John Wiley & Sons, Chichester.

Fanger, P. O. (1970), *Thermal Comfort*. Danish Technical Press, Copenhagen.

Fanger, P. O., Melikov, A. K. Hanzawa, H. and Ring, J. (1998), Air turbulence and sensation of draught. *Energy and Building*, 12, 21–39.

Ferziger, J. H. and Peric, M. (1996), *Computational Methods for Fluid Dynamics*. Springer, Berlin.

Fletcher, C. A. J. and Srinivas, K. (1992), *Computational Techniques for Fluid Dynamics*. Springer-Verlag, Berlin.

Germano, M., Piomelli, U., Moin, P. and Cabot, W. (1990), A dynamics subgrid-scale eddy viscosity model. Proceedings of the 1990 Summer Program, Center for Turbulence Research, Standford, CA.

Gladstone, C. and Woods, A. W. (2001), On buoyancy-driven natural ventilation of a room with a heated floor. *Journal of Fluid Mechanics*, 441, 293–314.

Hunt, G. R. and Linden, P. F. (2000), Multiple steady airflows and hysteresis when wind opposes buoyancy. *AIR Infiltration Review*, 21 1–3 (2).

Kato, S., Murakami, S., Mochida, A., Akabayashi, A. S.-I. and Tominaga, Y. (1992), Velocity-pressure field of cross ventilation with open windows analyzed by wind tunnel and numerical simulation. *Journal of Wind Engineering and Industrial Aerodynamics*, 44, 573–587.

Koestel, A. (1955), Paths of horizontally projected heated and chilled air jets. *Transactions ASHAE*, 61, 213–232.

Landahl, M. T. and Mollo-Christensen, E. (1992), *Turbulence and Random Processes in Fluid Mechanics*. Cambridge University Press, Cambridge.

Launder, B. E., Reece, G. J. and Rodi, W. (1975), Progress in the development of Reynolds stress turbulent closure. *Journal of Fluid Mechanics*, 68, 537–566.

Launder, B. E. and Spalding, D. B. (1974), The numerical calculation of turbulent flows. *Computer Methods in Applied Mechanics and Engineering*, 3, 269–289.

Li, Y., Sandberg, M. and Fuchs, L. (1992), Vertical temperature profiles in rooms ventilated by displacement: full-scale measurement and nodal modeling. *Indoor Air*, 2, 225–243.

Li, Y. and Delsante, A. (1996), Derivation of capture efficiency of residential kitchen range hoods in a confined space. *Building and Environment*, 31, 461–468.

Li, Y. and Delsante, A. (1998), On natural ventilation of a building with two openings. In the Proceedings of the 19th AIVC Conference: Ventilation Technologies in Urban Areas, Oslo, Norway, 28–30 September 1998.

Li, Y. and Delsante, A. (2001), Natural ventilation induced by combined wind and thermal forces. *Building and Environment*, 36, 59–71.

Lilly, D. K. (1966), On the application of the eddy viscosity concept in the inertial subrange of turbulence. NCAR Manuscript, 123.

List, E. J. (1982), Turbulent jets and plumes. *Annual Review of Fluid Mechanics*, 14, 189–212.

Mundt, E. (1995), Displacement ventilation systems—Convection flows and temperature gradients. *Building and Environment*, 30, 129–133.

Nielsen, P. V. (1974), Flow in air conditioned rooms – model experiments and numerical solutions of the flow equations. PhD thesis, Technical University of Denmark, Nordborg, Denmark.

Nielsen, P. V., Restivo, A. and Whitelaw, J. H. (1979), Buoyancy-affected flows in ventilated rooms: *Numerical Heat Transfer*, 2, 115–127.

Nitta, K. (1996), Study on the Variety of theoretical solutions of ventilation network. *Journal of Architecture, Planning, and Environmental Engineering* (Transactions of AIJ), 480, 31–38.

Patel, V. C., Rodi, W. and Scheuerer, G. (1985), Turbulence models for near-wall and low Reynolds number flows – A review. *AIAA Journal*, 23, 1308–1319.

Roache, P. J. (1972), *Computational Fluid Dynamics*. Hermosa Publishers, Albuquerque.

Sandberg, M. (1981), What is ventilation efficiency. *Building and Environment*, 16, 123–135.

Spengler, J. D., Samet, J. M. and McCarthy J. F. (eds), (2001), *Indoor Air Quality Handbook*. McGraw–Hill, New York.

Tennekes, H. and Lumley, J. L. (1972), *A First Course in Turbulence*. MIT Press, Cambridge.

Tritton, D. J. (1988), *Physical Fluid Dynamics*. Oxford University Press, New York.

Turner, J. S. (1973), *Buoyancy Effects in Fluids*. Cambridge University Press, Cambridge.

Wilcox, D. C. (1993), *Turbulence Modeling for CFD*. DCW Industries, Inc., La Canada, CA.

Nomenclature

Symbols and units

A	area	m^2
A^*	effective opening area	m^2
Ar	Archimedes number	–
B	buoyancy flux	$m^4 s^{-3}$
c	concentration	$kg\ m^{-3}$ or $m^3 m^{-3}$
C_d	discharge coefficient	–
C_D	model constant	–
c_M	heat capacity of thermal mass	$J\ m^{-3} kg^{-1} K^{-1}$
c_p	specific heat capacity	$J\ m^{-3} kg^{-1} K^{-1}$
C_p	wind pressure coefficient	–
E	energy flux	$J\ m^{-3}$
E	total heat power in the room	W
$E(k)$	energy frequency spectrum distribution	–
E_v	pressure loss due to friction	Pa
F	flatness	–
F	shape factor	–
$\mathbf{F}_g$	body force vector	N
$\mathbf{F}_s$	surface force vector	N

g	gravity acceleration	m s^{-2}
G	pollutant generation rate	kg s^{-1}
Gr	Grashof number	–
h	convective heat transfer coefficient	W m^{-2}K^{-1}
G	pollutant generation rate	kg s^{-1}
Gr	Grashof number	–
h	convective heat transfer coefficient	W m^{-2}K^{-1}
h	convective heat transfer coefficient	W m^{-2}K^{-1}
H	enthalpy	J
H	room height or vertical distance between two openings	m
H	vertical distance between two openings	m
h_f	convective heat transfer coefficient	W m^{-2}K^{-1}
h_r	radiative heat transfer coefficient	W m^{-2}K^{-1}
k	thermal conductivity	W m^{-1}K^{-1}
k	turbulence kinetic energy	m^{2}s^{-2}
L	characteristic length scale	m
l	mixing length	m
l_k	Kolmogorov length scale	m
M	mass	kg
m	mass flow rate	kg s^{-1}
n	air change rate	h^{-1}
N	number of samples	–
Nu	Nusselt number	–
p	pressure	Pa
Pl	Plank number	–
Pr	Prantl number	–
Q	heat added to the system from surrounding	W
q	heat flux	W m^{-2}
q	mass flow rate	kg m^{-3}
q	volumetric flow rate or ventilation rate	m^{3}s^{-1}
Ra	Rayleigh number	–
Re	Reynolds number	–
$R_{uv}(r)$	spatial correlation	–
$R_{uv}(\tau)$	autocorrelation	–
S	skewness	–
S_{ij}	mean strain-rate tensor	s^{-1}
t	time	s
t_c	overall convection time scale	s
t_d	overall viscous diffusion time scale	s
t_k	Kolmogorov time scale	s
t_{tc}	thermal convection time scale	s
t_{td}	thermal diffusion time scale	s
T	temperature	K
U	characteristic velocity	m s^{-1}
$\hat{U}$	internal energy	J
u	x-component velocity or velocity	m s^{-1}

u_c	convection velocity scale	m s^{-1}
$\mathbf{v}$	velocity vector	m s^{-1}
V	volume	m^3
v	y-component velocity or velocity	m s^{-1}
V	volume	m^3
v	y-component velocity or velocity	m s^{-1}
V_e	ventilation rate	m^{3}s^{-1}
W	work done by the surrounding to the system	W
w	z-component velocity	m s^{-1}
z	vertical height	m

Greek symbols

α	thermal air change parameter	m^{3}s^{-1}
α	thermal diffusivity	–
α	thermal expansion coefficient	K^{-1}
β	phase shift	rad (h)
β	thermal expansion coefficient	K^{-1}
ε	dissipation rate of turbulence kinetic energy	W kg^{-1}
ε	emissivity	–
γ	wind air change parameter	m^{3}s^{-1}
κ	Karman constant	–
λ	convective heat transfer number	–
λ	Taylor micro scale	m
μ	dynamic viscosity	Pa.s
μ_t	dynamic eddy viscosity	Pa.s
ν	kinematic viscosity	m^{2}s^{-1}
ν_t	kinematic eddy viscosity	m^{2}s^{-1}
θ	temperature	K
ρ	density	kg m^{-3}
σ	Stefan-Boltzmann constant	W m$^{-2} \cdot$ K^{-4}
τ	time constant	s h^{-1}
τ_{ij}	Reynolds-stress tensor	pa
u_τ	friction velocity	m s^{-1}
$\hat{\Phi}$	potential energy	J
Δp	pressure difference	Pa
Λ	Taylor integral length scale	m

Subscripts

1	opening 1
2	opening 2
b	bottom opening
e	exit
f	floor
i	indoor

m opening "m"
M thermal mass
o outdoor
r radiation
s stack pressure; supply
t top opening
w wind pressure

Superscripts

a air
r *r*-th moment
rms root mean square

Chapter 2

Ventilation and indoor environmental quality

Bjarne Olesen, Philo Bluyssen and
Claude-Alain Roulet

2.1 Introduction

The well-being of people depends largely on their health and comfort as well
as on the safety level of conditions under which they perform their main
activities of living, working and transportation. These activities take place
in an enclosed space in which people spend more than 90 per cent of their
time (Jenkins *et al.*, 1990), and in more than 40 per cent of the enclosed
spaces, people suffer from health-, comfort- and safety-related complaints
and illnesses (see Table 2.1; Dorgan Associates, 1993). Improving health,
comfort and safety of the population in such spaces has consequently great
potential for economic and societal benefits through increased productivity,
reduced sick leave and medical costs, and reduction of the number of casu-
alties in accidents, as well as through the prevention of liabilities (Dorgan
Associates, 1993; Bonnefoy *et al.*, 2004).

Health, comfort and safety issues are particularly pronounced in the area
of social housing, where deterioration of the existing building stock and
the need for renovation is a priority. Across the European Union countries, social housing numbers over fifty-five million dwellings; much of this
housing is of poor quality, creating problems of ill-health and insecurity
for the occupants and ongoing maintenance problems for the owners. At
the moment, more than 170 million people live in mass housing areas
constructed in the post-war period in Europe – 25 countries.

Early findings of a recent pan-European housing survey by WHO clearly
indicate a link between present-day housing conditions, including the immediate environment, and human health and well-being. These data confirm
that the indoor dwelling characteristics that most affect human health are
connected to thermal comfort, lighting, moisture, mould and noise.

Moisture and mould problems are estimated to occur in 15–30 per cent
of the European housing stock (Samson *et al.*, 1994). It is only in recent
years that a link between the effects of indoor fungi on allergic reactions
and respiratory infections has become widely known. Approximately 20
per cent of the European population are allergic to mites and fungi, and
the prevalence of asthma and allergies in domestic buildings is increasing

Table 2.1 Established relations between physical parameters and health and comfort problems (Bonnefoy *et al.* 2004)

Physical parameter	Comfort	Human health: determined links
Temperature	50% (highly) dissatisfied with thermal comfort	Respiratory diseases, cold and throat illness, multiple allergies
Light	25% people (highly) dissatisfied (daylight)	(Trends of) depression, chronic anxiety, household accidents
Noise	25% people annoyed (e.g. traffic, neighbours)	Hypertension, (trends of) depression, fatigue, accidents
Moisture and mould	25% of dwellings: mould growth in > 1 room	Respiratory diseases, asthma, allergies
Indoor air quality (general)	8% of dwellings: smells, dampness 10% dissatisfied	Fatigue, (trends of) depression, anxiety, respiratory diseases

(Jantunen *et al.*, 1998; Institute of Medicine, 2000). In Europe, asthma affects one of seven children, and the rates of asthma among children in Western Europe are ten-fold those in Eastern Europe.

Indoor environmental quality (IEQ) complaints are also related to sickness absence rates of office workers due to the sick building syndrome (SBS) and building-related illnesses (BRI). Losses in work productivity and performance have a direct financial impact on businesses (Fisk, 2000a).

In this chapter, the latest knowledge on human requirements and ventilation will be presented and discussed, mainly with respect to an acceptable indoor air quality (IAQ) and thermal comfort.

2.2 Indoor environmental quality

From the occupant point of view, the ideal situation is an indoor environment that satisfies all occupants (i.e. they have no complaints) and does not unnecessarily increase the risk or severity of illness or injury. Both the satisfaction of people (comfort) and health status are influenced by numerous factors: general well-being, mental drive, job satisfaction, technical competence, career achievements, home/work interface, relationship with others, personal circumstances, organizational matters, etc. and last but not least environmental factors, such as

- IAQ: comprising odour, indoor air pollution, fresh air supply, etc.;
- thermal comfort: moisture, air velocity, temperature;
- acoustical quality: noise from outside, indoors, vibrations;
- visual or lighting quality: view, illuminance, luminance ratios, reflection;
- aesthetic quality.

Although there is rich scientific literature and the reports of several national experiences on this subject, a uniform set of criteria for the countries of Europe has not yet been defined.

Currently, in several standards and guidelines, human indoor environmental requirements for spaces are expressed by physical and chemical indicators (temperature, Decibel, Lux, CO concentration, etc.) (CEN, 1998; ASHRAE, 2004a; ISO, 2005). Although required levels in those standards and guidelines are met, it can be concluded from several studies that the IEQ as experienced by occupants is not always acceptable and sometimes is even unhealthy, causing health and comfort problems (Bluyssen *et al.*, 1995). This mismatch is due to several reasons:

- the relationship between objective measurement and human assessment is not known for all physical/chemical parameters. No consensus model is available for air quality. For light, recent findings show that brightness of the surroundings is the key element and not only the illuminance (Light & Health Research Foundation, 2002).
- even if established models for separate subjective issues exist [e.g. thermal comfort (Fanger, 1972) and noise], the holistic effects of all separate physical/chemical factors are still largely unknown.

Besides the physical/chemical indicators, other indicators are being used such as the percentage of dissatisfied occupants, productivity numbers (Clements-Croome, 2002), sick leave, estimated life expectations (Carrothers *et al.*, 1999) and even the number of deaths related to a BRI. However, the determination and use of these indicators has not been documented in guidelines or standards.

In several spaces (cars, space industry and buildings), health complaints and comfort problems are strongly related to the available methods of ventilation. This relation has been shown, for example, by the increased risk of infectious disease transmission (recirculated air), sources in heating, ventilating and air-conditioning (HVAC) systems causing an overall distribution of unwanted pollutants (Bluyssen *et al.*, 2003), and stagnant zones and draught (insufficient ventilation effectiveness). Complaints are in general related to air quality, thermal comfort and noise parameters.

2.3   Indoor air quality

2.3.1   Human perception of air quality

Besides direct measurement of the contents of indoor air, which comprises thousands of compounds, human subjects are used as measuring instruments when exposed to odours and irritants. The basic biological principles of the perception mechanisms are fairly well understood whereas for the information processes at higher centres of the brain, this is less clear (ECA, 1999). How these basic processes relate to the more complex psychological responses to odorant/irritant stimulation, such as perceived air quality, annoyance and symptom reporting, is uncertain.

Many airborne substances are complex stimuli. They are combinations of many chemicals that can interact at one or several levels before or during perception: chemical or physical interaction in the gas mixture, interaction of molecules at the receptor surfaces (olfactory and trigeminal systems), peripheral interaction in the nervous system and, finally, interaction in the central nervous system. Therefore, effects such as masking, neutralization and counteraction are not surprising in gas mixtures (Berglund, 1976).

The attributes that can be measured in this way are the same as for all other sensory modalities:

- detection (the limit value for absolute detection)
- intensity (odour intensity, sensory irritation intensity)
- quality (value judgement such as hedonic tone or acceptability).

Detection

The classical threshold theory assumes the existence of a momentary absolute sensory threshold. However, in real life, there is no fixed odour or irritation threshold of absolute detection for a particular individual or a particular pollutant but rather a gradual transition from total absence to definitely confirmed sensory detection (Garriga-Trillo, 1985). Therefore, in the theory of signal detect ability (Engen, 1972), the same repeated signal is assumed to have a defined distribution, and thus each sensory evaluation by a subject is executed on a probability basis. Berglund and Lindvall (1979) have used this signal detection approach to test a few single compounds and a few building investigations.

In the classical methods, the threshold level is defined as the level at which 50 per cent of a given population will detect the odour. One of these methods is the threshold method, which is standardized in many countries for the evaluation of outdoor air (CEN, 1994). In this threshold method, an air sample is diluted stepwise (for each step, by a factor of 2) with clean (odour-free) air to determine the dilution at which 50 per cent of a panel of eight persons can no longer distinguish the diluted air from odour-free air. This number of dilutions, expressed in odour units per m^3 air of 20°C (o.u. m^{-3}), is the numerical value for the odour concentration of the original air sample.

Some measurements using the classical threshold level method have been made on indoor air, ventilation systems and building materials (Berglund and Lindvall, 1979; Bluyssen and Walpot, 1993).

The absolute detection threshold varies widely with chemical substances, as is shown by the large spread in literature-reported odour thresholds for single compounds (Devos *et al.*, 1990). This is caused, for example, by the procedure used by the purity of chemical substance, the equipment used and the sample of subjects.

Recognition threshold values (concentration at which a certain chemical is recognized) are usually measured in the same way as detection levels. Both use either the method of limits or the method of constant stimuli (ECA, 1999).

In the method of limits, the chemical substance is presented in alternating ascending and descending series, starting at different points to avoid having the subject fall into a routine. The subject is asked to report whether the sample can be detected or not.

The method of constant stimulus is based on the assumption that the momentary individual threshold value varies from time to time and that this variation has a normal distribution. The chemical substance is usually presented in a random selection of concentrations.

For both methods, no training is required, although subjects may be selected on their sensitivity to the chemical substances tested.

Intensity

The intensity of odours or irritants can be ascertained by several methods: equal-intensity matching, magnitude estimation and direct scaling methods (ECA, 1999). The latter is the most common in IAQ studies and uses, for example, visual, semantic scales (e.g. no odour, weak odour, moderate odour, strong odour, very strong odour, over-powering odour). With equal-intensity matching, the subject matches the intensity of, for example, two different odorants.

Magnitude estimation techniques generate magnitude estimates of intensity resulting from direct numerical estimations by subjects. The perceived intensity of an odour is established by rating the intensity of that odour on a magnitude scale, using or not using reference odours. The American Standards for Testing and Measurements (ASTM) technique (ASTM, 1981) uses, for example, samples of 1-butanol vapour presented at varying concentrations, and the Master Scale unit method (Berglund and Lindvall, 1979) uses five concentrations of pyridine, which are jointly measured with indoor air samples.

The assessment of decipol levels of the air in office buildings using trained panels [European Audit project (Bluyssen *et al.*, 1996)] is an example of magnitude estimation with memory references. The same method but instead having the references (with numerical values) nearby to compare is an example of magnitude estimation with several references. This method was applied in several other European projects as well [Database and

MATHIS (de Oliveira Fernandes and Clausen, 1997; de Oliveira Fernandes, 2001), AIRLESS (Bluyssen *et al.*, 2001)].

Quality

A value judgement of IAQ can be given in several ways. One can make a classification (e.g. yes/no), as used in ASHRAE 62-1989 (ASHRAE, 1996) (whether the air is acceptable or not), resulting in a percentage of dissatisfied, or one can use a list of descriptors to describe a chemical substance. The latter is used mainly in the food and perfume industry, from which many classification systems of odours have been developed.

For evaluation of the acceptability of an air sample (percentage of dissatisfied persons), several methods have been applied. Besides the yes/no classification ('acceptable' or 'not acceptable'), the continuous acceptability scale (Gunnarsen and Fanger, 1992) is used. The middle of the scale is indicated as the transition between just acceptable and just not acceptable. With both methods, however, large panels (up to 100 persons depending on the statistical relevance required) of untrained persons are required.

Two units, olf and decipol, were introduced to quantify sensory source emissions and perceived air quality (Fanger, 1988). This theory is based on the assumption that the pollutants in buildings all have the same relation between exposure and response after one factor normalization based on human bioeffluents. Emission rates are measured in olf, where one olf is defined as the emission rate causing the same level of dissatisfaction as bioeffluents from one seated person at any airflow. Concentration or 'perceived air quality' is measured in decipol. One decipol is defined as the concentration of pollution causing the same level of dissatisfaction as emissions from a standard person diluted by a clean airflow of 10 ls^{-1}. In this context, 'perceived air quality' means the dissatisfaction or acceptability with IAQ.

With these units, the so-called decipol method was developed. The decipol method comprises a panel of ten or more persons who are trained to evaluate the perceived air quality in decipol, or an untrained panel of at least fifty persons (Gunnarsen and Bluyssen, 1994). A method to train a panel to evaluate perceived air quality in decipol has been developed (Bluyssen, 1990 and 1991). It uses a reference gas, a scale, special equipment, selection and training procedures (Bluyssen, 1998). Research indicates, however, that this method does not evaluate the acceptability but the intensity of the air sample.

Besides the questionable use of the term 'perceived air quality', which can involve many dimensions other than dissatisfaction and acceptability (e.g. odour intensity, stuffiness, perceived dryness, degree of unpleasantness) (ECA, 1999), the main item that is discussed with this method is the assumption that all pollutants have the same relation between exposure and response, i.e. that the calculated olf values from separate sources can simply be added.

According to Stevens' law, perceived odour intensity of one single compound increases as a power function of concentration (Stevens, 1957):

$$R = C(S - S_0)^n$$

where R is the perceived odour intensity, S is the stimulus concentration and C, S_0 and n are constants.

The perceived intensity of a mixture of two compounds may in theory be as strong as the sum of the perceived intensities of the unmixed compounds (complete addition), more intense than the sum of its compounds (hyper-addition), or less intense than the sum of its compounds (hypo-addition). There are three kinds of hypo-addition:

- partial addition: the mixture is perceived as more intense than the stronger compound perceived alone;
- compromise addition: the mixture is perceived as more intense than one compound perceived alone, but less intense than the other; and
- compensation addition: the mixture is perceived as weaker than both the stronger and the weaker compounds.

According to Berglund (1976), stimulation is proportional to the number of molecules, as long as just one type of molecule is present. The odour interaction for mixtures of constituent odorants is governed by a strongly attenuating function, hypo-addition (Berglund *et al.*, 1976). The concentration of numerous compounds may be less important to the perceived air quality than the addition or subtraction of a few specific compounds to the gas mixture (Berglund and Lindvall, 1990).

Reviewing previous addition studies of perceived air quality (in decipol), it was observed that for the majority of the comparisons between predicted (by using the addition assumption) and measured pollution loads of combinations of sources, the predicted pollution loads are frequently higher than the measured pollution loads (Bluyssen and Cornelissen, 1999), thus implying hypo-addition as well. The same was found for the perceived air qualities.

Olfactory adaptation, which is similar to visual adaptation to light, makes it even more difficult to predict or model perceived odour intensity or perceived air quality. With continuous exposure, the perceived odour intensity will decrease with time and the odour threshold will increase with time. Recovery or re-adaptation will occur within less than a minute after removal from odour (VDI, 1986).

In conclusion, modeling of perceived quality or intensity of the indoor air is not possible yet, based on single compounds, although some predictions have been made with mixtures of several compounds. The indoor air comprises thousands of compounds, of which some are odorous, some are not, and others are irritants. Besides the combined odours and irritant effects of the thousands of compounds, the qualitative character is even more complex. The 'sensory' print is in general different from the 'chemical' print.

2.3.2 Instruments for measuring IAQ

The development of instruments, an artificial nose or an electronic nose, that can evaluate the air quality in the same way as the human nose does, is an ongoing activity. Many attempts have been made, some successful for the purpose for which they are designed, others not. The reason is not only related to the still incomplete knowledge of the perception mechanism (information processes in the brain), but also to the fact that the nose is able to detect very low concentrations.

In Table 2.2 for a number of compounds that are emitted by the human body, the lowest odour detection level that could be found in literature is presented (Bluyssen, 1990). From this table, it can be seen that the human nose is able to detect certain compounds at ppt level.

Furthermore, the results of Cain and Cometto-Muniz (1993) indicate that complex chemical environments may enable chemosensory and particularly irritative detection, when single volatile organic compounds (VOCs) lie far

Table 2.2 Odour detection levels for some compounds emitted by the human body (Bluyssen, 1990)

Compound	Molecular weight	Structure	Odour detection level	
			$\mu g\,m^{-3}$	ppb[a]
Acetaldehyde	44	CH_3CHO	0.2	0.111
Benzaldehyde	106	C_6H_5CHO	0.8	0.185
Butyric acid	88	$CH_3CH_2CH(OCCH_3)$ $CO_2C_2H_5$	1	0.278
Coumarin	146	$C_9H_6O_2$	0.007	0.0012
Dimethylsulphide	62	$(CH_3)_2S$	2.5	0.986
Dimethyldisulphide	94	CH_3SSCH_3	0.1	0.026
n-Decanal	156	$CH_3(CH_2)_8CHO$	0.25	0.039
Ethanethiol or ethylmercaptan	62	C_2H_5SH	0.1	0.039
Hydrogensulphide (inorganic)	34	H_2S	0.7	0.503
Methylmercaptan or methanethiol	48	CH_3SH	0.04	0.020
Phenylacetic acid	136	$(C_6H_5CH_2CO)_2O$	0.03	0.0054

[a] ppb $= 24.45 \times \mu g\,m^{-3}$ molecular weight.

below their individual thresholds. This means that in gas mixtures the nose may detect even far below single thresholds, i.e. below the ppt–ppb range.

It should be borne in mind that the use of human subjects to evaluate perceived air quality, the so-called sensory evaluation of air, is only one way of measuring the air quality. Compounds such as carbon monoxide cannot be smelled by a human being and can be nevertheless health threatening. Such compounds should therefore be measured in another way.

The methods and instruments available for measuring indoor air compounds can be divided into two groups (Bluyssen, 1996):

- those that require an extraction step before making a physical or chemical measurement (e.g. chromatography);
- those that make a direct physical measurement of a certain property of the sample (e.g. non-dispersive infrared spectrometry).

Chromatography

Chromatography is a separation technique, in which an inert gas or liquid (mobile phase) flows at a constant rate in one direction through the stationary phase, a solid with a large surface-to-volume ratio or a high-boiling liquid on a solid support. The sample may be a gas or a liquid, but it must be soluble in the mobile phase. Gas chromatography is used for separation of volatile, relatively non-polar materials or members of homologous series; liquid chromatography is used for separation in particular of those materials with low volatility and labile or unstable compounds; and thin layer and column chromatography for separation of inorganic or organic materials, and low-molecular-weight species up to high chain-length polymers.

Spectrometry

Spectrometry or photometric methods make use of discrete energy levels of molecules and the emission or absorption of radiation which usually accompanies changes by a molecule from one energy level to another. They are generally based on the measurement of transmittance or absorbance of a solution of an absorbing salt, compound or reaction product of the substance to be determined. They include absorption spectroscopy, emission spectroscopy, laser spectroscopy, photo-acoustic techniques and X-ray analysis. In photometry, it is necessary to decide upon the spectral levels to be used in the determination. In general, it is desirable to use a filter or monochromator setting such that the isolated spectral portion is in the region of the absorption maximum. A monochromator is a device or instrument that, with an appropriate energy source, may be used to provide a continuous calibrated series of electromagnetic energy bands of determinable wavelength or frequency range.

Mass spectrometry and flame ionization

Mass spectrometry and flame ionization can be placed under the category ionization methods. In mass spectrometry, a substance is analysed by forming ions and then sorting the ions by mass in electric or magnetic fields. Positive ions are produced in the ion source by electron bombardment or an electric discharge.

A flame ionization detector (FID) makes use of the principle that very few ions are present in the flame produced by burning pure hydrogen or hydrogen dilute with an inert gas. The introduction of mere traces of organic matter into such a flame produces a large amount of ionization. The response of the detector is roughly proportional to the carbon content of the solute. The response to most organic compounds on a molar basis increases with molecular weight.

Chemical sensors

Chemical sensors for gas molecules may in principle monitor physisorption, chemisorption, surface defect, grain boundary or bulk defect reactions (Gardner and Bartlett, 1992). Several chemical sensors are available: mass-sensitive sensors, conducting polymers and semiconductors. Mass-sensitive sensors include quartz resonators, piezoelectric sensors or surface acoustic wave (SAW) sensors (Elma et al., 1989; Nakamoto et al., 1990; Bruckman et al., 1994). The basis is a quartz resonator coated with a sensing membrane that works as a chemical sensor.

With conducting polymers, a wide range of aromatic and hetero-aromatic monomers undergo electrochemical oxidation to yield adherent films of conducting polymer under suitable conditions (Gardner et al., 1990). The conductivity of the polymer film is altered on exposure to different gases.

The principle with semiconductor sensors is based on the change of the electrical characteristics of the semiconductor when the gas to be measured is absorbed. The change of the number of free load carriers or the change of polarization of the bounded load carriers is then measured (Bruckman et al., 1994).

Several commercial instruments are available. Some comprise conducting polymers, others tin oxide gas sensors (thick or film devices) or metal oxide semiconductors, and combinations. However, none of them can yet evaluate IAQ as the nose does.

The indoor environment comprises thousands of chemical compounds in low concentrations, of which not all can be measured and interpreted by currently available equipment. The nose can detect very low concentrations (ptt range) and interpret all at the same time.

2.3.3 Sources of pollution

Categories

The possible sources of indoor air pollution can be categorized into:

- outdoor sources: traffic, industry.
- occupant-related activities and products: tobacco smoke, equipment (laser printers and other office equipment), consumer products (cleaning, hygienic, personal care products).
- building materials and furnishings: insulation, plywood, paint, furniture (particle board), floor/wall covering, etc.
- ventilation systems.

In the European Audit project ('European Audit Project to Optimise Indoor Quality and Energy Consumption in Office Buildings'), nine European countries were audited during the heating season of 1993–1994 (Bluyssen et al., 1996). In this audit, besides the normal measurements such as questionnaire, physical/chemical analysis of air, etc., panels of persons, trained to evaluate the perceived air quality, were used to measure the air quality in pre-selected spaces of office buildings as well as the outdoor and supply air. From this investigation, it was concluded that the main pollution sources were the materials, furnishings and activities in the offices and the ventilation system in the buildings.

The main groups of pollutants found in the indoor air are chemical and biological pollutants. Among the chemical group, one can distinguish gases and vapours (inorganic and organic) and particulate matter. And among the biological group micro-organisms belong: mould, fungi, pollens, mites, spores, allergens, bacteria, airborne infections, droplet nuclei, house dust and animal dung.

Ventilation systems

For ventilation systems, a separate study was performed named AIRLESS (Bluyssen et al., 2001). Experiments were performed to investigate why, when and how the components of HVAC systems pollute or are the reason for pollution. Different combinations of temperature, relative humidity (RH), airflow and pollution in passing air were investigated. Measurements of perceived air quality, particles, chemical compounds (such as VVOCs and aldehydes) and biological compounds were selected for each component. The most polluting components of HVAC systems were studied in the laboratory and in the field. The perceived air quality was in most cases measured with a trained sensory panel, according to the protocol developed for the AIRLESS project (Bluyssen, 1998).

The conclusion was that main sources and reasons for pollution in a ventilation system may vary considerably, depending on the type of construction, use and maintenance of the system. In normal comfort ventilation systems, the filters and the ducts seem to be the most common sources of pollution, especially odours. If humidifiers and rotating heat exchangers are used, they are also suspected of being significant pollution sources, especially if not constructed and maintained properly. The pollution load caused by the heating and cooling coils seems to be less notable. The effect of airflow on the pollution effect of HVAC system components seems to be less important.

Pollution sources in ventilation systems are presented in Chapter 6.

2.3.4 Recommendations for improving IAQ

From the AIRLESS project, a number of recommendations were extracted to achieve an optimum balance between IAQ and energy consumption (Bluyssen, 2004). A major part of these recommendations were related to IAQ.

Some general recommendations, not related to the ventilation system itself, were as follows:

1 Apply source control: A good way to reduce the airflow rate (and energy use) while increasing IAQ is to reduce the pollutant source strength indoors in combination with a reduction of the ventilation rate. Relevant actions are as follows:

   - avoid smoking indoors.
   - use building and furnishing materials that do not emit pollutants.
   - evacuate contaminants close to the sources with local exhaust (hoods) (e.g. printers and copy machines).

2 For efficient and controlled ventilation, the building envelope should be airtight, so that all the air passes through the (natural or mechanical) ventilation system: To reduce energy use whilst maintaining adequate ventilation, the airflow rate should be controlled and therefore pass through control devices. Computer simulations indicate that the tightness of the building envelope can have an effect on energy demand up to a factor 2.

A number of recommendations were formulated for the HVAC system and its components, as follows.

Commissioning and regular servicing of the air-handling system

- keep heat exchangers, ducts, etc. clean.
- change filters on time.

Recirculation

Do not use recirculation, supply only fresh air: pollution is also recirculated in this way. Tune system settings with operation strategies: make sure that the system is restarted early enough before air is required in order to purge the building from contaminants accumulated during off time before occupants arrive.

Heat recovery

- use heat recovery, but only if the envelope is airtight, there is parasitic (external) recirculation and no leakage in the ventilation unit (internal recirculation).
- rotating heat exchangers should not be installed where some recirculation of odours cannot be accepted. If you do use them, select a wheel equipped with a purging sector, and install it with the purging sector on the hot side of the wheel. Avoid using hygroscopic wheels.
- in office buildings audited within the HOPE project, the perceived air quality in winter in the fourteen buildings equipped with rotating heat exchangers was, on average, significantly worse than in the thirty-six buildings without heat recovery or with other airtight heat recovery systems.

Humidification

The basic strategy of not humidifying the air should be mentioned. In most European climates, humidification is not necessary. Furthermore, a lower RH generally results in a better-perceived air quality (lower odour intensity). However, an excessively low RH (< 30 per cent) can cause complaints such as dry air, eye irritation, dry/irritated throat, dry/irritated skin and nose. And excessive humidity favours mould growth. Therefore, a RH between 40 and 60 per cent is considered to be optimal.

Air filtering

Common filter techniques (bag filters) are such that filtering leads to an increasing film of dust collected on the filter. The fact that fresh outdoor air is transported through dirt accumulated on the filter is asking for problems. Therefore, filters should be changed or cleaned often enough to prevent the filter from being a source of pollution.

- change the filter, depending on the situation, traffic and other loads, once in 3–12 months, but in general every six months for highly polluted areas (urban) and once per year for low-polluting areas (countryside). Change the filters only when the system is turned off.
- inspect the filter regularly visually for penetration, deposit at the bottom of the system below the filter, leakage of the connections of the filter to the system, wetness and possible mould growth. Checking whether the filter emits any odours with ones own nose (in off condition) is advised one per month.

Air ducts

Start with clean, oil-free ducts, and then through efficient filtering avoid dirt that enters the system. Quality management on the construction site should be a principal focus. Oil residuals are the dominating sources in new ducts. Growth of micro-organisms, dust/debris accumulated in the ducts during construction at the worksite and organic dust accumulated during the operation period can be sources as well.

Heating and cooling coils

Heating and cooling coils without condensed or stagnating water in the pans are components that make a small contribution to the overall perceived air pollution. On the other hand, cooling coils with condensed water in the pans are microbiological reservoirs and amplification sites that may be major sources of odour in the inlet air.

- droplets catcher downstream of the cooling coil: Cooling coils may also release water droplets from condensation. These droplets wet the devices located downstream, such as filters or acoustic dampers, which then will be a good place for mould growth.
- increase/decrease the set point for cooling/heating as much as possible (with respect to comfort for occupant). The air quality is perceived as being better when the air is cooler.

2.3.5 Recommended ventilation rates

Earlier, most standards and guidelines for required ventilation rates were given as ventilation per person. Both laboratory and field studies have, however, shown that people and their activity (smoking, activity level), building and furnishings (floor covering, paint, furniture, cleaning, electronic equipment, etc.) and ventilation systems (filters, humidifiers, ducts, etc.) may also contribute. Even the outside air may be a source to indoor air.

Both people and buildings are taken into account in newer standards for the required ventilation rates in buildings. In all the standards more than one procedure is included. They all include a prescriptive method, where the minimum ventilation rates can be found in a table listing values for different types of space, as well as an analytical procedure for calculation of the required ventilation rate. By means of the analytical procedure, the ventilation rates can be calculated on the basis of type of pollutant, emission rates and acceptable concentration. All the proposed standards deal also with the health issue and not only the comfort issue.

Prescriptive procedure

For the prescriptive method, a minimum ventilation rate per person and a minimum ventilation rate per square metre floor area are required. The two ventilation rates are then added. The person-related ventilation rate should take care of pollution emitted from the person (odour), and the ventilation rate based on the person's activity and the floor area should cover emissions from the building, furnishings, HVAC system, etc.

The design outdoor airflow required in the breathing zone of the occupied space or spaces in a zone, i.e. the breathing zone outdoor airflow (V_{bz}), is determined in accordance with the equation:

$$V_{bz} = R_p P_z + R_a A_z \tag{2.1}$$

where

A_z = zone floor area: the net occupied floor area of the zone (m^2).

P_z = zone population: the greatest number of people expected to occupy the zone during typical usage. Note: If P_z cannot be accurately predicted during design, it may be an estimated value based on the zone floor area and the default Occupant Density listed in Table 2.3.

R_p = outdoor airflow rate required per person: these values are based on adapted occupants.

R_a = outdoor airflow rate required per unit area.

Table 2.3 summarizes the required ventilation rates from recent standards such as prEN15251 (CEN, 2005), ASHRAE 62.1 (ASHRAE, 2004a) and CR 1752 (CEN, 1998). There are, however, quite big differences between the European recommendations and those listed by ASHRAE. One major reason is that ASHRAE requirements are minimum code requirements, where the basis for design is adapted people, whereas the European recommendations are based on unadapted people.

Table 2.3 Smoking-free spaces in commercial buildings according to ASHRAE 62.1, CR 1752 and EN15251

Type of building/space	Occupancy person/m²	Category CEN	Minimum ventilation rate, i.e. for occupants only l s⁻¹ person		Additional ventilation for building (add only one) l s⁻¹ m⁻²			Total l s⁻¹ m⁻²	
			ASHRAE R_p	CEN	CEN low-polluting building	CEN not low-polluting building	ASHRAE R_a	CEN low polluting building	ASHRAE
Single office (cellular office)	0.1	A	2.5	10	1.0	2.0	0.3	2	0.55
		B		7	0.7	1.4		1.4	
		C		4	0.4	0.8		0.8	
Land-scaped office	0.07	A	2.5	10	1.0	2.0	0.3	1.7	0.48
		B		7	0.7	1.4		1.2	
		C		4	0.4	0.8		0.7	
Conference room	0.5	A	2.5	10	1.0	2.0	0.3	6	1.55
		B		7	0.7	1.4		4.2	
		C		4	0.4	0.8		2.4	
Auditorium	1.5	A	3.8	10	1.0	2.0	0.3	16	6
		B		7	0.7	1.4		11.2	
		C		4	0.4	0.8		6.4	
Cafeteria / Restaurant	0.7	A	3.8	10	1.0	2.0	0.9	8	1.17
		B		7	0.7	1.4		5.6	
		C		4	0.4	0.8		3.2	
Classroom	0.5	A	3.8	10	1.0	2.0	0.3	6	2.2
		B		7	0.7	1.4		4.2	
		C		4	0.4	0.8		2.4	
Kinder-garten	0.5	A	5.0	12	1.0	2.0	0.9	7	3.4
		B		8.4	0.7	1.4		4.9	
		C		4.8	0.4	0.8		2.8	
Department store	0.15	A	3.8	14.7	2.0	3.0	0.6	4.1	1.17
		B		10	1.4	2.1		2.9	
		C		6	0.8	1.2		1.7	

Analytical procedure

All of the listed standards have also an analytical procedure, either in the standard text or in an informative appendix. In this procedure, the required ventilation rate is calculated on a comfort basis (perceived odour and/or irritation) as well as on a health basis. The highest calculated value, which in most cases will be the comfort value, is then used as the required minimum ventilation rate. The basis for the calculation is in all standards based on a mass balance calculation.

The required ventilation rate is calculated as

$$Q = \frac{G}{(C_i - C_0) \cdot E_v} (\text{ls}^{-1})$$ (2.2)

where

> G = total emission rate (mgs^{-1})
> C_i = concentration limit (mgl^{-1})
> C_0 = concentration in outside air (mgl^{-1})
> E_v = ventilation effectiveness

In all the standards, however, knowledge concerning emission rates (G) and concentration limits (C_i), from a health point of view, is very limited. Within the next few years, knowledge will increase and data will be available from ongoing research projects and from testing by manufacturers of building materials and furnishings.

2.4   Thermal comfort

2.4.1   Heat balance of the human body

Existing methods for the evaluation of the general thermal state of the body, both in comfort and in heat- or cold-stress considerations, are based on an analysis of the heat balance for the human body:

$$S = M - W - C - R - E_{sk} - C_{res} - E_{res} - K(\text{Wm}^{-2})$$ (2.3)

where

> S = heat storage in body;
> M = metabolic heat production;
> W = external work;
> C = heat loss by convection;
> R = heat loss by radiation;
> E_{sk} = evaporative heat loss from skin;

C_{res} = convective heat loss from respiration;
E_{res} = evaporative heat loss from respiration;
K = heat loss by conduction.

The factors influencing this heat balance are as follows: activity level (metabolic rate, met or Wm^{-2}); thermal resistance of clothing I_{cl} (clo or m^2KW^{-1}); evaporative resistance of clothing R_e (m^2PaW); air temperature t_a (°C); mean radiant temperature t_r (°C); air speed v_{ar} (ms^{-1}); partial water vapour pressure p_a (Pa).

These parameters must be in balance so that the combined influence will result in a thermal storage equal to zero, or else the working time has to be limited to avoid too much strain on the body. A negative thermal storage indicates that the environment is too cool and vice versa. To provide comfort, the mean skin temperature also has to be within certain limits and the evaporative heat loss must be low. In existing standards, guidelines or handbooks, different methods are used to evaluate the general thermal state of the body in moderate environments, cold environments and hot environments; but all are based on the above heat balance and listed factors.

Besides the general thermal state of the body, a person may find the thermal environment unacceptable or intolerable if local influences on the body from asymmetric radiation, air velocities, vertical air temperature differences or contact with hot or cold surfaces (floors, machinery, tools, etc.) are experienced.

2.4.2 Requirements for thermal comfort

The main standards for comfortable thermal environment are ASHRAE 55-2004 (ASHRAE, 2004b) and ISO EN 7730-2005 (ISO, 2005). After the last revision, these two standards are very similar, using the same evaluation methods and recommended criteria. The evaluation is based on the methods and criteria for

- general thermal comfort [predicted mean vote (PMV)/predicted percentage of dissatisfied (PPD) index or operative temperature]
- local thermal discomfort (draft, radiant asymmetry, vertical air temperature differences, floor surface temperatures).

Thermal comfort is defined as 'that condition of mind which expresses satisfaction with the thermal environment'. Dissatisfaction may be caused by warm or cool discomfort of the body as a whole as expressed by the PMV and PPD indices or may be caused by an unwanted cooling (or heating) of one particular part of the body.

Because of individual differences, it is impossible to specify a thermal environment that will satisfy everybody. There will always be a percentage of dissatisfied occupants. But it is possible to specify environments predicted to be acceptable by a certain percentage of the occupants.

Because of local or national priorities, technical developments and climatic regions, in some cases a higher thermal quality (fewer dissatisfied) or lower quality (more dissatisfied) may be sufficient.

The following sections present the methods and give recommended criteria.

2.4.3   General thermal comfort in conditioned spaces

ISO EN 7730 standardizes the PMV–PPD index as the method for evaluation of moderate thermal environments. To quantify the degree of comfort, the PMV index gives a value on a 7-point thermal sensation scale: $+3$ hot, $+2$ warm, $+1$ slightly warm, 0 neutral, -1 slightly cool, -2 cool, -3 cold. An equation in the standard calculates the PMV index based on the six factors (clothing, activity, air and mean radiant temperature, air speed and humidity).

The PMV index can be determined when the activity (metabolic rate) and the clothing (thermal resistance) are estimated, and the following environmental parameters are measured: air temperature, mean radiant temperature, relative air velocity and partial water vapour pressure [see ISO EN 7726 (ISO, 1998)].

The PMV index is derived for steady-state conditions but can be applied with good approximation during minor fluctuations of one or more of the variables, provided that time-weighted averages of the variables during the previous 1 h period are applied. Because the PMV index assumes that all evaporation from the skin is transported through the clothing to the environment, this method is not applicable to hot environments. It can be used within a range of PMV index of -2 to $+2$, i.e. thermal environments where sweating is minimal.

Furthermore, it is recommended to use the PMV index when the six main parameters are within the following ranges:

$$M = 46\text{--}232\,\mathrm{Wm}^{-2}\,(0.8\text{--}4\,\mathrm{met});$$
$$I_{\mathrm{clo}} = 0\text{--}0.310\,\mathrm{m}^2\cdot{}^\circ\mathrm{CW}\,(0\text{--}2\mathrm{clo});$$
$$t_{\mathrm{a}} = 10\text{--}30^\circ\mathrm{C};$$
$$t_{\mathrm{r}} = 10\text{--}40^\circ\mathrm{C};$$
$$v_{\mathrm{ar}} = 0\text{--}1\,\mathrm{ms}^{-1};$$
$$p_{\mathrm{a}} = 0\text{--}2,700\,\mathrm{Pa}.$$

The metabolic rate can be estimated using ISO EN 8996 (ISO, 2004) and the thermal resistance of clothing can be estimated using ISO EN 9920 (ISO, 2006), taking into account the type of work and the time of year. For varying metabolic rates, it is recommended to estimate a time-weighted average during the previous 1 h period. For sedentary people, the insulation of a chair must also be taken into account.

The PMV index can be used to check whether a given thermal environment complies with the comfort criteria specified and to establish requirements for different levels of acceptability. Whereas some existing standards specify only one level of comfort (ASHRAE 55-04), others (ISO EN 7730, CR 1752, prEN15251) recommend three categories as summarized in Table 2.4. Each category prescribes a maximum percentage of dissatisfied for the body as a whole (PPD) and for each of the four types of local discomfort. Some requirements are hard to meet in practice whereas others are quite easily met. The different percentages express a balance struck between the aim of providing few dissatisfied and what is practically obtainable using existing technology.

The three categories in Table 2.4 apply to spaces where persons are exposed to the same thermal environment. It is an advantage if some kind of individual control of the thermal environment can be established for each person in a space. Individual control of the local air temperature, mean radiant temperature or air velocity may contribute to balance the rather large differences between individual requirements and therefore provide fewer dissatisfied.

By setting PMV = 0, an equation is established that predicts combinations of activity, clothing and environmental parameters which will provide a thermally neutral sensation. Figure 2.1 shows lines of PMV equal to 0 and shaded areas indicating the range corresponding to a certain maximum PPD value.

The PMV index predicts the mean value of the thermal votes of a large group of people exposed to the same environment. But individual votes are scattered around this mean value, and it is useful to predict the number of people likely to feel uncomfortably warm or cool. The PPD index establishes a quantitative prediction of the number of thermally dissatisfied people. The PPD predicts the percentage of a large group of people likely to feel too warm or cool, i.e. voting hot (+3), warm (+2), cool (−2) or cold (−3) on the 7-point thermal sensation scale.

When the PMV value has been determined, the PPD can be found from Figure 2.2 or determined from the equation:

$$PPD = 100 - 95 \, e^{\left(-0.03353 \cdot PMV^4 - 0.2179 \cdot PMV^2\right)} \tag{2.4}$$

2.4.4 Metabolic rate

All assessments of thermal environments require an estimate of metabolic heat production of the occupants. ISO EN 8996 presents three types

Table 2.4 Three categories of thermal comfort

Category	Thermal state of the body as a whole		Local discomfort				
	Predicted percentage of dissatisfied (PPD %)	Predicted mean vote (PMV)	Percentage of dissatisfied due to draught (DR %)	Percentage of dissatisfied due to vertical air temperature difference(%)	Percentage of dissatisfied due to warm or cool floor (%)	Percentage of dissatisfied due to radiant asymmetry (%)	
A	<6	−0.2 < PMV < +0.2	<15	<3	<10	<5	
B	<10	−0.5 < PMV < +0.5	<20	<5	<10	<5	
C	<15	−0.7 < PMV < +0.7	<25	<10	<15	<10	

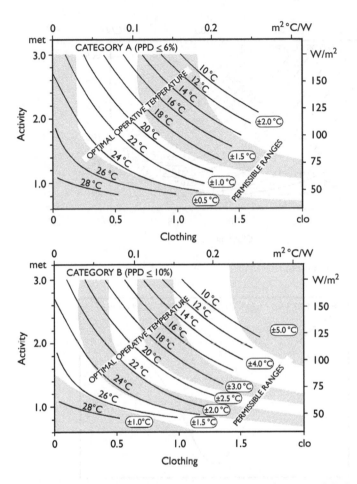

Figure 2.1 The optimum operative temperature as a function of clothing and activity for the three categories of the thermal environment.

The three diagrams show also the range around the optimum temperature for the three categories. The air velocity in the space is assumed $< 0.1\,\mathrm{ms^{-1}}$. The relative air velocity, v_{ar}, caused by body movement is estimated to be zero for a metabolic rate, M, less than 1 met and $v_{ar} = 0.3\,(M - 1)$ for $M > 1$ met. The diagrams are determined for a relative humidity of 50 per cent, but the humidity has only a slight influence on the optimum and permissible temperature ranges.

of methods. The first is by use of tables, where estimates are provided based on a description of the activity. These range from a general description (e.g. light, heavy, etc.) to methods of summating components of tasks (e.g. basal metabolic rate + posture component + movement component, etc.). An example of activity levels is given in Table 2.5.

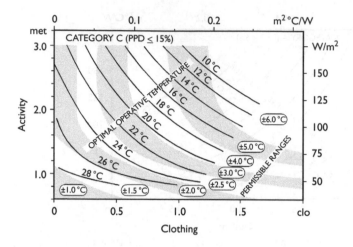

Figure 2.1 (Continued)

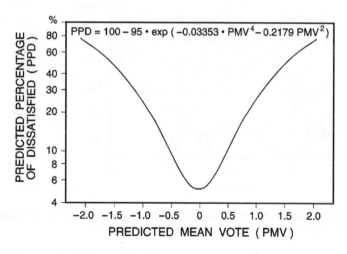

Figure 2.2 Predicted percentage of dissatisfied (PPD) as a function of predicted mean vote (PMV).

2.4.5 Clothing

ISO EN 9920 provides a large database of thermal insulation values that have been measured on a standing thermal manikin. One set of tables gives the insulation values for a large number of ensembles (Table 2.6).

Another set of tables in ISO EN 9920 gives insulation values for individual garments (Table 2.7), based on which the insulation for a whole ensemble can be estimated.

Table 2.5 Metabolic rates

Activity	Metabolic rates	
	$(W\,m^{-2})$	met
Reclining	46	0.8
Seated, relaxed	58	1.0
Sedentary activity (office, dwelling, school, laboratory)	70	1.2
Standing light activity (shopping, laboratory, light industry)	93	1.6
Standing, medium activity (shop assistant, domestic work, machine work)	116	2.0
Walking on the level:		
$2\,km\,h^{-1}$	110	1.9
$3\,km\,h^{-1}$	140	2.4
$4\,km\,h^{-1}$	165	2.8
$5\,km\,h^{-1}$	200	3.4

The insulation of an ensemble, I_{cl}, may be estimated as the sum of the individual garment insulation values:

$$I_{cl} = \sum I_{clu}$$

The data on evaporative resistance are not so extensive. A few data are given in the standard, and a method to calculate the evaporative resistance based on the thermal insulation is also given. For the insulation of chairs, typically 0.1–0.4 clo should be added.

2.4.6 Operative temperature range

For a given conditioned space, there exists an optimum operative tempera-ture corresponding to PMV $= 0$, depending on the activity and the clothing of the occupants. Figure 2.1 shows the optimum operative temperature and the permissible temperature range as a function of clothing and activity for each of the three PPD categories. The optimum operative temperature is the same for the three categories, whereas the permissible range around the optimum operative temperature varies.

The operative temperature at all locations within the occupied zone of a space should at all times be within the permissible range. This means that the permissible range should cover both spatial and temporary variations, including fluctuations caused by the control system.

Figure 2.1 applies for an RH of 50 per cent; however, in moderate environments, the air humidity has only a modest impact on the thermal

Table 2.6 Thermal insulation for clothing ensembles

Work clothing	I_{cl}		Daily wear clothing	I_{cl}	
	clo	$m^2 \cdot K\,W^{-1}$		clo	$m^2 \cdot K\,W^{-1}$
Underpants, boiler suit, socks, shoes	0.70	0.110	Panties, T-shirt, shorts, light socks, sandals	0.30	0.050
Underpants, shirt, boiler suit, socks, shoes	0.80	0.125	Underpants, shirt with short sleeves, light trousers, light socks, shoes	0.50	0.080
Underpants, shirt, trousers, smock, socks, shoes	0.90	0.140	Panties, petticoat, stockings, dress, shoes	0.70	0.105
Underwear with short sleeves and legs, shirt, trousers, jacket, socks, shoes	1,00	0.155	Underwear, shirt, trousers, socks, shoes	0.70	0.110
Underwear with long legs and sleeves, Thermojacket, socks, shoes	1.20	1.85	Panties, shirt, trousers, jacket, socks, shoes	1.00	0.155
Underwear with short sleeves and legs, shirt, trousers, jacket, heavy quilted outer jacket and overalls, socks, shoes, cap, gloves	1.40	0.220	Panties, stockings, blouse, long shirt, jacket, shoes	1.10	0.170
Underwear with short sleeves and legs, shirt, trousers, jacket, heavy quilted outer jacket and overalls, socks, shoes	2.00	0.310	Underwear with long sleeves and legs, shirt, trousers, V-neck sweater, jacket, socks, shoes	1.30	0.200
Underwear with long sleeves and legs, thermojacket and trousers, Parka with heavy quitting, overalls with heave quilting, socks, shoes, cap, gloves	2.55	0.395	Underwear with short sleeves and legs, shirt, trousers, vest, jacket, coat, socks, shoes	1.50	0.230

Table 2.7 Thermal insulation for individual garments

Garment description	Thermal insulation clo
Underwear	
Panties	0.03
Underpants with long legs	0.10
T-shirt	0.09
Shirts-Blouses	
Short sleeves	0.15
Normal, long sleeves	0.25
Trousers	
Shorts	0.06
Normal	0.25
Dresses-Skirts	
Light skirts (summer)	0.15
Heavy skirt (winter)	0.25
Winter dress, long sleeves	0.40
Sweaters	
Thin sweater	0.20
Thick sweater	0.35
Jackets	
Light, summer jacket	0.25
Jacket	0.35
High-insulative, fibre-pelt	
Boiler suit	0.90
Trousers	0.35
Jacket	0.40
Outdoor clothing	
Coat	0.60
Parka	0.70
Sundries	
Socks	0.02
Thick, long socks	0.10
Shoes (thick soled)	0.04
Boots	0.10

sensation. Typically a 10 per cent higher RH is felt equally as warm as a 0.3°C higher operative temperature. The number of dissatisfied persons in Table 2.4 is not additive. Some of the same people experiencing general thermal comfort (PMV–PPD) may be the same as the people experiencing local thermal discomfort. In practice, a higher or lower number of dissatisfied persons may be found, when using subjective questionnaires in field investigations [ISO 10551 (ISO, 2001)].

Table 2.8 gives examples of recommended operative temperatures in some typical spaces.

Table 2.8 Example criteria for operative temperature and mean air velocity for typical spaces

Type of build-ing/space	Clothing Cooling season (summer) clo	Clothing Heating season (winter) clo	Activity met	Category	Operative temperature Cooling season (summer)°C	Operative temperature Heating season (winter)°C	Mean air velocity Cooling season (summer) m s⁻¹	Mean air velocity Heating season (winter) m s⁻¹
Office	0.5	1.0	1.2	A	24.5 ± 0.5	22.0 ± 1.0	0.18	0.15
				B	24.5 ± 1.5	22.0 ± 2.0	0.22	0.18
				C	24.5 ± 2.5	22.0 ± 3.0	0.25	0.21
Cafeteria / restaurant	0.5	1.0	1.4	A	23.5 ± 1.0	20.0 ± 1.0	0.16	0.13
				B	23.5 ± 2.0	20.0 ± 2.5	0.20	0.16
				C	23.5 ± 2.5	20.0 ± 3.5	0.24	0.19
Department store	0.5	1.0	1.6	A	23.0 ± 1.0	19.0 ± 1.5	0.16	0.13
				B	23.0 ± 2.0	19.0 ± 3.0	0.20	0.15
				C	23.0 ± 3.0	19.0 ± 4.0	0.23	0.18

Relative humidity is assumed to be 60% for 'summer' (cooling season) and 40 % for 'winter' (heating season).

2.4.7 Local thermal comfort

Besides the general thermal state of the body, a person may find the thermal environment unacceptable if local influences on the body from asymmetric temperature radiation, draught, vertical air temperature differences or contact with hot or cold surfaces (floors, machinery, tools, etc.) are experienced. The data for local thermal discomfort are mainly based on studies with people under low activity levels (1.2 met). For higher activities, it can be expected that people are less sensitive to local thermal discomfort. The given relations between dissatisfied and local discomfort parameters are from EN ISO 7730-2005, ASHRAE 55-2004, and CR 1752 and EN15251.

Draught – local air velocities

One of the most critical factors is draught. Many people at low activity level (seated–standing) are very sensitive to air velocities, and therefore, draught is a very common cause for occupant complaints in ventilated and air-conditioned spaces. Fluctuations of the air velocity have a significant influence on a person's sensation of draught. The fluctuations may be expressed either by the standard deviation of the air velocity or by the turbulence intensity T_u, which is equal to standard deviation (σ_{v_a}) divided by the mean air velocity, v_a. The percentage of people feeling draught (draught rating, DR) may be estimated from the equation:

$$DR = (34 - t_a)(v_a - 0.05)^{0.62}(3.14 + 0.37 \cdot \sigma_{v_a}) \qquad (2.5)$$

where

v_a = mean air velocity (3 min) (ms^{-1})
σ_{v_a} = standard deviation of air velocity (3 min) (ms^{-1})
t_a = air temperature (°C)

For $v_a < 0.05$ ms^{-1}, insert $v_a = 0.05$ ms^{-1}.
For DR > 100 per cent, use DR = 100 per cent.

The model applies to people at light, mainly sedentary activity with a thermal sensation for the whole body close to neutral. The sensation of draught is lower at activities higher than sedentary and for people feeling warmer than neutral.

For people at higher activity levels and/or at ambient temperatures above the comfort range, an increased air velocity may improve the general thermal comfort. This influence is taken into account by using the PMV equation. Also high local velocities (spot cooling) may decrease discomfort from high activity and/or high ambient temperatures.

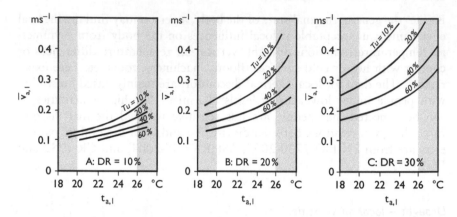

Figure 2.3 Mean air velocity as a function of local air temperature and turbulence intensity for the three categories of the thermal environment.

The turbulence intensity may vary between 30 and 60 per cent in spaces with mixing flow air distribution. In spaces with displacement ventilation or without mechanical ventilation, the turbulence intensity may be lower.

Based on the above equations, the diagrams in Figure 2.3 were made.

Increased air velocity

Standards 55-2004 and ISO7730-2005 include a diagram to estimate the air speed required to offset an increase in temperature (Figure 2.4). This shows that the requirement for personal control to increase the air speed is essential for accepting the higher speed. Therefore, it may not be appropriate to offset a temperature increase by increasing the air speed within a centrally controlled air system.

Vertical air temperature difference

A high vertical air temperature difference between head and ankles may cause discomfort. The recommended criteria in Table 2.9 for the vertical air temperature difference are valid only for people with low activity level like sedentary and standing. For higher activities, as is often found in industry, people are less sensitive to this temperature difference.

Warm or cool floors

If the floor is too warm or too cool, the occupants may feel uncomfortable because of warm or cool feet. For people wearing light indoor shoes, the recommended floor temperatures are summarized in Table 2.10a. It is

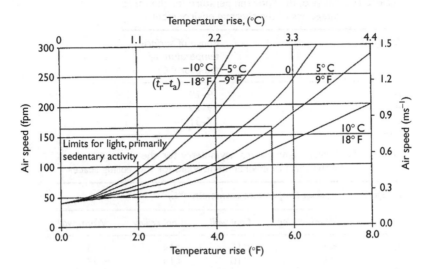

Figure 2.4 Air speed required offsetting increased temperature.

The air speed increases in the amount necessary to maintain the same total heat transfer from the skin. This figure applies to increases in temperature above those allowed in the summer comfort zone with both t_r and t_a increasing equally.

Table 2.9 Vertical air temperature difference between head and ankles (1.1 and 0.1 m above the floor) for the three categories of the thermal environment

Category	Vertical air temperature difference (°C)
A	<2
B	<3
C	<4

the temperature of the floor rather than the material of the floor covering which is important for comfort.

Radiant temperature asymmetry

Radiant asymmetry may also cause discomfort. People are most sensitive to radiant asymmetry caused by warm ceilings or cool walls (windows). Recommended criteria are summarized in Table 2.10b. These data apply to sedentary people and low ceiling height. A study with a high ceiling-mounted (9 m) gas-fired infrared heater (Langkilde *et al.*, 1985) showed a higher acceptable temperature asymmetry. For seated and standing people, a temperature asymmetry of 10–14 K resulted in less than 5 per cent dissatisfied.

Table 2.10a Range of the floor temperature for the three categories of the thermal environment

Category	Range of surface temperature of the floor (°C)
A	19–29
B	19–29
C	17–31

Table 2.10b Radiant temperature asymmetry for the three categories of the thermal environment. Valid for low ceiling spaces

Category	Radiant temperature asymmetry (°C)			
	Warm ceiling	Cool wall	Cool ceiling	Warm wall
A	<5	<10	<14	<23
B	<5	<10	<14	<23
C	<7	<13	<18	<35

2.4.8 Adaptation

The above-mentioned requirements are based largely on laboratory studies with test subjects mainly from Europe and North America. But studies with Asian and African subjects (Fanger, 1973; Tanabe *et al.*, 1987; de Dear *et al.*, 1991) under laboratory test conditions have found similar results for general thermal comfort. Several extensive field studies from 160 buildings all over the world summarized by de Dear and Brager (1998) show that in buildings with HVAC systems, the PMV model works well (Figure 2.5a). The studies show also that in naturally ventilated buildings (free running, no mechanical cooling), people seem to adapt (behaviourally, psychologically) and can accept higher indoor temperatures than those predicted by the PMV model (Figure 2.5b). Whether people will still have the same level of performance at the higher temperatures is a further consideration.

Based on this study, ASHRAE55-2004 includes a figure similar to Figure 2.6, where lines for 90 and 80 per cent comfort ranges are included. It is important to emphasize that the climatic data used are those of the monthly average outside air temperature. For some cities, the range for the maximum monthly average outside temperature is shown. For most European cities, the upper limit is 26–27°C, which is similar to the recommended limits based on the PMV–PPD index. For warmer environments, however, it will be acceptable, according to this model, to have indoor temperatures that are a few degrees higher.

McCartney and Nicol (2002) arrived at similar conclusions from experiments in European buildings. From these experiments, adaptive comfort

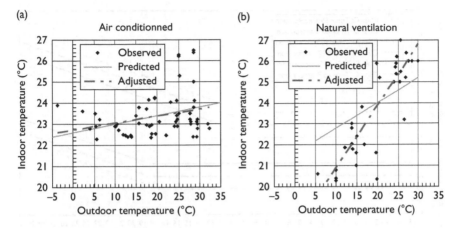

Figure 2.5 (a) Observed and predicted indoor comfort temperatures from ASHRAE RP-884 database for air-conditioned buildings. (b) Observed and predicted indoor comfort temperatures from RP-884 database for naturally ventilated buildings. Reprinted from de Dear and Brager (2002), with permission from Elsevier.

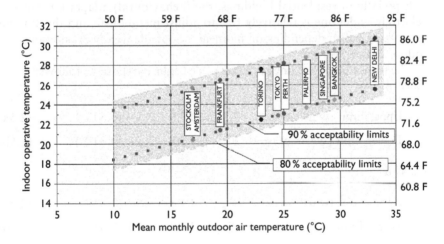

Figure 2.6 Acceptable operative temperature ranges for naturally conditioned spaces according to ASHRAE 55-2004. Range shown for different climatic areas.

models were developed, as shown in Figure 2.7. This concept has been adapted in CEN EN15251 (2005). A running weekly mean outdoor temperature is used here instead of a monthly average. The diagrams are valid for office buildings, where occupants have an activity in the range of 1.2– 1.6 met. Clothing can be varied from 0.5 to 1.0 clo. There must be access to operable windows.

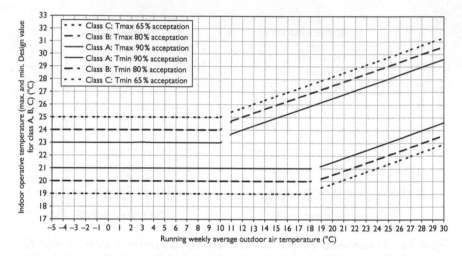

Figure 2.7 Design values for the indoor operative temperature for buildings without mechanical cooling systems.

Especially in residential buildings, the (behavioural) adaptation range is relatively wide: one is relatively free to adjust metabolism and the amount of clothing worn, depending on momentary outside weather conditions and indoor temperatures.

The formulae of the lines in Figure 2.7 (right part) are as follows:

Class A

$$\text{Upper limit (warm season)}: T_{i\,max} = 17.8 + 2.5 + (0.31\,T_o) \tag{2.5a}$$

$$\text{Lower limit (warm season)}: T_{i\,min} = 17.8 - 2.5 + (0.31\,T_o) \tag{2.5b}$$

Class B

$$\text{Upper limit (warm season)}: T_{i\,max} = 17.8 + 3.5 + (0.31\,T_o) \tag{2.5c}$$

$$\text{Lower limit (warm season)}: T_{i\,min} = 17.8 - 3.5 + (0.31\,T_o) \tag{2.5d}$$

Class C

$$\text{Upper limit (warm season)}: T_{i\,max} = 17.8 + 4.2 + (0.31\,T_o) \tag{2.5e}$$

$$\text{Lower limit (warm season)}: T_{i\,min} = 17.8 - 4.2 + (0.31\,T_o) \tag{2.5f}$$

where T_i = acceptable indoor temperature (°C)
T_o = mean monthly outdoor temperature (°C)

Heating (cold) season: Below an outdoor temperature of 10°C, the *upper* limits are the same as for mechanically cooled buildings.

2.5 Indoor environment and performance

The effects of IAQ on productivity became an issue only in the last decade, as a result of extensive research and an understanding of the strong connections between factors such as ventilation, air-conditioning, indoor pollutants and adverse effects on health and comfort. The complexity of a real environment makes it very difficult to evaluate the impact of a single parameter on human performance, mostly because many of them are present at the same time and, as a consequence, act together on each individual. In addition, worker motivation affects the relationship between performance and environmental conditions (e.g. highly motivated workers are less likely to have reduced performance in an unfavourable environment; however they may become more tired and that may also affect performance).

One way of evaluating the performance is the use of self-reported performance. This was used to study the self-evaluation of the influence of the environment, job satisfaction and job stress on performance (Roelofsen, 2002). The study was performed among 170 people in 6 offices. The self-reported performance was made on a 9-point scale. Based on the data, the following equation for self-reported performance (WEP) could be established:

$$WEP = 6.739 - 0.419E - 0.164JD - 0.048JS \qquad (2.6)$$

where

E = dissatisfaction with environment

JD = job satisfaction

JS = job stress

It is clear that the indoor environment was evaluated as having the greatest influence on performance: much higher than job satisfaction and job stress.

A common approach to evaluate the influence of climatic factors on human performance could be to measure the extent to which SBS symptoms occur, as these are known to cause distraction from work or even short-term absenteeism. However, this link is not well established yet and must be better understood and recognized. A possible mechanism may be described as follows: (a) inadequate ventilation or superfluous emissions from different sources increase the concentration of pollutants, which negatively affect perceived air quality; (b) reduced air quality negatively affects the central nervous system, increasing SBS symptoms such as headache, difficulty in concentration, tiredness; (c) these symptoms will cause distraction from work and decreased work ability, i.e. productivity loss. Nevertheless, indoor pollution may also exacerbate the sensation of dryness and irritation of the eyes. As a consequence, a higher blinking rate and watery eyes will negatively affect visual skills and decrease the performance of visually demanding work.

There is limited information in the literature showing a direct relationship between SBS symptoms and worker productivity. Analysing the data of a British Office Environment Survey (Raw et al., 1990), Raw found that people reporting more than two symptoms on the SBS list are likely to have reduced performance ratings, and a linear relationship exists between SBS and self-estimated productivity.

Based on their data, Fisk and Rosenfeld (1997) estimated an average decrement in self-reported productivity of 2 per cent. Raw and his colleagues emphasized that the responses evaluated on a 9-grade subjective scale reflect the responder's belief, regardless of whether that belief is correct, and the actual productivity was not assessed. Mucous and work-related symptoms were also found to affect self-reported productivity (Hall et al., 1991), but no further validation on the accuracy of self-reports related to actual productivity loss was made by other field investigations. Measured data in a field experiment (Nunes et al., 1993) indicate a relationship between SBS symptoms and worker performance. As part of an SBS study of 3 weeks, in which the outdoor air supply was experimentally varied, forty-seven employees undertook two computerized neurobehavioral tests at their workplace. The workers presenting with more SBS symptoms were found to respond 7 per cent longer in a continuous performance task and to have a 30 per cent higher error rate in a symbol-digit substitution test. As correlations were found also with temperature but not for the measured pollutants, it is more likely that the effects observed were due not only to air quality factors.

There is substantial evidence that poorly perceived IAQ is likely to have a negative effect on work performance. This effect was demonstrated first by Wargocki et al. (1999) when they exposed impartial female subjects in a realistic office environment to the emissions from a carpet. The study showed that by improving perceived air quality, the SBS symptoms were reduced and the performance of typical office tasks increased. These findings were later confirmed by several other independent investigations conducted in Denmark using different ventilation rates (Lagercrantz et al., 2000; Wargocki et al., 2000a; Bakó-Biró et al., 2004) using various types of pollution sources and different subjects. Based on these results, an overall relation between ventilation rate per person and performance was established (Figure 2.8). The quantitative relationships were developed based on these results and show that for every 10 per cent increment in the percentage of dissatisfied in the range 15–68 per cent, a c. 1 per cent decrement in performance of text-typing can be expected (Wargocki et al., 2000b).

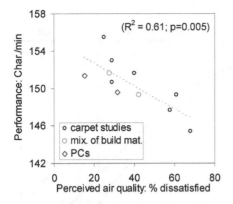

Figure 2.8 Performance of text-typing as a function of perceived air quality expressed in per cent dissatisfied, based on the results of laboratory studies, using typical indoor pollution sources such as carpets, linoleum, books and papers on wooden bookshelves, sealant and personal computers (Wargocki et al., 2000b; Bakó-Biró et al., 2004).

It is natural to ask whether such an improvement in the air quality level to obtain only a few per cent increment on the productivity side will justify any investment to improve the IAQ, especially when there are no obvious complaints, and knowing that thermal conditions even within the thermal comfort zone according to Wyon (1996) may reduce performance by 5–15 per cent. Seppänen and Fisk (2005) compiled the results from studies relating the indoor thermal temperature to performance and the results are shown in Figure 2.9.

Details concerning clothing and activity were not listed for all studies included in Figure 2.9, so the temperatures cannot easily be related to the corresponding comfort zone. The authors conclude that the nature of this association is that productivity improves as thermal conditions approach a predicted thermal comfort zone. Assuming sedentary work and normal winter clothing, 1.0 clo, the corresponding PMV index values have been added to the figure from Seppanen and Fisk.

The salaries of workers in typical office buildings exceed the building energy and maintenance cost by a factor of 100 approximately. The same applies for the salaries and annual construction or rental costs (Skåret, 1992; Fisk, 2000b). Thus, even a 1 per cent increase in productivity should be sufficient to cover any expenses related to doubling of energy or maintenance costs or other large investments involving construction costs or rent.

In view of the fact that a good IAQ also reduces the prevalence of SBS symptoms, Fisk and Rosenfeld (1997) estimated that considerable gains and savings may result in health care costs, involving billions of dollars

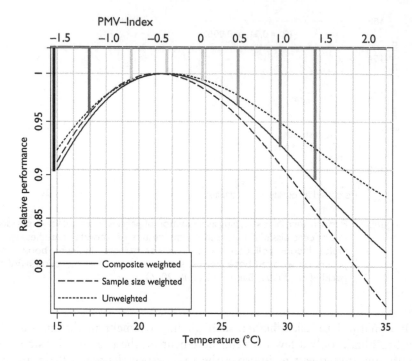

Figure 2.9 Relation between indoor room temperature and performance from several published studies based on Seppänen and Fisk (2005).

nationwide in the US (Fisk and Rosenfeld, 1997; Fisk, 2000b). In another study, Milton *et al.* (2000) investigated the sick leaves of 3,270 employees in 40 buildings. For the employees in offices, the risk of short sick leave was a factor 1.53 higher at a ventilation rate of $12\,ls^{-1}$ per person compared with a ventilation rate of $24\,ls^{-1}$ per person.

The review by Wyon (1996) on the published literature showed that the payback time for general upgrading of currently unhealthy office buildings (representing 40 per cent of the building stock) would be as low as 1.6 years if only a 0.5 per cent increase in the overall productivity is achieved. Moreover, the cost–benefit simulation made by Djukanovic *et al.* (2002) showed that the annual increase in productivity was worth at least ten times as much as the increase in annual energy and maintenance costs, when improving the perceived air quality in office buildings, specifying a payback time of no more than 4 months due to the productivity gains achieved.

The results of recent studies (Wargocki *et al.*, 2003 and 2004; Tham, 2004) show that improving IAQ in real buildings has in fact greater effect on the actual performance of office work in the field (up to 9 per cent) than would be predicted from the field laboratory experiments mentioned above. Based on the current knowledge regarding IAQ and performance of human work, it seems that it is worth investing fundamental resources to improve the quality of indoor air, next to other environmental factors in real buildings, that will definitely lead to an improved work performance among the occupants, an improvement that is not necessarily measured in terms of characters typed or number of units added.

References

ASHRAE (1996) *Ventilation for Acceptable Indoor Air Quality*, Appendix C: Air quality guidelines – informative, Atlanta, GA, American Society of Heating, Refrigerating and Air-Conditioning Engineers (ASHRAE Standard 62-1989R, proposed).

ASHRAE (2004a) *Ventilation for Acceptable Indoor Air Quality*, Atlanta, GA, American Society of Heating, Refrigerating and Air-Conditioning Engineers (ASHRAE Standard 62.1-2004).

ASHRAE (2004b) *Thermal Environment Conditions for Human Occupancy*, Atlanta, GA, American Society of Heating, Refrigerating and Air-Conditioning Engineers (ASHRAE Standard 55-2004).

ASTM (1981) *Standard Practices for Referencing Supra Threshold Odour Intensity*, Annual book for ASTM standards E544-75, ASTM, USA (re-approved 1981), pp. 32–44.

Bakó-Biró, Z., Wargocki, P., Weschler, C. J. and Fanger, P. O. (2004) 'Effects of pollution from personal computers on perceived air quality, SBS symptoms and productivity in offices', *Indoor Air*, 14(3): 178–187.

Berglund, B. (1976) 'Psychological processing of odour mixtures', *Psychological Review*, 83(6): 432–441.

Berglund, B. and Lindvall, T. (1979) 'Olfactory evaluation of indoor air quality'. In: *Proceedings of Indoor Climate '78*, Copenhagen, Danish Building Research Institute, pp. 141–157.

Berglund, B. and Lindvall, T. (1990) 'Sensory criteria for healthy buildings'. In: *Proceedings of Indoor Air '90*, Toronto, Canada, Vol. 5, pp. 65–78.

Berglund, B., Berglund, U. and Lindvall, T. (1976) 'Psychological processing of odour mixtures', *Psychological Review*, 83: 432–441.

Bluyssen, P. M. (1990) *Air Quality Evaluated by a Trained Panel*, Ph.D. study, Laboratory of Heating and Air Conditioning, Technical University of Denmark.

Bluyssen, P. M. (1991) 'Air quality evaluated with the human nose', *Air Infiltration Review*, 12(4): 5–9.

Bluyssen, P. M. (1996) *Methods and Sensors to Detect Indoor Air Pollutants Perceived by the Nose*, TNO-report 96-BBI-R0873, TNO, Delft, The Netherlands.

Bluyssen, P. M. (1998) *Protocol for Sensory Evaluation of Perceived Air Pollution with Trained Panels*, Internal project document 1.10, October, Delft, The Netherlands.

Bluyssen, P. M. (2004) *A Clean and Energy-Efficient Heating, Ventilating and Air-Conditioning System: Recommendations and Advice*, February 2004, ISBN 90-5986-009-8, TNO Building and Construction Research, Delft, The Netherlands.

Bluyssen, P. M. and Cornelissen, H. J. M. (1999) 'Addition of sensory pollution loads – simple or not, that is the question'. In: *Design, Construction and Operation of Healthy Buildings*, ASHRAE, Atlanta, USA, pp. 161–168.

Bluyssen, P. M. and Walpot, J. (1993) 'Sensory evaluation of perceived air quality: a comparison of the threshold and the decipol method'. In: *Proceedings of Indoor Air '93*, Finland, Vol. 1, p. 65.

Bluyssen, P. M., de Oliveira Fernandes, E., Fanger, P. O., Clausen, G., Roulet, C. A., Bernhard, C. A. and Valbjörn, O. (1995) *European Audit Project to Optimise Indoor Air Quality and Energy Consumption in Office Buildings*, Final report, Contract JOU2-CT92-002, March, Delft, The Netherlands.

Bluyssen, P. M., de Oliveira Fernandes, E., Groes, L., Clausen, G., Fanger, P. O., Valbjrn, O., Bernhard, C. A. and Roulet, C. A. (1996) 'European Audit project to optimize indoor air quality and energy consumption in office buildings', *Indoor Air*, 6(4): 221.

Bluyssen, P. M., Seppänen, O., de Oliveira Fernandes, E., Clausen, G., Büller, B., Molina, J. L. and Roulet, C. A. (2001) 'AIRLESS: A European project to optimise air quality and energy consumption of HVAC-systems'. In: *Proceedings of CLIMA 2000*, September, Naples, Italy.

Bluyssen, P. M., Cox, C., Seppänen, O., de Oliveira Fernandes, E., Clausen, G., Müller, B. and Roulet, C. A. (2003) 'Why, when and how do HVAC-systems pollute the indoor environment and what to do about it' (Ref No. 01/708), *Building and Environment*, 38(2): 209–225.

Bonnefoy, X. R., Annesi-Maesano, I., Aznar, L. M., Braubachi, M., Croxford, B., Davidson, M., Ezratty, V., Fredouille, J., Gonzalez-Gross, M., van Kamp, I., Maschke, C., Mesbah, M., Moissonnier, B., Monolbaev, K., Moore, R., Nicol, S., Niemann, H., Nygren, C., Ormandy, D., Röbbel, N. and Rudnai, P. (2004) 'Review of evidence on housing and health', Fourth Ministerial Conference on Environment and Health, Budapest, Hungary, 23–25 June 2004.

Bruckman, H. W. L. *et al.* (1994) *Kunsstof CO-sensor*, TNO-industrie, TNO-rapport 0795/U94, Delft, The Netherlands.

Cain, W. S. and Cometto-Muniz, J. E. (1993) 'Irritation and odour: symptoms of indoor air pollution'. In: *Proceedings of Indoor Air '93*, Vol. 1, pp. 21–31.

Carrothers, T. J., Graham, J. D. and Evans, J. (1999) 'Putting a value on health effects of air pollution', *IEQ Strategies-Managing Risk*, 3(10).

CEN (1994) *Dynamic Olfactometry to Determine the Odour Threshold*, Draft European preliminary standard, CEN TC264/WG2, Brussels, Belgium.

CEN (1998) *Ventilation for Buildings – Design Criteria for the Indoor Environment*, Brussels, Belgium (CR 1752).

CEN (2005) *Criteria for the Indoor Environment Including Thermal, Indoor Air Quality, Light, and Noise*, Brussels, Belgium (prEN 15251).

Clements-Croome, D. (2002) (ed.) *Creating the Productive Workplace*, London and New York, E&FN Spon, Conference on Indoor Air Quality and Climate, Helsinki, Finland, Helsinki University of Technology, Vol. 1, pp. 53–58.

de Dear, R. and Brager, G. S. (1998) 'Developing an adaptive model of thermal comfort and preference', *ASHRAE Transactions*, 104(1a): 145–167.

de Dear, R. J. and Brager, G. S. (2002) 'Thermal comfort in naturally venti-lated buildings: revisions to ASHRAE Standard 55', *Energy and Buildings*, 34: 549–561.

de Dear, R., Leow, K. and Ameen, A. (1991) 'Thermal comfort in the humid tropics-part1', *ASHRAE Transactions*, 97(1): 874–879.

de Oliveira Fernandes, E. (2001) MATHIS Publishable Final Report, Joule III Programme, EC, Porto, Portugal, March.

de Oliveira Fernandes, E. and Clausen, G. (1997) *European Database on Indoor Air Pollution Sources*, Final Report, Porto, Portugal, February.

Devos, M., Patte, F., Ronault, J., Laffort, P. and van Gemert, L. J. (1990) *Standardised Human Olfactory Thresholds*, IRL, New York.

Djukanovic, R., Wargocki, P. and Fanger, P. O. (2002) 'Cost-benefit analysis of improved air quality in an office building'. In: *Proceedings of Indoor Air 2002*, Monterey, The 9th International Conference on Indoor Air Quality and Climate, Vol. 1, pp. 808–813.

Dorgan Associates (1993) *Productivity and Indoor Environmental Quality Study*, Alexandria, VA, National management institute.

ECA (1999) *Sensory Evaluation of Indoor Air Quality*, Report no. 20, European Collaborative Action, Indoor air quality & its impact on man, EUR18676EN, Italy.

Elma, K., Yokoyama, M., Nakamoto, T. and Moriizumi, T. (1989) 'Odour-sensing system using a quartz-resonator sensor array and neural network pattern recognition', *Sensors and Actuators*, 18: 291–296.

Engen, T. (1972) 'Psychophysics I: discrimination and detection'. In: Kling, J. W., Riggs, L. A. (eds) *Woodworth and Schlosberg's Experimental Psychology*, Vol. 1: Sensation and perception, New York, Holt, Rinehart and Winston, pp. 11–46.

Fanger, P. O. (1972) *Thermal Comfort, Analysis and Applications in Environmental Engineering*, McGrawhill, ISBN 0-89874-446-6.

Fanger, P. O. (1973) 'The variability of man's preferred ambient temperature from day to day', *Archives of Science and Physiology*, 27(4): A403–A407.

Fanger, P. O. (1988) 'Introduction of the olf and the decipol units indoors and outdoors', *Energy and Buildings*, 12: 1–6.

Fisk, W. and Rosenfeld, A. (1997) 'Estimates of improved productivity and health from better indoor environments', *International Journal of Indoor Air Quality and Climate*, 7(3): 158–172.

Fisk, W. J. (2000a) 'Review of health and productivity gains from better IEQ'. In: *Proceedings of Healthy Buildings 2000*, Helsinki, Finland, Vol. 4, pp. 22–34.

Fisk, W. J. (2000b) 'Estimates of potential nationwide productivity and health benefits from better indoor environment'. In: John D. Spengler, John F. McCarthy, Jonathan M. Samet (eds) *Indoor Air Quality Handbook*, McGraw-Hill, NY, pp. 4.1–4.32.

Gunnarsen, L. and Fanger, P. O. (1992) 'Adaptation to indoor air pollution', *Environment International*, 18: 43–54.

Gunnarsen, L. and Bluyssen, P. M. (1994) 'Sensory measurements using trained and untrained panels'. In: *Proceedings of Healthy Buildings '94*, Budapest, Hungary, Vol. 2, pp. 533–538.

Gardner, J. W. and Bartlett, P. N. (eds) (1992) *Sensors and Sensory Systems for an Electronic Nose*, NATO ASI series, Vol. 212, NATO.

Gardner, J. W., Bartlett, P. N., Dodd, G. H. and Shurmer, H. V. (1990) 'The design of an artificial olfactory system', In: Schild, D. (ed.) *Chemosensory Information Processing*, Vol. H39, Berlin, Heidelberg, Springer-Verlag, pp. 131–173.

Garriga-Trillo, A. (1985) 'Funcion psicofisica y medida de la sensibilidad olfativa', unpublished thesis, Universidad Autonoma de Madrid.

Hall, H. I., Leaderer, B. P., Cain, W. S. and Fidler, A. T. (1991) 'Influence of building-related symptoms on self-reported productivity'. In: *Proceedings of Healthy Buildings IAQ '91*, Washington, DC, USA, ASHRAE, pp. 33–35.

Institute of Medicine, Committee on the assessment of asthma and indoor air (2000) *Cleaning the Air. Asthma and Indoor Exposures*, Washington DC, National Academy Press, p. 438.

ISO (1998), International Organization for Standardization, *Instruments for Measuring Physical Quantities* (EN ISO 7726).

ISO (2001), International Organization for Standardization, *Assessment of the Influence of the Thermal Environment Using Subjective Judgement Scales* (EN ISO 10551).

ISO (2004), International Organization for Standardization, *Ergonomics – Determination of Metabolic Heat Production* (EN ISO 8996).

ISO (2005), International Organization for Standardization, *Moderate Thermal Environments – Determination of the PMV and PPD Indices and Specification of the Conditions for Thermal Comfort* (EN ISO 7730).

ISO (2006), International Organization for Standardization, *Estimation of the Thermal Insulation and Evaporative Resistance of a Clothing Ensemble* (EN ISO 9920).

Jantunen, M. J., Hänninen, O., Katsouyanni, K., Knöppel, H., Kuenzli, N., Lebret, E., Maroni, M., Saarela, K., Sram, R. and Zmirou, D. (1998) 'Air pollution exposure in European cities: The "Expolis" study'. *JEAEE* 8(4): 495–518.

Jenkins, P. L., Philips, T. J. and Mulberg, E. J. (1990) 'Activity patterns of Californians: use of and proximity to indoor pollutant sources'. In: *Proceedings of Indoor Air '90*, Toronto, Vol. 2, pp. 465–470.

Lagercrantz, L., Wistrand, M., Willén, U., Wargocki, P., Witterseh, T. and Sundell, J. (2000) 'Negative impact of air pollution on productivity: previous Danish findings repeated in new Swedish test room'. In: *Proceedings of Healthy Buildings 2000*, Espoo, Finland, Vol. 1, pp. 653–658.

Langkilde, G., Gunnarsen, L. and Mortensen, N. (1985) 'Comfort limits during infrared radiant heating of industrial spaces'. In: *Proceedings of CLIMA 2000*, Copenhagen.

Light & Health Research Foundation (2002) *Proceedings of Symposium Healthy Lighting*, November 2002, Eindhoven, The Netherlands (Prof. Dr S. Daan, University of Groningen, 'The physiology of the non imaging system: the eye and the brain'; Dr D-J. Dijk, Centre for Chronobiology, University of Surrey, Guildford, UK, 'Biological effects of ocular light'; Dr G.C. Brainard, Jefferson Medical College, Philadelphia, USA, 'Biological effects and the administration of ocular light'; L. Zonneveldt, TNO-Building and Construction Research; Eindhoven, 'Applications of healthy lighting in the working place'; Dr M.S. Rea, Lighting Research Centre, Troy, NY, USA, 'The future of healthy lighting').

McCartney, K. J. and Nicol, J. F. (2002) 'Developing an adaptive control algorithm for Europe', *Energy and Buildings*, 34: 623–635.

Milton, K., Glenross, P. and Walters, M. (2000) 'Risk of sick-leave associated with outdoor air supply rate, humidification, and occupant complaint, *Indoor Air*, 10: 211–221.

Nakamoto, T., Fukunishi, K. and Moriizumi, T. (1990) 'Identification capability of odour sensor using quartz-resonator array and neural-network pattern recognition', *Sensors and Actuators*, B1: 473–476.

Nunes, F., Menzies, R., Tamblyn, R. M., Boehm, E. and Letz, R. (1993) 'The effect of varying level of outdoor air supply on neurobehavioural performance function during a study of sick building syndrome (SBS)'. In: *Proceedings of Indoor Air '93*, The 6th International Conference on Indoor Air Quality and Climate, Helsinki, Finland, Helsinki University of Technology, Vol. 1, pp. 53–58.

Raw, G. J., Roys, M. S. and Leaman, A. (1990) 'Further findings from the office environment survey: productivity'. In: *Proceedings of Indoor Air '90*, The 5th International Conference on Indoor Air Quality and Climate, Ottawa, Canada Mortgage and Housing Corporation, Vol. 1, pp. 231–236.

Roelofsen, P. (2002). 'The impact of office environments on employee performance: the design of the workplace as a strategy for productivity enhancement', *Journal of Facilities Management*, 1(3): 247–264.

Samson, R. A., Flannigan, B., Flannigan, M., Verhoeff, A. P., Adan, O. C. G. and Hoekstra, E. S. (1994) 'Health implications of fungi in indoor environments', *Air Quality Monographs*, Vol. 2, Elsevier, Amsterdam, pp. 602.

Seppänen, O. and Fisk, W. J. (2005) 'Some quantitative relations between indoor environmental quality and work performance or health'. In: *Proceedings of 9th International Conference on Indoor Air Quality and Climate*, Beijing, September 2005.

Skåret, J. E. (1992) *Indoor Environment and Economics*, Project no. N 6405, Oslo, The Norwegian Institute of Building Research (NBI-Byggforsk) (in Norwegian).

Stevens, S. S. (1957) 'On the psycho-physical law', *Psychological Review*, 64: 153–181.

Sundell, J. (1999) *Indoor Environment and Health*, National Institute of Public Health, Stockholm, Sweden.

Tanabe, S., Kimura, K., Hara, T. and Akimoto, T. (1987) 'Effects of air movement on thermal comfort during the summer season', *Proceedings of Indoor Air '87*, Berlin, pp. 496–500.

Tham, K. W. (2004) 'Effects of temperature and outdoor air supply rate on the performance of call center operators in the tropics', *Indoor Air*, 14(s7): 119–125.

VDI (1986) *Olfactometry, Odour Threshold Determination, Fundamentals*, Düsseldorf, VDI-Richtlinien, VDI 3881, Blatt 1/part 1.

Wargocki, P., Wyon, D. P., Baik, Y. K., Clausen, G. and Fanger, P. O. (1999) 'Perceived air quality, Sick Building Syndrome (SBS) symptoms and productivity in an office with two different pollution loads', *Indoor Air*, 9(3): 165–179.

Wargocki, P., Wyon, D., Sundell, J., Clausen, G. and Fanger, P. O. (2000a) 'The effects of outdoor air supply rate in an office on perceived air quality, sick building syndrome (SBS) symptoms and productivity', *Indoor Air*, 10: 222–236.

Wargocki, P., Wyon, D. P. and Fanger, P. O. (2000b) 'Pollution source control and ventilation improve health, comfort and productivity'. In: *Proceedings of Cold Climate HVAC 2000*, Sapporo, Japan, pp. 445–450.

Wargocki, P., Wyon, D. P. and Fanger, P. O. (2003) 'Call-centre operator performance with new and used filters at two outdoor air supply rates'. In: *Proceedings of Healthy Buildings 2003*, Singapore, The 7th International Conference Energy-Efficient Healthy Buildings, Vol. 3, pp. 213–218.

Wargocki, P., Wyon, D. and Fanger, P. O. (2004) The performance and subjective responses of call-center operators with new and used supply air filters at two outdoor air supply rates, *Indoor Air*, 14(s8): 7–16.

Wyon, D. P. (1996) 'Indoor environmental effects on productivity'. In: *Proceedings of IAQ '96*, Paths to Better Building Environments, USA, ASHRAE, pp. 5–15.

Chapter 3

Energy implications of indoor environment control

Claude-Alain Roulet

3.1 Introduction

The main purpose of buildings is to provide a comfortable living environment for their occupants despite a sometimes uncomfortable external environment (extreme temperatures, wind, rain, noise, solar radiation, etc.). This includes, among others, thermal, visual and acoustic comfort as well as indoor air quality. Energy is required to control the indoor climate and indoor air quality, as well as lighting and to provide services in the building.

Except during the 1950s and 1960s, it has always been considered important that an excessive use of energy should be avoided in the construction and the management of a building, sometimes even at the cost of user comfort.

Except in very mild climates, the largest amounts of energy are used to ensure a comfortable indoor climate, either by heating or by cooling. Therefore, indoor environment control has the largest impact on the energy use of buildings.

Since the Rio de Janeiro conference, there have been more and more incentives to save energy and lower the impact of buildings on the environment. Therefore, there is no excuse for the building sector not to adopt a sustainable development policy.

This chapter provides methods to assess the energy flows in buildings and in ventilation systems. It presents guidelines for improving the energy performance of buildings without compromising the indoor environment quality, in particular indoor air quality. It shows that, in many cases, it is even possible to improve both indoor environment quality and energy performance. These guidelines address both the building as a whole and the HVAC systems. The chapter ends with an estimation of the energy impacts of various ventilation strategies, assessed by detailed simulations of a particular building as a case study.

3.2   Energy flow in buildings

Energy is used to control the indoor climate and indoor air quality. The main uses of energy in buildings are for

- heating
- cooling
- drying and humidifying
- ventilation
- hot water supply
- lighting
- cooking, washing or producing goods and services
- electrical appliances.

The amount of energy consumed for each purpose depends on many factors such as climate, local habits, energy policies and cost.

It is often suspected that energy savings result in poorer indoor environment quality or, in another context, that a high comfort level is the result of high technology and high energy consumption. This is actually not true. It is now generally admitted among building scientists that high-quality energy services do not necessarily incur a high energy use and that good environment quality can be obtained with a reasonable amount of energy and power and with a low environmental impact. Section 3.5 provides more information on this issue.

The thermal balance of a building is schematically presented in Figure 3.1.

Depending on the indoor–outdoor temperature difference, the building has heat gains or loss by transmission through the envelope and by ventilation, internal heat gains from lighting, electrical appliances, metabolic activity, etc., and passive solar heat gains, mostly through windows or, in some cases, specific passive solar devices such as greenhouses or transparent

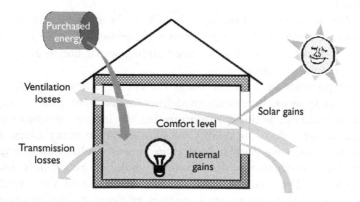

Figure 3.1 Heat balance of a building (Roulet, 2004).

insulation. External power is used, when needed, for maintaining thermal comfort at a comfortable level. A part if it is used for ventilation and, in some cases, for air conditioning.

3.2.1 Energy for heating

During the cold season, the internal environment should be heated to keep it at a comfortable temperature. At each instant, the heating power that should be provided by the heating plant, Φ_h, equals to the heat power transferred to the outdoor environment (also called heat loss), Φ_l, minus the free gains (passive solar gains, Φ_s, and internal gains, Φ_i), to which the heat stored during this instant in the building fabric, P_{hs}, should be added:

$$\Phi_h = \Phi_l - \Phi_s - \Phi_i + \Phi_{hs} \tag{3.1}$$

Φ_{hs} is negative if heat is transferred from the building fabric to the internal environment.

Heat losses include transmission heat loss through the building envelope and ventilation heat loss, i.e. the heat contained in the air leaving the building. These losses are, as a first approximation, proportional to the indoor–outdoor temperature difference. Transmission heat loss is reduced by thermal insulation, and ventilation heat loss is controlled by adjusting the airflow rate at the required level and can be reduced by heat recovery, using a heat exchanger to extract heat from the exhaust air and provide this heat to another medium, most often to supply air.

Passive solar heat gain is the result of solar radiation entering the building through various ways:

- directly through the windows exposed to direct and diffuse solar radiation;
- indirectly by heating attached sunspaces, a part the heat generated into these sunspaces being transferred to the internal environment by transmission and convection;
- indirectly through the opaque envelope elements. These gains are generally a small part of the loss of common opaque elements, because most of the heat generated at the external surface by solar radiation is immediately lost to the external environment. In envelope elements with transparent insulation, however, the solar radiation is transformed into heat behind the transparent insulation, and most of this heat can reach the indoor environment;
- there are also more sophisticated passive or hybrid solar systems such as the Trombe wall, which is a massive wall with a window in front of it and possibly ventilation openings between the internal environment and the air layer contained between the wall and the glazing.

Internal heat gains are supplied by the occupants (about 100 W/person), lighting and other appliances that are not part of the heating system.

Passive solar gains can be controlled by movable solar protections. Internal heat gains however are not much controlled: appliances are not switched off and occupants do not necessarily leave the building when it is too warm inside. When these gains are larger than the losses, several processes help in maintaining the indoor temperature at a comfortable level:

- the heating control system reduces the heating power or switches it off.
- the occupant increases the heat loss, mainly by opening the windows, and reduces the heat gains by using the moveable solar protections.
- the internal temperature increase also increases the heat loss.
- some heat is stored in the building fabric. This stored heat is larger if the fabric is massive and in good thermal contact with the indoor environment.
- increase ventilation rate to the building.

The energy used for heating is the integral of the equation above, taken over periods when heating is actually needed (negative heating power is taken as zero):

$$Q_h = \frac{1}{\eta_h} \int_{>0} (\Phi_l - \Phi_s - \Phi_i + \Phi_{hs}) \, dt \qquad (3.2)$$

where η_h is the global efficiency of the heating system.

Simplified models for calculating the seasonal or annual energy use for heating, such as the international standard EN ISO 13790 (CEN, 2004), integrate separately each term of the sum, and introduce a *'utilisation factor'*, η, for the passive solar and internal gains to take into account the building storage capacity and the rejected gains.

$$Q_h = Q_l - \eta_h (Q_s + Q_i) \qquad (3.3)$$

This utilisation factor is determined empirically by comparing the results of detailed simulations according to Equation (3.2) with simplified calculations according to Equation (3.3). The utilisation factor used in the international standard EN ISO 13790 depends on a building time constant, τ, and on the ratio γ between the gains and the loss.

The time constant characterises the internal thermal inertia of the heated space and is calculated by:

$$\tau = \frac{C}{H} \qquad (3.4)$$

where

- C is the internal heat capacity of the building that may be roughly calculated as the mass of the building fabric in contact with indoor air multiplied by a conventional specific heat capacity of $1000\,\mathrm{J/(kg \cdot K)}$;
- H is the heat loss coefficient of the building, that is the heating power required to maintain an indoor–outdoor temperature difference of $1\,\mathrm{K}$. This includes ventilation and transmission heat loss coefficient.

The ratio γ_{gl} between the gain and the loss is defined as

$$\gamma_{gl} = \frac{Q_g}{Q_l} \tag{3.5}$$

The utilisation factor is calculated from the empirical formulas:

$$\text{if } \gamma_{gl} \neq 1\colon \eta = \frac{1 - \gamma_{gl}^a}{1 - \gamma_{gl}^{a+1}} \qquad \text{if } \gamma_{gl} = 1\colon \eta = \frac{a}{a+1} \tag{3.6}$$

where a is a numerical parameter depending on the time constant, τ, defined by

$$a = a_0 + \frac{\tau}{\tau_0} \tag{3.7}$$

Values of a_0 and τ_0 are given in Table 3.1.

Figures 3.2 and 3.3 illustrate utilisation factors for monthly calculation periods and for various time constants for type I and type II buildings.

The utilisation factor is defined independently of the heating system characteristics, assuming perfect temperature control and infinite flexibility. A slowly responding heating system and a less-than-perfect control system can significantly affect the use of gains.

The building time constant can be increased either by increasing the accessible internal mass or by decreasing its losses (improving its insulation or recovering heat on ventilation). As it can be seen in Figure 3.3, heavy,

Table 3.1 Values of the numerical parameter a_0 and reference time constant τ_0, for a calculation period of one month ((CEN, 2004))

Type of building	a_0	$\tau_0(h)$
Continuously heated buildings (more than 12 h per day) such as residential buildings, hotels, hospitals, homes and penitentiary buildings	1	15
Building heated during day-time only (less than 12 h per day) such as education, office and assembly buildings and shops	0.8	70

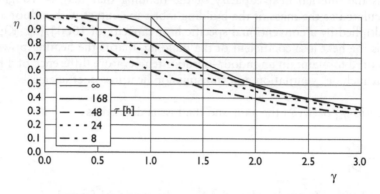

Figure 3.2 Utilisation factor for 8 h, 1 day, 2 days, 1 week and infinite time constants, valid for monthly calculation period of continuously heated buildings (CEN, 2004).

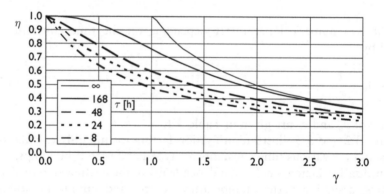

Figure 3.3 Utilisation factor for 8 h, 1 day, 2 days, 1 week and infinite time constants, valid for monthly calculation period for buildings heated during the day only (CEN, 2004).

well-insulated buildings are close to the theoretical maximum for the utilisation of gains, whereas buildings with a small time constant cannot use a large part (one-third up to the half) of the free gains.

3.2.2 Energy for cooling

The aim of the mechanical cooling system is to extract excessive heat gains from the building to maintain a comfortable temperature. The heat gains are essentially the same as in Equation (3.1), but in most cases, when cooling, the outdoor temperature is larger than the indoor temperature. Therefore, there are additional heat gains by transmission through the envelope and ventilation.

$$\Phi_c = \Phi_l + \Phi_s + \Phi_i - \Phi_{hs} \tag{3.8}$$

As heat gains are the main part of the cooling energy requirement, and as these heat gains strongly vary with time, detailed calculation models are used in most cases to calculate the energy need for cooling, using:

$$Q_c = \frac{1}{\eta_c} \int_{>0} (\Phi_l + \Phi_s + \Phi_i - \Phi_{hs}) \, dt \tag{3.9}$$

The integral is over periods when cooling is actually needed.

However, a simplified, monthly-based calculation method is proposed in the last version of (CEN, 2004). This method is similar to Equation (3.3) for heating, but the roles of gains and losses are reversed:

$$Q_c = (Q_s + Q_i) - \eta_c Q_l \tag{3.10}$$

Where η_c is calculated according to Equation (3.6) using the values of a_0 and τ_0 for continuously heated buildings.

3.2.3 Energy for air conditioning

The buildings are primarily ventilated for the purpose of removing the pollutants generated within them. The air leaving the buildings has the characteristics (temperature, humidity, chemical composition) of the indoor air. It is replaced at the same mass airflow rate by air coming from outdoors, which has its own characteristics. Therefore, the building gives or takes heat, water vapour and other chemical components to or from the air entering the building to reach the characteristics of the indoor air.

This needs energy. Figure 3.4 shows several characteristics of humid air. The curves show the water content of air as a function of its temperature, for various relative humidity. The water content cannot be larger than that shown by the saturation curve. Air with a relative humidity φ has a water vapour concentration φ times ($\varphi < 1$) that of the saturated air.

Energy is needed, on the one hand, to heat or cool the air and on the other hand to evaporate water in it or to condense water for drying it. The *enthalpy*[1] of a gas mixture is, as a first approximation, the sum of the enthalpies of its components. Taking as a reference dry air at 0°C, the increase of specific enthalpy (in J/m^3) of the mixture air–water vapour at temperature θ and specific humidity ω is:

$$h = c_{da}\theta + (L + c_w\theta) \frac{\omega}{1 - \omega} \tag{3.11}$$

where

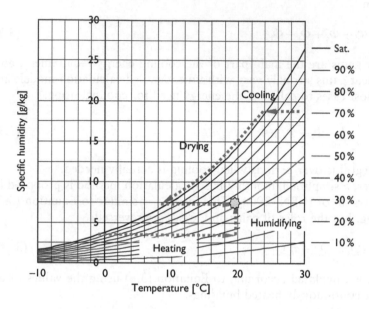

Figure 3.4 Psychrometric chart with constant relative humidity curves and constant enthalpy lines.

c_{da} is the specific heat capacity of dry air [about $1,006\,J/(kg{\cdot}K)$]
c_w is the specific heat capacity of water vapour [about $1,805\,J/(kg{\cdot}K)$]
L is the latent heat of evaporation, i.e. the heat required for evaporating
   1 kg water, about 2,501,000 J/kg.
$x = \omega/(1 - \omega)$ is the humidity ratio, i.e. the mass of water vapour per
   kg of dry air.

Figure 3.4 illustrates the paths of temperature and specific humidity of air for two processes, both ending at 20°C and 50 per cent relative humidity:

• starting from outdoor air in winter, at −1°C and 80 per cent relative humidity, heating and humidifying it in order to get 50 per cent relative humidity at 20°C.
• cooling and drying summer outdoor from 30°C and 70 per cent relative humidity down to 20°C and 50 per cent relative humidity. Note that, for drying the air, it should first be cooled down at the dew point temperature corresponding to the required specific humidity, then reheated to the required indoor temperature. It should also be noticed that it is impossible to cool down the air below its dew point without drying it.

The heat required for heating and humidifying the air may be brought from free sources such as solar radiation or metabolic activity of occupants. Lighting and electrical appliances that are not part of the heating system may also give heat to the indoor air. Water vapour may be added to indoor air by plants and occupants and their activities such as cooking and drying laundry. However, this 'free' energy may not suffice, and, in this case, the complement is provided by a heating system. In this case, about 0.34 Wh is needed to heat or cool $1\,m^3$ of air by 1°C, as long as its specific humidity does not change or if the air is humidified by 'free sources'. This value is therefore used in models calculating the energy for heating buildings.

Hot and humid outdoor air cools down and eventually dries at the contact of cold surfaces, on which excess water vapour may condense. If these surfaces are not cooled, such as the building fabric or furniture, their temperature will rise and cooling stops after a while. However, the air temperature rises more slowly if the air is in contact with massive structures that were cooled down before, for example by strong airing during the cool night.

Mechanical cooling is needed to keep the surfaces in contact with the air cold and to get continuous air drying and cooling. Starting from warm, humid air, it is first cooled down when passing through a refrigerated heat exchanger (horizontal 'cooling' line in Figure 3.4) until it reaches its dew point. Then it is dried by losing the water that condenses on the heat exchanger ('drying curve') until it reaches the required specific humidity, at a new, lower dew point. It should then be heated at the required temperature.

Numerical values for this process are given in Table 3.2. The largest change in enthalpy is when drying, because 2,500 J should be withdrawn from the heat exchanger to condense each gram of water.

The energy required to reheat the dried, cold air can be provided by various means:

- the indoor environment, from heat loads and solar gains. This way is common in tropical climates. It saves investing in a heating system, and heating energy is free. It has however the disadvantage of blowing cold air in the occupied spaces, often leading to draughts. The temperature control is obtained by varying the supply airflow rate.
- heat provided to a warm heat exchanger by a separate heating system. This is expensive both in investment and running cost and should no more be used.
- heat provided to a warm heat exchanger by the chiller. The heat pump used to cool down the chilled water must also be cooled and therefore provides cooling water at temperatures higher than indoor temperature. This water or a part of it can be circulated into the warm heat exchanger without any running cost. The investment is limited to pipes connecting the chiller condenser to the warm heat exchanger and to a control valve.

Table 3.2 Humidity ratio and specific enthalpy of warm, humid air cooled down and dried
       as shown in Figure 3.4

Temperature θ (°C)	Relative humidity φ(%)	Humidity ratio x (g/kg)	Specific enthalpy h [J/(kg·K)]	Enthalpy increase Δh [J/(kg·K)]
30.0	70	18.8	78,756	
23.9	100	18.8	71,815	−6,941
9.3	100	7.3	27,571	−44,244
20.0	50	7.3	38,724	11,153

3.2.4 Energy for ventilation

The energy to move the air is the product of a force by a displacement. The force is the pressure, Δp, exerted on the section area, A, of the duct, and the displacement is the path, l, of the air during a time interval, Δt. But $A\,l$ is the volume of air displaced during Δt. Hence

$$Q_m = \Delta p \, A \, l = \Delta p \, V \qquad (3.12)$$

The power is energy divided by the time used to deliver the energy. Dividing the above equation by Δt, we get

$$\Phi_m = \frac{Q_m}{\Delta t} = \frac{\Delta p V}{\Delta t} = \Delta p \frac{V}{\Delta t} = \Delta p \dot{V} \qquad (3.13)$$

In natural ventilation, this energy is provided by wind and stack effect. In mechanical ventilation systems, this energy is provided by the fan. The mechanical power delivered by a fan is the product of the volume airflow rate $\dot{V}$ delivered by the fan, by the pressure differential Δp, across the fan. The mechanical power required to move the air through a ductwork is also the product of the volume airflow rate through the ductwork, by the pressure difference between the main supply and main exhaust ducts. As the pressure difference is proportional to the square of the airflow rate, the mechanical power for ensuring a given airflow rate into a ductwork is proportional to the cube of the airflow rate! Increasing the airflow rate in a room by 10 per cent requires 33 per cent more fan power, and doubling the airflow rate requires a power eight times larger if the ductwork is not changed.

The electrical power used by the fan motor is larger than Φ_m, because of the motor and fan losses. The energy losses of fans are shared between the elements of the chain linking the electrical network to the aeraulic ductwork *(Figure 3.5)*.

Typical values for the specific ventilation energy, i.e. the energy used to move one cubic meter of air in a ventilation network or the power required to ensure $1\,m^3/h$ airflow rate are given in Table 3.3. It should be noticed that there is a factor close to 2 between typical and energy efficient systems.

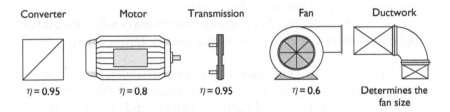

Converter	Motor	Transmission	Fan	Ductwork
$\eta = 0.95$	$\eta = 0.8$	$\eta = 0.95$	$\eta = 0.6$	Determines the fan size

Figure 3.5 Approximate figures for the efficiencies of various elements needed to move the air in the ductwork.

Table 3.3 Energy for moving one cubic meter of air in typical and energy efficient buildings (SIA, 2006)

Use of ventilated room	Typical systems ($Wh\,m^{-3}$)	Energy-efficient systems ($Wh\,m^{-3}$)
Dwelling	0.42	0.17
Office and meeting rooms	0.56	0.35
Shop	0.35	0.22
Shopping center	0.90	0.56
Classroom, auditorium	0.35	0.22
Hospital room	0.35	0.22
Hotel room	0.56	0.35
Restaurant	0.35	0.22
Professional kitchen	0.42	0.28
Workshop	0.35	0.22
Toilets	0.14	0.08
Couloirs, archives	0.28	0.17
Garage (private)	0.14	0.0^a
Garage (public)	0.28	0.0^a
Computer room (exhaust air only)	0.14	0.08

[a] Natural ventilation.

3.3   Assessing energy flows

3.3.1   Assessing energy use of buildings

Energy assessments of buildings are carried out for various purposes, such as

- monitoring of the energy efficiency of the building including heating, ventilation and air conditioning systems;
- helping in planning retrofit measures, through prediction of energy savings that would result from various actions;
- judging compliance with building regulations expressed in terms of a limitation on energy use or a related quantity;
- transparency in commercial operations through the certification and/or display of a level of energy performance (energy certification).

The energy use of building may be assessed basically by two ways: calculation using building models and measurement of the energy use. A combination of both ways could provide more confidence in the results or allow estimating missing data. Therefore, a recent European standard (CEN, 2006) proposes and describes two principal options for energy rating of buildings, the first being calculated, the second being based on measurements. These options are the asset rating and the operational rating.

The asset rating is based on calculations of the energy used by the building for heating, cooling, ventilation, hot water and lighting, with standard input data related to climate and occupancy. This rating provides an assessment of the energy efficiency of the building under standardised conditions that enables a comparison to be made between different buildings within climatic main regions and with identical or at least similar activities. A rating calculated using actual climatic and occupancy data can be used to compare two buildings having different climates or different uses, to compare retrofit scenarios, to optimise energy performance etc.

The operational rating is based on measurement of energy use. The operational rating measures the in-use performance of a building, including all deviations between theoretical properties in calculations and realised properties (air-infiltration, heat transfer, generation efficiency, control etc.) and is influenced by the way the building is maintained and operated. It assists those trying to improve the efficiency of building operation and allows displaying the actual energy performance of a building. It can also provide useful feedback to the owners, occupiers and designers of new buildings if assessed a few years after occupation and compared to the calculated asset rating, for the same set of energy end uses.

The operational rating includes all energy uses under actual conditions, whereas the asset rating may include only some uses, and for standard conditions. Therefore, these ratings cannot be directly compared.

In both cases, the boundaries of the building under study should be clearly defined and the same boundaries should be used for all calculations and measurements. It is the energy flows through these boundaries that will be assessed.

3.3.2 Assessing heat flow rates in ventilation systems

The heat power transported by an airflow is the product of its mass flow rate $\dot{m}$ by its enthalpy, h:

$$\Phi = \dot{m}h = \rho \dot{V}h \tag{3.14}$$

The airflow rate can be measured as described in Chapter 8. The density depends on the temperature and humidity, as does the enthalpy (Equation 3.11). Therefore, in addition to airflow rate measurements, only temperature and humidity should be measured to assess the heat power.

In addition to this, the airflow has some kinetic energy as given by Equation (3.13).

In parts of the ventilation systems where the air is neither humidified nor dried naturally or artificially, the enthalpy does not change, and heat transferred from one place to the other is measured by the airflow rate and the temperature:

$$\Phi = \dot{m}c_{\mathrm{da}}\theta = \rho \dot{V}c_{\mathrm{da}}\theta \tag{3.15}$$

3.3.3 Assessing the fan power efficiency

The efficiency of the whole air moving system shown in Figure 3.5 can be assessed by measuring on one hand the consumption of electrical energy and on the other hand the kinetic energy given to the air in duct.

The fan efficiency is the ratio of useful power, i.e. Φ_{m}, to the electrical power consumed by the fan motor, Φ_{e}:

$$\eta_{\mathrm{f}} = \frac{\Phi_{\mathrm{m}}}{\Phi_{\mathrm{e}}} = \frac{\dot{V}\Delta p}{\Phi_{\mathrm{e}}} \tag{3.16}$$

Measuring the airflow rate, the pressure differential and the electrical power used by the fan allows the determination of the fan efficiency.

Poor fan efficiency not only spills costly electrical energy but also hinders efficient cooling. The cooling power of the air blown by the fan is:

$$\Phi_{\mathrm{cool}} = \rho c\dot{V}\Delta\theta \tag{3.17}$$

where

    ρ is the density of air
    c is the heat capacity of air
    $\Delta\theta$ is the temperature difference between exhaust air and supply air.

The kinetic energy given to the air by the fan is, sooner or later, degraded into heat by viscosity and friction on the surfaces of ducts, room walls and furniture. As the fan motor is usually in the airflow, its heat loss is also delivered to the air. Therefore, nearly all the energy given to the fan ends as heat in the indoor air. This corresponds to a heating power equal to the electrical power consumed by the fan motor, Φ_e. Hence

$$\Phi_h = \Phi_e = \frac{\dot{V}\Delta p}{\eta_f} \tag{3.18}$$

For air conditioning, the air temperature increase resulting from the fan heat gain is

$$\Delta\theta_h = \frac{\Phi_e}{\rho c \dot{V}} = \frac{\Delta p}{\eta_f \rho c} \tag{3.19}$$

should be as small as possible. Therefore, the fan efficiency should be large, and the pressure differential should be kept as small as possible.

To measure the fan efficiency, the following quantities should be assessed:

- airflow rate through the fan
- pressure differential across the fan
- electric power used by the fan motor.

Tracer gas measurements, as described in Chapter 6, allow for the measurement of airflow rates through both supply and exhaust fans. Depending on the units used when interpreting the results, these measurements may provide either volume airflow rates, $\dot{V}$, or mass airflow rates, $\dot{m}$. These are related by

$$\dot{m} = \rho\dot{V} \tag{3.20}$$

where ρ is for the density of the air, which can be calculated by

$$\rho = \frac{\overline{M}p}{RT} \cong 3.46 \times 10^{-3}\frac{p\,[\text{Pa}]}{T\,[\text{K}]}\,[\text{kg/m}^3] \tag{3.21}$$

where

    p is the atmospheric pressure (at average 101,300 Pa at sea level);
    T is the absolute temperature;

$\overline{M}$ is the average molar mass of the air mixture (about 28.8 g/mole);
$R = 8.31396$ J/(mole·K) is the molar gas constant.

Then, if mass airflow rates are given from tracer gas measurements, the airflow through the fans should first be converted into volume airflow rate.

The pressure differential is assessed with a differential manometer able to read about 200 Pa (20 mm water column).

The two ports of this manometer are fitted with two small pipes, which other ends are placed on each side of the fan (Figure 3.6). Care should be taken to avoid too much dynamic pressure on these ends. It is advised to install these ends perpendicular to the airflow, preferably close to a wall of the air duct and at locations where the air velocity is about the same on both sides of the fan. If the pressure varies significantly when moving one of the pressure taps, the dynamic pressure has an effect, and the pressure taps should be moved to a best place.

On most AHUs, a differential pressure switch is installed to check the function of the fan. This switch is connected by two pipes to taps installed in the ducts before and after the fan. These taps can be used to connect the differential manometer, but the safety switch should be either disabled or short-circuited. Otherwise, the fan motor will stop as soon as the pipes of the pressure switch are disconnected.

The electrical power used by the fan motor is measured with a wattmeter. This measure may not be easy and should be performed by a professional.

In principle, the rms voltage, U, between phase and neutral point, and the rms current, I, running into each motor coil are measured for each phase

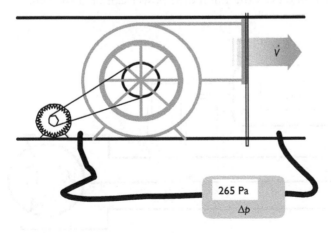

Figure 3.6 Installation of the differential manometer to measure the pressure differential across the fan.

together with the phase shift, ϕ, between voltage and current. The power is then:

$$\Phi_e = \sum_{j=1}^{3} U_j I_j \cos(\phi_j) \qquad (3.22)$$

the sum being related to all three phases.

In very small AHUs, the fan motor is single phase, and

$$\Phi_e = UI \cos(\phi) \qquad (3.23)$$

The measure of the current requires installing an ampere-meter in the circuit. An easy way is to use clamp-on ampere meters. A measuring clamp is installed around each wire leading to the motor. This clamp contains a transformer that gives a current proportional to the current running through the closed clamp.

If the three-phase motor is wired as shown in Figure 3.7, as it is usually the case, the current measured in each wire is then passing in two coils in parallel.

Then, if the three coils are identical, the power is:

$$\Phi_e = \sqrt{3} U_{UV} I_W \cos(\phi) \qquad (3.24)$$

where U, V and W denote any of the three phases: U_{UV} is the inter-phase voltage and I_W the current running into one of the three wires.

In some AHUs, the fan is controlled by a variable frequency controller. Such devices are often equipped with a screen on which the frequency, the voltage, the current and even the fan motor power can be displayed.

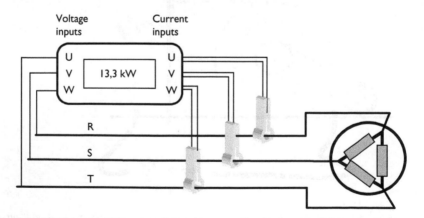

Figure 3.7 Schematics of electrical power measurement on a three-phase motor.

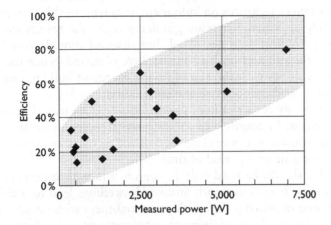

Figure 3.8 Fan efficiencies as a function of actual fan motor power.

Finally, the fan power efficiency is calculated according to

$$\eta_f = \frac{\dot{V}\Delta p}{\Phi_e} \qquad (3.25)$$

where

$\dot{V}$ is the airflow rate through the fan;
Δp is the pressure differential across the fan;
Φ_e is the electrical power used by the fan.

The fan efficiencies were measured on several fans of various units in Switzerland. Results are shown in figures below. Figure 3.8 shows a general improvement of the fan efficiency when fan power increases. However, the dispersion is huge, and large differences can be observed for each power class. For example, the efficiency of 1,600 W fans varies from 40 to nearly 70 per cent and that of small fans from 13 to 32 per cent. All motors were running at about 63 per cent of their nominal capacity.

3.4 Energy and indoor environment quality

3.4.1 Introduction

Directly following the oil cost crisis in the 1970s, measures were hastily taken in many buildings to reduce their energy use. These measures were planned with only two objectives: energy efficiency and return on invest-ment. The effects of these measures on indoor environment, health or com-fort were completely neglected. Therefore, in many cases, the results were dramatic. Not only comfort was decreased, but cases of mould growth, increased indoor pollution, and health hazards were observed. Since then, there seems to be a conflict with the aim of saving energy in buildings and the aim of creating a good indoor environment quality.

Of course, some energy conservation opportunities (ECOs) may degrade the indoor environment. Measures such as low internal temperature or too low ventilation rate should therefore either be avoided or taken only in case of emergency and for a limited period of time.

Some other ECOs should be used only in conjunction with others. For example, retrofitting windows in poorly insulated dwellings lead to a risk of mould growth, and improving the envelope air tightness without taking care of ensuring and controlling a minimum ventilation rate may decrease the indoor air quality.

In buildings, energy is required, among others, for purposes given in the first column of Table 3.4. This table also proposes known ways to save energy and presents some effects of these energy saving measures on comfort or indoor environment quality. It can readily be seen that there are many cases where ECOs, when well designed and executed, improve the indoor environment quality.

Several recommendations, resulting from experience and recent surveys performed within European projects (Bluyssen *et al.*, 1995; Roulet *et al.*, 2005b) are given below.

3.4.2 Design intentions

The building is (or at least should be) designed and constructed first to bring a good indoor environment to its occupants. There could be other objectives, such as:

- prestige, image
- low cost
- energy saving
- real estate business, speculation.

However, the occupants should nevertheless be given the highest pri-ority. The experts in the HOPE project[2] propose the following definition for a high quality (HQ) building: (Bluyssen *et al.*, 2003): 'A healthy and energy-efficient building does not cause or aggravate illnesses in the build-ing occupants, assures a high level of comfort to the building's occupants in the performance of the designated activities for which the building has been

Table 3.4 Functions of the building requiring energy, together with some ways to save energy and effects of these energy saving measures on comfort

Energy required for	Ways to save energy	Impact on indoor environment
Compensation of transmission heat loss in winter	Better, thicker insulation, low emissivity-coated multiple glazing	Improves comfort Improves health by preventing mould growth
Compensation of ventilation heat loss in winter	Lower ventilation rate Limit the ventilation rate to the required level	May result in low IAQ Less drafts, less noise, good IAQ
	Use heat recovery on exhaust air	Generally improves IAQ in winter
Winter heating in general	Improve solar gains with larger, well-located windows. Improve the use of gains by better insulation and good thermal inertia	If windows are poor: cold surfaces. Overheating if poor solar protections If well designed: good visual contact with outdoor environment, excellent summer and winter comfort
Elimination of heat gains during warm season	Use passive cooling Use efficient, well-commissioned and maintained systems Higher internal temperature	Very comfortable in appropriate climates and buildings. Better IAQ and comfort Should be kept within comfort zone
Internal temperature control	Comfortable set-point temperature, improved control	Avoids over- and under-heating
Humidification	Switch it off	No effect in many cases
Lighting	Use daylighting.	Comfortable light, with limited heat gains when well controlled
	Use efficient artificial lighting	Comfort depends on the quality of light. Limited heat gains

intended and designed, and minimises the use of non-renewable energy, taking into account available technology including life cycle energy costs.' Indeed, sustainable development requires that HQ buildings should be designed, built and maintained taking account of environmental, economical and social stakes.

Healthy, comfortable and energy efficient buildings are the result of a conscious design keeping constantly these three objectives in mind. It is not by chance that most of the 16 apartment buildings and seven office buildings fulfilling at best the HOPE criteria for these objectives were designed that way.

Essential recommendations that could be given to reach these objectives are:

- prefer passive methods to active ones wherever possible;
- think about the user comfort, needs and behaviour;
- adapt the building to its environment and climate.

Passive and active ways to get HQ buildings

Passive ways are architectural and constructive measures that naturally provide a better indoor environment quality without or with much less energy use. Examples are:

- improving winter thermal comfort with thermal insulation, passive solar gains, thermal inertia and controlled natural ventilation;[3]
- improving summer thermal comfort with thermal insulation, solar protections, thermal inertia and appropriate natural ventilation;
- ensuring indoor air quality by using low-emitting materials and controlled natural ventilation;
- providing controlled daylighting;
- protecting from outdoor noise with acoustical insulation, adjusting the reverberation time for a comfortable indoor acoustics.

Passive means are often cheap, use very little or no energy and are much less susceptible to break down than active means. However, they often depend on meteorological conditions and therefore cannot always fulfil the objectives. They should be adapted to the location and therefore need creativity and additional studies from the architect, and a design error may have dramatic consequences.

Active (or technological) ways allow reaching the objectives by mechanical means, using energy for complementing the passive ways or even for compensating poor building performance. Examples are:

- heating boilers and radiators for winter comfort;
- artificial cooling by air conditioning or radiant panels for summer comfort;
- mechanical ventilation;
- artificial lighting;
- actively diffusing background music or noise to cover the ambient noise.

Active ways, when appropriately designed, built and maintained, are perfectly adapted to the needs. The architect does not have to take much care of them, because these are designed and applied by specialised engineers according to known technology. Flexible and relatively independent on meteorological conditions, they allow correcting architectural errors. However, the required technology is often expensive, requires a higher maintenance and uses much energy. It may break down, and when it does, the resulting situation may be dramatic. Furthermore, the fact that they allow correcting architectural 'errors' can also be considered as a disadvantage.

Passive ways are preferred, but cannot always fulfil the comfort objectives. Therefore, the appropriate strategy is to use them as much as reasonably possible and to compensate for their insufficiencies with active systems, which will then be smaller. This strategy often allows more freedom in choosing the type and location of active systems.

Taking account of the user

The occupant of a building expects that the building provides an acceptable indoor environment, according to his wishes. The occupant likes to have a control on this environment and even needs such a control to adapt it to his needs (Figure 3.9).

Well-being of occupants can be assessed with personal questionnaires as follows. Occupants are asked if they have had two or more episodes of several symptoms, and if they feel better on days out of the office. A symptom that does disappear when out of the building is assumed to be building-related. The list of symptoms includes those commonly related to sick building syndrome, i.e., in office buildings: dryness of the eyes, itchy or watery eyes, blocked or stuffy nose, runny nose, dry throat, lethargy or tiredness, headaches, dry, itching or irritated skin. In homes, additional symptoms are sneezing and breathing difficulties. From these replies, a

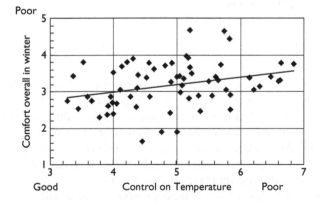

Figure 3.9 Perceived comfort correlates with the perceived control of the occupants over their environment (Johner *et al.*, 2005). Scale for both axes goes from 1 = satisfactory to 7 = unsatisfactory.

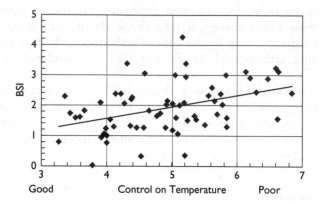

Figure 3.10 Building Symptom Index correlates with the perceived control of the occupants on their environment (Johner *et al.*, 2005).

building symptom index (BSI) is calculated to get the average number of building-related symptoms per occupant (see Figure 3.10).

The control an occupant has over his environment not only affects his perceived comfort but is linked in some way with his well-being, as measured by the BSI.

Therefore, the building design as well as the system must take into account the user's needs and wishes and allow the user to adapt its environmental conditions to his needs as much as possible.

Where the design does not allow the occupant to adapt his environment, he finds another way: bringing in heaters, opening the window in winter instead if he cannot put the thermostat down, using tape or paper to close draughty ventilation openings, etc.

Adaptation to the environment

The outdoor environmental characteristics (temperature, solar radiation, wind, dust, pollution, noise, etc.) change with the location of the building. Therefore, a design that is well adapted in a place may be completely unsuitable to another one: Bedouin tents, igloos, tropical huts, all well adapted to their environment, cannot be used elsewhere. This is also valid for contemporary building design: it is of course possible to compensate for environmental changes using active techniques, but this often decreases the indoor environment quality and increases the energy use.

A building that is well adapted to its climate protects its occupants against the extreme conditions observed outdoors without creating uncomfortable internal conditions. According to (Chatelet *et al.*, 1998), the internal climate in a free-running building (i.e. without any heating or cooling system

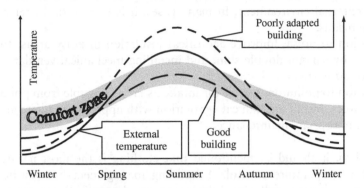

Figure 3.11 Evolution of temperatures in a free-running building and its environment throughout the year (Roulet, 2004).

running) should be at least as comfortable as the outdoor climate. This strategy is explained below and in Figure 3.11.

Because of changes in their clothing, occupants require different temperatures in order to be comfortable (the so-called 'comfort temperatures') during summer or winter. Therefore, the comfort 'zone' (the range of comfortable temperatures) is higher in summer than in winter.

A well-adapted building (curve A in Figure 3.11) has a good thermal insulation, appropriate passive solar gains (including moveable and efficient shading systems) and adaptive ventilation devices. In summer, it is protected against solar radiation and designed for passive cooling. In winter, it uses solar gain to increase the internal temperature. The result is a building that, in most temperate climates, provides comfort without energy sources other than the sun during most of the year. The energy use for heating is strongly reduced as a result of a shorter heating season. Cooling is not required as long as the internal heat load stays within reasonable limits and, in warmer climates, cooling energy is kept reasonable.

On the other hand, a poorly adapted building is not well insulated and protected against solar radiation. It is designed neither for an efficient use of solar energy nor for passive cooling. Its free-floating internal temperature is then too low in winter and too high in summer. Expensive and energy-consuming systems have to be installed in order to compensate for this misfit between the building and its surrounding climate. Such poorly adapted buildings will require heating in winter and cooling in summer and are the cause of the belief that the use of large amounts of energy is necessary for comfort.

Adaptation of the building to the environment includes the following:

• adaptation to climate: Appropriate thermal insulation, solar protections and ventilation openings. An acceptable architecture should ensure that the building is, without any heating or cooling, at least as comfortable as the external environment. In many cases, it is easy to do better by passive means.

- adaptation to noise: Improve acoustical insulation in noisy areas, for example by using a double skin, and installing mechanical ventilation with sound barriers.
- adaptation to pollution: Locate air intake as far as possible from pollution sources, install mechanical ventilation with appropriate filters and ensure appropriate maintenance.

Nevertheless, it should be mentioned that clothing is the most natural first step for temperature control. A building management has justified air conditioning because full casual dress was mandatory in the company. Allowing clothing adaptation in buildings certainly improves comfort and may save much energy!

3.4.3   Ensuring thermal comfort

Thermal insulation

A good thermal insulation not only reduces heating and cooling energy use but also improves comfort by reducing the unpleasant effects of cold or warm internal surfaces. In temperate climates, where buildings require essentially cooling, a good thermal insulation reduces the heat gains from outdoor air and solar radiation. However, it should be combined with passive cooling, to evacuate internal heat gains. It was observed that HQ buildings are better insulated (lower U-values for roofs, walls and glazing) than other ones (Roulet *et al.*, 2003).

Table 3.5 Rating of thermal environment and air movement in winter in apartment buildings with and without supply air heating in Finland (Roulet et *al.*, 2005a) (scale from 1 = satisfactory to 7 = unsatisfactory; averages over 5 and 6 buildings)

	Heated supply air		Unheated supply air		
	Mean	σ	Mean	σ	P[4]
Thermal comfort	2.65	0.37	3.43	0.33	0.04%
Thermal sensation	3.96	0.22	4.18	0.27	3.27%
Temperature stability	3.10	0.32	3.56	0.44	1.10%
Feeling draughts	3.82	0.21	4.34	0.30	0.07%
Overall comfort in winter	2.57	0.33	2.76	0.39	13.06%

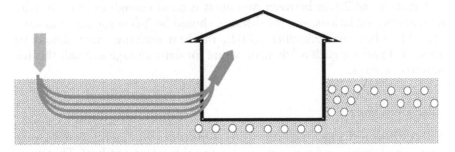

Figure 3.12 Principle of a ground heat exchanger. Circles represent possible ducts locations.

Preheated ventilation

Use of preheated supply air improves thermal comfort in cold climate (Table 3.5). This can be achieved by placing air inlets close to the radiators, by installing ground air exchangers ('Canadian wells') or by installing mechanical air supply with heat recovery.

In ground-to-air heat exchangers, air is preheated or precooled in a network of underground ducts (Figure 3.12) before being supplied to the building. The ground temperature is colder than outdoor air in the warm season and warmer than outdoor air in the cold season.

The guidelines below should be followed to get good results:

- place cleaning openings at least at each turn;
- ensure draining of the whole network: water should not stay in the ducts;
- ensure, if necessary, enough thermal insulation between the building and the ducts. Heat for the air should not come from the building;
- ensure a good thermal contact between the external duct surface and the soil. A thin air layer could strongly reduce the heat exchange efficiency.

Duct's material should fulfil the following requirements (Fraefel *et al.*, 2000):

- resist to ground pressure
- resist to soil acidity
- air-and gas-tight (water vapour, radon)
- smooth, to make cleaning easy
- easy to install, low cost.

The length of the ducts should not be larger than 30 m, to avoid too large pressure drop. For the same reason and to ensure a good heat exchange, the air velocity in the ducts should be between 2 and 3 m/s.

A distance of 20 cm between two ducts is good enough to smooth daily temperature variations, but the distance should be 2–3 m for seasonal storage. According to Hollmuller (2002), the heat exchange area should be about $1/15 \, m^2$ for each m^3/h airflow rate for daily storage and half this for seasonal storage.

Solar protections

Solar radiation changes with time and is sometimes very useful (passive solar gains, daylighting) and sometime uncomfortable (glare, too hot). Therefore, in order to increase comfort and decrease energy use, windows as well as other envelope elements collecting solar radiation (sunspaces, transparent insulation) should be equipped with mobile, controllable solar protections, in order to control light and solar gains. An automatic control with possibility for individual override improves comfort (Guillemin and Molteni, 2002).

Solar radiation is partly reflected by the solar protection, but a large part is absorbed in it and transformed into heat, which increases the temperature of the solar protection. If this protection is outdoors, this heat is released to outdoor air and does not enter the building. If the solar protection is inside, this heat is delivered to indoor air, thus heating the building (Figure 3.13).

Therefore, good solar protections for mid-season and summer are external, because only these can significantly reduce the gains to avoid overheating. Internal solar shading might be necessary to control lighting conditions (glare). Use wind proof external solar protection, if necessary use solar protection behind a second skin. In this case, the space between the two skins should be well ventilated, with openings at least every second floor.

In winter, internal shadings are preferred, because they avoid glare and allow solar heat gains. Overhangs or claustras are not ideal, because they

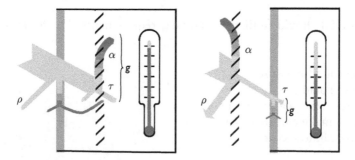

Figure 3.13 A large part of the solar radiation is absorbed by the solar protection and transformed into heat. If the solar protection is inside, this heat increases the internal heat load.

do not allow any control: they either are not sufficient on some sunny days or reduce daylight on overcast days. Vegetation with deciduous leaves may help by bringing a seasonal solar protection. However, it also shadow the building the whole day long, even when more daylight is necessary.

The use of sun reflective glazing decreases daylighting, hence increases artificial lighting and heat load. The sun remains the most efficient light source: only $1\,\mathrm{W/m^2}$ heating for 100 lux.

Within the HOPE project, it was observed that few of the audited buildings, in particular among office buildings, have efficient solar protection. In many cases, external solar protections are not allowed, either for architectural reasons (!) or because of too high wind pressure in high-rise buildings. Expected comfort problems are, too hot on sunny days, possible glare, increasing cooling energy use, increased cooling load, encouraging the use of mechanical cooling.

In buildings with natural ventilation, the risk of having too high temperature increases in buildings without solar protections. If all buildings are taken into account (both uncooled and cooled), there are still significant differences ($P < 5\%$) in the perceived temperature (Figure 3.14), thermal comfort and temperature stability between buildings with and without solar protections. All these differences are in favour of buildings with solar protections.

Passive solar heating

Solar energy enters freely in the building, mainly through the windows, thus contributing to heating the building. In order to maximise the use of these gains during the cold season without suffering from overheating, the following conditions should be met (Figure 3.15):

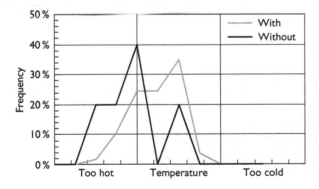

Figure 3.14 Vote on the perceived temperature in summer in buildings with and without solar protections.

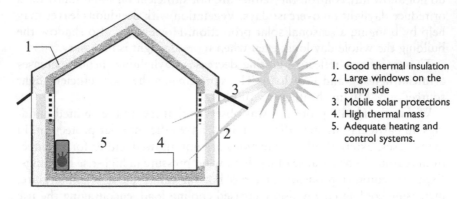

Figure 3.15 Requirements for passive solar heating.

1. Avoid excessive heat loss with a good thermal insulation and low-e coated double glazing.
2. Design large collecting areas on the sunny side of the building (from south-east to south-west in Northern hemisphere). Collecting areas are mostly windows but can also be sunspaces or walls with transparent insulation.
3. Equip these collecting areas with mobile, controllable solar protections (see above).
4. Install a heating control that cuts the heating system down as soon as (or even before than) solar gains suffice to keep a comfortable internal temperature, and switch it on as soon as necessary.
5. Design the building with a large thermal mass (heavy construction) allows to store heat, thus avoiding overheating during sunny day and maintaining a mild indoor climate during the night.

Passive solar gains can be useful to improve comfort at low cost during the heating season but may be uncomfortable when poorly controlled, especially in mid- or hot season.

Overheating in summer

The building may be at risk of being too hot in summer for various reasons:

- located in a warm climate
- large glazed area without solar protection
- neither passive nor active cooling
- too high internal heat load.

Expected comfort problems are transpiration, reduced performance and productivity, increase of human errors. The risk of reduced air quality

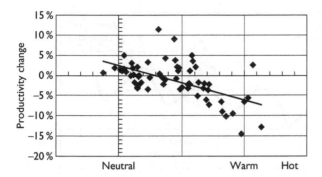

Figure 3.16 Perceived productivity and temperature in summer (Roulet et al., 2005a).

(smells, humidity) also exists. Hot buildings increase the cooling load or encourage mechanical cooling.

A high perceived temperature in summer decreases the perceived productivity (Figure 3.16). This relationship was not observed in winter. The negative slope in Figure 3.16 has a probability smaller than 10^{-5} to be zero or positive.

Not surprising also is that the perceived global comfort in Summer is better in office buildings with mechanical cooling [observed mean vote is 0.75 on the ISO 7730 .(ISO, 1993) scale] than in buildings without mechanical cooling (mean vote is +1.17). This difference is very significant ($P = 10^{-6}$). However, the difference between the perception of thermal comfort between these two groups of buildings is smaller: votes are both close to 4 (3.93 and 4.29) on a scale from 1 'comfortable' to 7 'uncomfortable'. The difference in average votes on this question is also less significant ($P = 0.5$ per cent). This confirms (see Chapter 2) that occupants of buildings without mechanical cooling are more tolerant than those in fully conditioned buildings (de Dear and Brager, 2002). The productivity is perceived as unchanged in cooled buildings and decreased by 3 per cent in the other buildings. This difference is not large but significant ($P = 0.01$ per cent).

Reducing inside air temperature by improved insulation and solar protections, together with passive cooling using night ventilation (see below) is sufficient to prevent discomfort and energy problems in many cases. Mechanical cooling improves thermal comfort but increases energy use.

Passive cooling

Passive cooling through nighttime ventilation is a cheap and energy-efficient way to keep the indoor environment within a comfortable temperature range in most temperate climates, in particular in continental climates, and higher altitudes in sub-tropical areas. In well-adapted buildings, it can

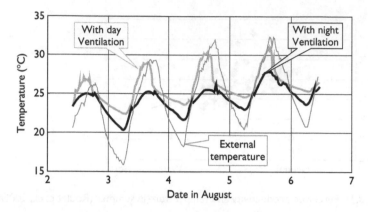

Figure 3.17 Temperatures in two identical office rooms. One is aired as usual, only during the day. The other one is aired mainly at night (Van der Maas *et al.*, 1994).

ensure a comfortable indoor climate in summer without artificial cooling, provided that internal heat load is not too large.

Figure 3.17 shows the evolution of internal and external temperatures in two identical office spaces (40 m³) of the LESO,[4] which has been ventilated following two different strategies:

1. The usual strategy in office buildings, with ventilation during the day but not at night; and
2. The passive cooling strategy with natural ventilation at night.

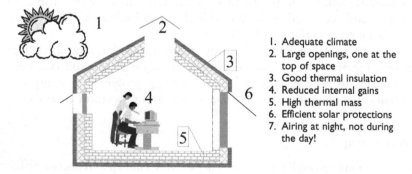

1. Adequate climate
2. Large openings, one at the top of space
3. Good thermal insulation
4. Reduced internal gains
5. High thermal mass
6. Efficient solar protections
7. Airing at night, not during the day!

Figure 3.18 Requirements for an efficient passive cooling.

The office spaces have considerable thermal inertia and external, moveable solar blinds. The natural night ventilation rate corresponds to about 10 building air volumes per hour. One person occupies the office during 8 h per day, often with a personal computer running. This experiment, along with many others, shows that summer comfort can be greatly improved at low cost using passive cooling.

Principles of passive cooling are compatible with those of passive solar heating. As shown in Figure 3.18, they are:

1. Passive cooling strategy can be applied only in climates where the daily average outdoor temperature is within comfort limits, and where there is a significant temperature swing between night and day.
2. Design large openings for natural ventilation at night, one of them being at the top of the ventilated space. This one should preferably be larger than all the other ones. These openings should be safe enough to remain open at night (protected against rain, insects, unwanted intrusions).
3. Avoid heat gains by using good thermal insulation, efficient shading systems (external and movable), and minimum ventilation rate during hot hours.
4. Avoid internal heat load by promoting daylighting and energy efficient appliances.
5. Store the remaining heat gains in the building structure. For this, the heavy building structure should be in direct contact with the indoor environment.
6. Cool the building structure down with a large ventilation rate when the external temperature is lower than the internal temperature. Large ventilation rates are easily obtained by natural ventilation through windows and doors.

If passive cooling is not possible, mechanically driven night ventilation is an alternative that should be considered. It is however more expensive and energy intensive to get the large airflow rates required for night cooling by mechanical means. This issue was addressed within the European project HybVent.[5]

The LESO building is a passive solar office building (Figures 3.19 and 3.20). Its total energy performance index is less than $60 \, \text{kWh} \, \text{m}^2$ per annum, including all appliances. This massive building is very well thermally insulated (20 cm insulation thickness, low-emissivity coated double glazing) and has large passive solar gains controlled by external movable solar protections. For passive cooling, there are safe openings at the bottom and the top of the building, so that a large airflow rate may

Figure 3.19 The LESO building, in which measurements shown in Figure 3.17 were performed (photo: EPFL). This is the south façade, with large windows and special daylight devices.

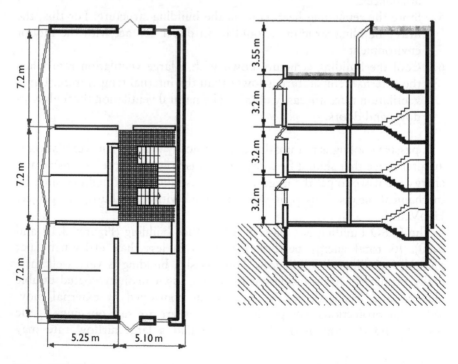

Figure 3.20 Horizontal and vertical cuts of the LESO building.

cool down the staircase during the night, as well as all offices that have their door open to the staircase. It is also possible to ventilate each office room individually by leaving the windows ajar. This way is however less efficient.

3.5 Strategies for HVAC systems and components

3.5.1 Air and water as heat transfer media

Heat can be transferred from one place to the other by moving a heated fluid. Both air and water are used for this purpose. The amount of heat transferred that way is:

$$Q = cm\Delta\theta = c\rho V\Delta\theta \tag{3.26}$$

where

- c is the specific heat capacity of the fluid
- m is the transferred mass of fluid
- $\Delta\theta$ is the temperature difference of the fluid before and after having transferred the heat
- ρ is the density of the fluid
- V is the transferred volume of fluid.

The air is not good for heat transport, especially when compared with water. First, its specific heat capacity is about $1,000\,J/(kg \cdot K)$, while that of water is four times larger, i.e. $4,180\,J/(kg \cdot K)$. Second, its density is about $1.2\,kg/m^3$ while that of water is $1,000\,kg/m^3$. Therefore, one cubic meter of water transports 3,500 times more heat than one cubic meter of air, for the same temperature difference (Figure 3.21).

Moreover, for comfort reasons, air cannot be supplied in occupied space above 30°C for heating or below 18°C for cooling. These comfort requirement limit the temperature difference, hence the transferred amount of heat. On the contrary, hot water can be distributed in the radiators at much higher temperature, and chilled water may be cooled down to 0°C and distributed into radiant panels (see Section 3.2.2) at the dew point of the ambient air. However, the hot water temperature should be decreased and chilled water temperature increased to avoid too large heat loss in the distribution network. Nevertheless of the temperature differences are taken into account, water can transfer 7,000 times more heat for cooling and 15,000 times more heat for heating than air.

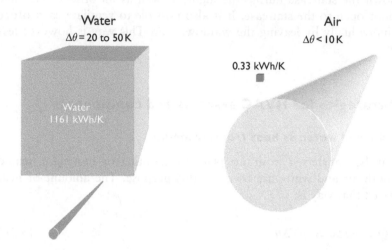

Figure 3.21 Comparing heat transfer capacities of water and air. The cubes represent the volume heat capacity, and the cylinders represent ducts able to transfer the same amount of heat.

The air velocity in ducts may be up to 20 m/s in high-pressure systems, but this should be avoided to reduce noise and energy for fans. It should ideally be about 2–5 m/s. Water velocity in pipes is limited to 1 m/s to avoid erosion. When everything is taken into account, the section of air ducts should be 2,000 times larger than that of water pipes, and their diameter 45 times larger (Figure 3.21).

A larger duct size implies also a larger external area, hence larger heat losses for the same insulation thickness. For all the reasons mentioned above, heat should be transported with water instead of air wherever possible, for example in hydronic heating and cooling systems.

3.5.2   Hydronic heating and cooling

Hydronic systems for heating are of common use, the room heat exchangers – called convectors or radiators depending on the convective and radiative parts of the heat exchange – being most often installed vertically on the walls, often below the windows. This cumulates the advantages of a hydronic system with the fast reaction to load changes, provided the installation of thermostatic valves on each unit.

Floor heating also becomes more and more popular. In this system, a network of small pipes is embedded in the floor rendering. The heating power is limited by the surface heat exchange coefficient, the floor area and the comfort limits: floor temperature in winter should be between 20 and 27°C

Figure 3.22 Auditorium cooled with radiant panels on the ceiling (photo: Energie Solaire SA, Sierre).

(Fanger, 1983). For this reason, it is comfortable only in well-insulated buildings, where the heating load remains limited. The advantages are a good comfort and low temperature heating allowing the use of high efficiency condensing boilers and renewable energy sources. The slow reaction to load changes is partly compensated by the auto-regulation resulting from the low heating water temperature.

Fan coil units or induction units allow heating and cooling the room air through a forced ventilation heat exchanger, in which hot or chilled water circulates. The room air is forced through the heat exchanger either by a fan or by ventilation air jets. These units can react quickly to change in heating or cooling loads and are relatively small for a given power. They are however noisy and need frequent cleaning and maintenance to keep a good indoor air quality.

Cooling can also be achieved by using radiant panels installed on walls or preferably as false ceiling (Figure 3.22). This method has several advantages (Roulet *et al.*, 1999b):

- the cooling power is rather high: about $10\,W/(m^2 K)$, so easily $100\,W/m^2$ with flat panels $10\,K$ below the ambience. This power may be even larger with finned panels that increase the convective part (creating cold draughts in extreme cases).

- this hydronic system makes neither noise nor vibration. This latter advantage is used to cool rooms with high-resolution electron microscopes.
- if the panels are flat, a large part of the heat is transferred by radiation, ensuring thermal comfort even at high air temperatures or with open windows.
- flat panels in normal use induce no draughts and good comfort, because occupants accept cold ceilings very well (Fanger *et al.*, 1985).

The disadvantage is however that the radiant panels cannot be cooled down below the dew point of the indoor air, because condensation will then drop from the panels. This first requires large panel areas, because the temperature difference between the panel and the ambience is limited, and second limits the use to dry climates or requires drying the indoor air.

The same system may be used for heating, provided that the heat load remains limited to about $25 \, \text{W/m}^2$, because the surface heat transfer coefficients is rather small for a warm ceiling, and occupants would complain if the ceiling is too warm (see Fanger *et al.*, 1980 and Chapter 2).

3.6 Heat recovery

3.6.1 Introduction

Ventilation in buildings – especially in large buildings and advanced low-energy and passive-solar houses – is becoming increasingly important for many reasons. One of them is the excellent standard of thermal insulation, which easily raises the contribution of ventilation losses – depending on the building's compactness and air change rate – to more than 50 per cent of total thermal loss. Another reason for the importance of ventilation is airtightness of buildings' envelopes, which avoids air infiltration heat loss but does not anymore provide sufficient ventilation. To cope with ventilation requirements with regard to hygiene and building physics, mechanical ventilation systems are of increasing use. In order to reduce energy consumption, ventilation systems with energy-efficient heat recovery systems are almost mandatory. More information on these systems is found in Section 6.5.6.

However, air-handling units may have parasitic shortcuts and leakage (Hanlo, 1991; Fischer and Heidt, 1997; Heidt *et al.*, 1998; Roulet *et al.*, 2001), which can decrease dramatically the efficiency of ventilation and heat recovery. Moreover, leakage in a building's envelope allows warm air to escape outdoors without passing through the heat recovery system. In addition, these units use electrical energy for fans, which may, in some cases, overpass the saved heat. The influence of these phenomena on the real energy saving is addressed in this paper.

3.6.2 Effect of leakage and shortcuts on heat recovery

Airflow rates, heat loss and heat recovery efficiency

Let us consider the air- and heat flows in the unit schematically represented in Figure 3.23. Outdoor air, o, enters the inlet grille, i, and is blown through the heat recovery system HR, where it is either heated or cooled. Then, after subsequent heating or cooling, rs, it enters the supply duct, s, to be distributed into the ventilated space. As the envelope is not perfectly airtight, the supply air may be mixed with infiltration air, inf. A part of the indoor air may also be lost by exfiltration (exf). The extract air, x, passes through the other part of the heat recovery system, re, where it is either cooled or heated. The air is then blown out to the atmosphere, a, through the exhaust duct, e.

If the exhaust and inlet grilles are not well situated, a part of the exhaust air may re-enter the inlet grille, resulting in an external recirculation rate, R_e. Leakage through the heat recovery system may also result in internal recirculation, from inlet to exhaust R_{ie}, or from extract to supply, R_{xs}.

In simplified methods to calculate heating (or cooling) demand of buildings, ventilation heat loss, Φ_v, is calculated by (CEN, 2004):

$$\Phi_V = c\dot{m}\left(\theta_x - \theta_o\right)\left(1 - \eta_G\right) \tag{3.27}$$

where

c is the heat capacity of air, i.e. $1,000\,\text{J}/(\text{kg}\cdot\text{K})$
$\dot{m}$ is the mass flow rate of outdoor air in kg/s
θ_x the temperature of extract air, which is considered as representative of the indoor air
θ_o the temperature of outdoor air and
η_G is the global efficiency of the heat recovery system.

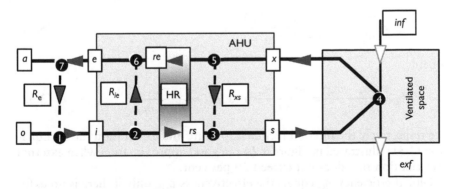

Figure 3.23 The simplified network representing the air handling unit and ducts. Arrows represent considered airflow rates.

This global efficiency, η_G, should consider the whole system, consisting of the ventilated building and its ventilation equipment. But, instead, often the nominal temperature efficiency of the heat recovery unit itself, ε_{HR}, is used. This efficiency is approximately

$$\varepsilon_{HR} \cong \frac{\theta_x - \theta_{re}}{\theta_x - \theta_o} \tag{3.28}$$

where the signification of subscripts can be seen in Figure 3.23. This replacement leads to optimistic results when the air-handling unit has recirculation or when the building is leaky.

Global heat recovery efficiency

If there were no heat recovery, the heat loss from the building, Φ_1, resulting from ventilation is the sum of extract heat flow and exfiltration heat loss, equal to the heat necessary to bring outdoor air to indoor climate conditions:

$$\Phi_1 = c\,(\dot{m}_x + \dot{m}_{exf})\,(\theta_x - \theta_o) = c\,(\dot{m}_s + \dot{m}_{inf})\,(\theta_x - \theta_o) \tag{3.29}$$

Neglecting latent heat, the recovered heat is

$$\Phi_r = c\dot{m}_{re}\,(\theta_x - \theta_{re}) = c\dot{m}_{rs}\,(\theta_{rs} - \theta_i) \tag{3.30}$$

as, in first approximation, all the heat taken from extract air is given to supply air. The global heat recovery efficiency of the system, η_G, can be calculated as a function of the fresh airflow, exfiltration, and recirculation rates, by taking account of mass conservation at the nodes of the system (Roulet *et al.*, 2001). The full relation is rather complex but, when there is no external recirculation, the global efficiency can be expressed as a function of exfiltration ratio and internal recirculation rates only:

$$\eta_G = \frac{\Phi_r}{\Phi_1} \cong \frac{(1 - \gamma_{exf})\,(1 - R_{xs})}{1 - R_{xs}\gamma_{exf}}\varepsilon_{HR} \tag{3.31}$$

where

$$R_{xs} = \frac{\dot{m}_s - \dot{m}_{rs}}{\dot{m}_x} = \frac{\dot{m}_x - \dot{m}_{re}}{\dot{m}_x} \text{ and } \gamma_{exf} = \frac{\dot{m}_{exf}}{\dot{m}_o + \dot{m}_{inf}} \tag{3.32}$$

are respectively the internal recirculation rate and exfiltration ratios. Equation (3.31) illustrated in Figure 3.24 is a good approximation when external recirculation rate does not exceed 20 per cent.

Global efficiency η_G equals the effectiveness ε_{HR} only if there is no exfiltration, and there is neither external- nor extract-to-supply recirculation. Otherwise, η_G is smaller than ε_{HR}.

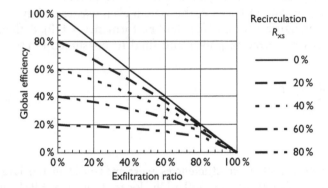

Figure 3.24 Global heat recovery efficiency as a function of exfiltration ratio γ_{exf} and internal recirculation rate R_{xs}. In this figure, $\varepsilon_{HR} = 100\%$, $R_{ie} = R_e = 0$.

The inlet to exhaust recirculation as well as the infiltration ratio has only a small effect on heat recovery efficiency but reduces the amount of fresh air supplied by the unit to the ventilated space. This recirculation obviously results in an increased consumption of electric energy for the fans, which is approximately proportional to the cube of the airflow rate, without delivering more fresh air. However, such parasitic recirculation is often not noticed and hence can lead to an undiscovered reduction of indoor air quality.

Specific net energy saving and coefficient of performance

A crucial issue is that HR-systems recover thermal energy but use electrical energy for the fans. As a useful figure to deal with this fact, we introduce the specific net energy saving per cubic meter of supplied outdoor air (*SNES* in Wh/m^3) averaged over a heating period, for which the mean outdoor temperature is $\bar{\theta}_o$. This figure is calculated by

$$SNES = \rho_o \frac{\eta_G \Phi_l + \Phi_{fan} \left(f_r - f_p \right)}{\dot{m}} \tag{3.33}$$

where

$\Phi_l = \dot{m}c \left(\bar{\theta}_x - \bar{\theta}_o \right)$ is the ventilation heat loss, based on average internal and external temperature during the heating season;

f_r is the part of the fan power recovered as heat in the supply air. This factor f_r is close to one for supply fans and zero for exhaust fans;

f_p is a production factor, accounting for the fact that the production of 1 kWh of electrical energy requires much more primary energy;

Only if *SNES* is positive, a net gain in thermal or primary energy is achieved by the HR-system. Otherwise, the system even wastes energy.

Another interesting figure is the coefficient of performance (COP), defined by the ratio of recovered heating power and final to consumed electrical power:

$$COP = \frac{\eta_G \Phi_1 + f_r \Phi_f}{\Phi_f}$$

(3.34)

Measurements

Airflow rates and heat exchanger efficiencies were measured in ten large units handling from 2,000 to 17,000 m^3/h located at the EPFL in Switzerland, and three small, wall-mounted room ventilation units handling 25–75 m^3/h, measured at the University of Siegen, in Germany. The measurement technique is described in Chapter 8 and in more detail in (Heidt and Rabenstein, 1994; Roulet et al., 1999a). All required airflow rates are determined using tracer gases and the assessment of fan efficiency as described in Section 3.3.3.

The temperature efficiency of the heat recovery system itself is simply calculated from temperature measurements upwind and downwind the heat exchanger in both supply and exhaust channels, using Equation (3.28). Figure 3.25 shows the global, actual heat recovery efficiency versus the theoretical one, represented by the nominal heat exchanger efficiency. Only two units are close to perfection, and two units perform very poorly.

Major leakage has been observed in several buildings. In three of them, infiltration represents a significant part of the outdoor air, and in four of

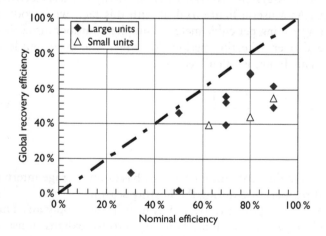

Figure 3.25 Global heat recovery efficiency versus nominal heat exchanger efficiency measured in several units.

them, most of the air leaves the building through the envelope instead of passing the heat recovery unit. Significant internal recirculation is observed in the three small units, and external recirculation above 20 per cent is measured in three large units. Leakage and shortcuts significantly affect heat recovery efficiencies, which drop from nominal values between 50 and 90 per cent down to actual values ranging between 5 and 69 per cent. On average, the heat recovery effectiveness ε_{HR} is 70 per cent, but the global, real efficiency is only 43 per cent. In the best case, an 80 per cent heat recovery effectiveness is reduced by 15 per cent down to a 69 per cent real efficiency.

The SNES and COP shown in Figure 3.26 are calculated with 16 K indoor–outdoor average temperature difference during 210 days, a recovery factor for fans, $f_r = 0.5$ (taking account that here are two fans in these units, one of them in the supply duct) and a production factor $f_p = 3.55$, which is the average for low-voltage electricity in Europe according to Frischtknecht *et al.*, 1994.

SNES can be very small or even negative. In the best case, it reaches $2.7\,\mathrm{Wh/m^3}$, corresponding to 8 K average temperature increase of fresh air. It should be also noticed that the COP can overpass 8 but might also be much smaller than expected, as it is often the case for air-to-air heat pumps.

Best net energy saving in the measured large units is 80,000 to 90,000 kWh per winter season, but another unit actually spills as much energy. Small measured units save between 80 and 350 kWh during an entire season. From energetic and economic aspects only, such ventilation units are disadvantageous and hard to recommend. Note that these results are obtained when the heat recovery is functioning.

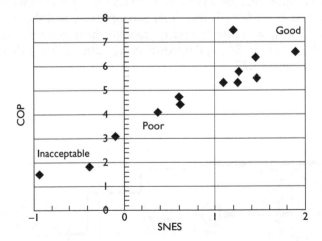

Figure 3.26 Coefficient of Performance (COP) versus Specific Net Energy Savings (SES) in the measured units. Only half of the measured units can be considered as actually efficient.

3.6.3 Conclusions

Heat recovery from extract air is often installed in advanced low energy buildings in order to ensure efficient ventilation at low energy cost. However, global efficiency of heat recovery depends significantly on air infiltration and exfiltration, which should be minimised during the heating period. Internal and external recirculation also decreases the efficiency of the heat recovery units. Moreover, electrical energy for fans is used in order to supply fresh air and to recover thermal energy from exhaust air.

Characteristic figures for the evaluation of ventilation units with heat recovery have been defined and measured using the tracer gas dilution method. The most important of them are global efficiency of heat recovery and SNES. For several examined ventilation units, energetic savings were small or even negative. Even if best technical performance is assumed (airtight building, $\varepsilon_{HR} = 90$ per cent, specific fan power equal to $0.2\,\mathrm{Wh/m^3}$), the economic viability of small ventilation units remains questionable. This, however, does not affect the other qualities of ventilation systems such as steady supply of fresh air with low concentrations of contaminants.

3.7 Effect of ventilation strategies on the energy use

3.7.1 Method

In order to estimate the effect of various ventilation strategies on energy use, a typical office building was simulated using TRNSYS 15 with two additional specific components: the Type 157 COMIS for TRNSYS (Weber *et al.*, 2002 developed by EMPA, Dübendorf, Switzerland), and the Type 168 HVAC module with absolute humidity control, heat recovery and cooling (developed by Sorane SA) (Jaboyedoff *et al.*, 2004). The HVAC system is outlined in Figure 3.27.

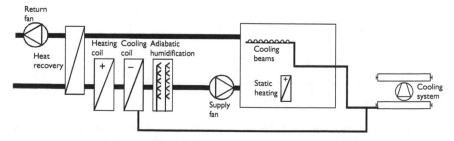

Figure 3.27 HVAC-system (Jaboyedoff *et al.*, 2004).

The three-floor building is of heavy construction and well insulated (U- values of walls, roof and floor equal or better than $0.35\,W/(m^2K)$, low-e glazing with $U = 1.4\,W/(m^2K)$ and 60 per cent total solar energy transmission. The office rooms are oriented south with 53 per cent glazed area. The indoor temperature can be controlled using air conditioning or hydronic heating and cooling.

The base case is simulated with the following parameters:

- outdoor air supply: 0.7–1.4 air changes per hour ($25–50\,m^3/h$ per person)
- no recirculation
- design room temperature for heating 20°C and off-set cooling system room of 26°C
- supply air temperature 20°C
- RH of supply air to room 30 per cent
- operating hours: during working hours (7–19 o'clock)
- pressure over system (supply and exhaust): 1,000 Pa (low velocities)
- fan efficiency of 0.5
- COP of 3.

Simulations were made for different climates: Northern (Oslo), Central (London and Zürich) and Southern Europe (Rome), and the following variants have been calculated:

- outdoor air supply: 15 and $35\,m^3/h$;
- RH of supply air to room: 50 per cent (basic) and no humidification;
- efficiency of heat recovery: 0 (no heat recovery), 0.75 and 0.85;
- infiltration 0.5 and 1.0 air changes per hour;
- setpoint for cooling (local): 24°C, 28°C and 50°C (no cooling);
- continuous ventilation (24 h on);
- natural ventilation instead of mechanical ventilation (openable windows).

For each variant, the effect of the following changes in design and operation has been calculated:

- with 50 per cent recirculation instead of no recirculation;
- without heating of air by coils;
- effect of ventilation strategies on heating demand;
- effect of air tightness on heating demand;
- without cooling of air by coils;
- lowering the set point for cooling;
- without droplet catcher behind cooling coil: 20 Pa lower pressure difference over system;

- with a more stage filter system instead of old F7 filter: Adding a F5 filter leads to a 150 Pa higher pressure difference over system.
- with new filter (F7) instead of old filter (F7): Changing used filters at a pressure drop of 180 Pa instead of 250 Pa leads to a reduction of the average pressure difference over the filter of 40 Pa and hence a 40 Pa lower pressure difference (average) over the system.
- no humidification;
- without heat recovery;
- with rotating wheel (RHE) with/without purging sector: 3 per cent difference in efficiency of RHE;
- with other heat exchanger: efficiency 0.75 instead of 0.85;
- effect of leakage on heat recovery: lower efficiency (e.g. 0.75 instead of 0.85);
- effect of increase of pressure difference caused by air velocity, length, curves, interior smoothness, and deposits in the ducts: 1,600 Pa versus 1,000 Pa.

3.7.2 Simulation results

The main results of these simulations are summarized below. These conclusions, in particular the numerical values, are valid for the building and the system simulated. However, the order of magnitudes and general tendencies are likely to be valid for more general situations.

Recirculation

With 50 per cent recirculation instead of no recirculation: The electric energy demand (cooling and fans) for ventilation decreases by about 40 per cent (27 per cent for Rome, 43 per cent for Zürich and 50 per cent for Oslo) if no recirculation is used, compared with 50 per cent recirculation. No heat recovery was used in these cases.

Heating

In all climates and for the same airflow rate, heating of air with heating coils in the air handling unit or using room hydronic heating does not change significantly the energy use.

The energy use for heating mainly depends on climate and internal gains. Except for heat recovery and time schedule of operation (working hours/24 h per day), ventilation strategies have a minor influence on heating energy demand.

The tightness of the building envelope has a big influence, up to a factor 2, on the heating energy need. When high infiltration occurs, humidity is also lowered.

Cooling

For the same airflow rate and same comfort conditions, more energy was required for cooling using air conditioning than with the hydronic cooling ceiling, depending on the location (Oslo, Zürich, Rome).

Lowering the set point for cooling from 26 to 24°C causes an augmentation of the cooling demand of the zones by a factor of 3–8, depending on the geographic location.

A droplet catcher behind a cooling coil results only in 20 Pa increase of the pressure difference over the total system, which results in a negligible effect for the energy demand. The droplet catcher is essential to avoid humidifying downstream filters or acoustic dampers.

Filtering the air

Using a two-stage filter system instead of an old F7 filter leads to an increase in electric power demand for fans by 10–15 per cent, depending on the pressure difference over the system. This leads to an increase in total electric power demand (fans and cooling) of 5–15 per cent, depending on the electric power demand for cooling.

Depending on the pressure difference over the system, earlier replacement of a filter results in a decrease in power demand for fans by 2–3 per cent. This results in a decrease of the total electric power demand (fans and cooling) of 1–2 per cent, depending on the electric power demand for cooling.

Humidification

In the northern (Oslo) and central locations (London/Zürich), the control of humidity supplied at 30 per cent relative humidity (against no humidification) generates an additional heating energy need of about 20–25 per cent for ventilation. For the southern climate (Rome), the increase is only 3 per cent. The increase of the total heating energy demand for a humidity of the supply air of 30 per cent is 5–10 per cent.

The control of humidity supplied at 50 per cent relative humidity (against no humidification) almost doubles the heating energy need for the ventilation, for all locations.

Heat recovery

Without heat recovery, the heating energy demand for ventilation is 67–137 per cent higher compared with a medium efficiency heat recovery (50 per cent).

A heat recovery with high efficiency (75 per cent), such as those achieved by well-installed rotating heat exchanger (RHE) leads to a reduction of the heating energy demand for ventilation by about 30 per cent compared with a medium efficiency (50 per cent) heat recovery. This means that the 3 per cent difference in efficiency (RHE with purging sector) has a negligible effect on the energy demand.

Infiltration or exfiltration through a leaky building envelope strongly reduces the efficiency of the heat recovery (see Section 3.6.2). With a heat recovery efficiency of 75 per cent, the heating energy demand for ventilation for cold and mild climate (Oslo, Zürich, London) was approximately 20 per cent higher than for an efficiency of 0.85. For warm climates (Rome), this was approximately 50 per cent higher, however with small absolute values (differences up to 2–3 MJ/m^2).

This results in a total heating energy demand for cold and mild climate (Oslo, Zürich, London) approximately 3–5 per cent higher with an efficiency of 0.75 than with 0.85. For warm climates (Rome), this number is approximately 15 per cent higher, however with low absolute values (differences up to 2–3 MJ/m^2).

Duct work

An increase of pressure difference caused by air velocity, length, curves, interior smoothness and deposits in the ducts from 1,000 to 1,600 Pa leads to an increase in electric power use of 60 per cent. The increase in total electric power depends on the geographic location and ranged from 25 to 55 per cent.

3.8 Summary

Energy is used in buildings for heating, cooling, ventilation, air conditioning, hot water, lighting, producing goods and services etc. The energy flows in buildings can be assessed either by calculation, using sophisticated or simplified mathematical models, or by measurements. Several models, summarized in Section 3.2 are standardised at international level.

Estimating various energy flows brings the necessary knowledge for improving energy performance. This improvement should however not be obtained at the cost of reduced comfort or decreased air quality. There are many ways to improve both the energy performance and indoor environment quality. It is recommended to apply first the passive ways, which are architectural measures that do not need purchased energy: thermal insulation, natural ventilation, daylighting, passive solar heating, passive ventilation cooling, etc. The aim is to adapt the building to its environment by a smart design. As these passive ways often depends on the meteorological conditions, active ways such as heating, cooling, air conditioning, artificial lighting, etc. are then used to compensate for the deficiencies of passive means.

It is also essential to give to the user the possibility to control his environment and to adapt it to his needs. This improves his well-being and productivity.

It should also be noticed that the air is not a good heat transferring medium and that water is about 7,000 times better for cooling and 15,000 times better for heating than air. In principle, ventilation systems should evacuate the contaminants and bring new air, but not transfer heat. Cooling and heating can be performed by hydronic systems.

If, nevertheless, air is used to transfer heat, heat recovery can be installed to greatly improve the energy performance. Care should be taken however that most of the air goes through the heat recovery device. For this, an airtight building envelope is paramount.

It was seen by simulations that recirculation decreases the energy use, but at the cost of increasing the indoor air pollution. Hydronic heating is not more energy efficient than air heating, but hydronic cooling uses less energy than air conditioning. Humidification needs much energy and should therefore be avoided wherever possible. Envelope air tightness improves the energy performance, because airflow rates can then be efficiently controlled. Improving the air filtering increases the electrical energy use by a few percent. Low-pressure ventilation network should be preferred. The energy use for fans is directly proportional to the pressure.

Notes

1. The enthalpy is a thermodynamic quantity equal to the internal energy of a system plus the product of its volume and pressure; enthalpy is the amount of energy in a system capable of doing mechanical work. The internal energy is the total kinetic and potential energy associated with the motions and relative positions of the molecules of an object, excluding the kinetic or potential energy of the object as a whole. An increase in internal energy results in a rise in temperature or a change in phase.
2. Reports and publications from the HOPE project can be found on http://HOPE.EPFL.ch
3. Natural ventilation can be controlled by installing (automatically or manually) adjustable vents in an airtight building envelope.
4. The LESO building is a passive solar office building at the EPFL, Lausanne, Switzerland.
5. Comprehensive reports on the HybVent project can be downloaded from http://hybvent. civil.auc.dk/

References

Bluyssen, P. M., Cox, C., Boschi, N., Maroni, M., Raw, G., Roulet, C.-A. and Foradini, F. (2003), European Project HOPE (Health Optimisation Protocol For Energy-Efficient Buildings). *Healthy Buildings*, p. 76–81.

Bluyssen, P. M., De Oliveira Fernandes, E., Groes, L., Clausen, G., Fanger, P.-O., Bernhard, V. O., C.-A. and Roulet, C.-A. (1995), European Audit Study in 56 Office Buildings: Conclusions and Recommendations. *Healthy Buildings*, p. 287–1292.

CEN – EN ISO 13790:2004 – Thermal performance of buildings – Calculation of energy use for heating, Brussels, Genève, CEN, ISO.

CEN – prEN 15203:2006 – Energy performance of buildings – Assessment of energy use and definition of ratings, Brussels, CEN.

Chatelet, A., Fernandez, P. and Lavigne, P. (1998), *Architecture Climatique*, v. 2: Aix-en-Provence, Edisud, 160 p.

de Dear, R. J. and Brager, G. S. (2002), Thermal comfort in naturally ventilated buildings: revisions to ASHRAE Standard 55. *Energy and Buildings*, 34, 549–561.

Fanger, P. O. (1983), Thermal comfort requirements: ICBEM international Conference on Building Energy Management, p. 6.1.

Fanger, P. O., Banhidi, L., Olesen, B. W. and Langkilde, G. (1980), Comfort limits for heated ceilings. *ASHRAE Transactions*, 86, 141–156.

Fanger, P. O., Ipsen, B. M., Langkilde, G., Olesen, B. W., Christensen, N. K. and Tanabe, S. (1985), Comfort limits for asymmetric thermal radiation. *Energy and Buildings*, 8, 225–236.

Fischer, T. and Heidt, F. D. (1997), Testing the ventilation efficiency of ventilation units with tracer gas methods: Second International Conference. *Buildings and the Environment*, p. 405–413.

Fraefel, R., Huber, H. and Trawnika, M. 2000, L'aération dans les bâtiments MINERGIE, CLIMA SUISSE, Olgastrasse 6, 8024 Zürich.

Frischtknecht, R., Hofstetter, P., Knoepfel, I., Dones, R. and Zollinger, E. (1994), *Oekoinventare für Energiesysteme*, Zurich, ETHZ.

Guillemin, A. and Molteni, S. (2002), An energy-efficient controller for shading devices self-adapting to the user wishes. *Building and Environment*, 37, 1091–1097.

Hanlo, A. R. (1991), Use of tracer gas to determine leakage in domestic heat recovery units (HRV): AIVC 12th Conference, 'Air Movement and Ventilation Control within Buildings', p. 19–28.

Heidt, F. D., Fischer, T. and Thiemann, A. (1998), Energetische Beurteilung dezentraler Raumlüftungsgeräte mit Wärmerückgewinnung – Methoden und Beispiele. *Heizung-Lüftung/Klima-Haustechnik*, 49, 52–62.

Heidt, F. D. and Rabenstein, R. (1994), MULTI-CAT – für Lüftungskenngrößen. *KI-Luft- und Kältetechnik*, 30, 119–124.

Hollmuller, P. (2002), Utilisation des échangeurs air/sol pour le chauffage et le rafraîchissment des bâtiments, Université de Genève, Genève, 120 p.

ISO, 1993, ISO-7730: Moderate thermal environments – Determination if the PMV and PPD indices and specifications of the conditions for thermal comfort, Geneva, ISO.

Jaboyedoff, P., Roulet, C.-A., Dorer, V., Weber, A. and Pfeiffer, A. (2004), Energy in air-handling units–results of the AIRLESS European Project. *Energy and Buildings*, 36, 391–399.

Johner, N., Roulet, C.-A., Oostra, A., Nicol, L. A. and Foradini, F. (2005), Correlations between SBS, perceived comfort, energy use and other building characteristics in European office and residential buildings. *Indoor Air*, Paper 306.

Roulet, C.-A. (2004), Qualité de l'environnement intérieur et santé dans les bâtiments: Lausanne, PPUR, 368 p.

Roulet, C.-A., Cox, C., Fernandes, E. d. O. and Müller, B. (2005a), *Health, Comfort, and Energy Performance in Buildings* – Guidelines to achieve them all Lausanne, EPFL, LESO, p. 64.

Roulet, C.-A., Cox, C. and Foradini, F. (2005b), Creating healthy and energy-efficient buildings: lessons learned from the HOPE project. *Indoor Air*, Paper 308.

Roulet, C.-A., Flourentzou, F. and Greuter, G. (2003), Multicriteria Analysis Methodology of Health, IEQ and Energy Use for Sustainable Buildings: Healthy Buildings 2003.

Roulet, C.-A., Foradini, F. and Deschamps, L. (1999a), Measurement of air flow rates and ventilation efficiency in air handling units. *Indoor Air*, p. 1–6.

Roulet, C.-A., Heidt, F. D., Foradini, F. and Pibiri, M.-C. (2001), Real heat recovery with air handling units. *Energy and Buildings*, 33, 495–502.

Roulet, C.-A., Rossy, J.-P. and Roulet, Y. (1999b), Using large radiant panels for indoor climate conditioning. *Energy and Buildings*, 30, 121–126.

Van der Maas, J., Flourentzou, F., Rodriguez, J.-A. and Jaboyedoff, P. (1994), Passive Cooling by Night Ventilation. *EPIC*, p. 646–651.

Weber, A., Koschenz, M., Hiller, M. and Holst, S. (2002), TRNFLOW: Integration of COMIS into TRNSYS TYPE 56: EPIC 2002 Lyon.

Nomenclature

Symbols and units

A		area, cross section m^2
a	numerical parameter	–
C	thermal capacity	J/K
c	specific heat capacity	J/(kg · K)
f	factor	–
h	enthalpy	J/kg
H	heat loss coefficient	W/K
l	distance, length	m
I	Electric current	A
L	latent heat	J
m	mass	kg
$\dot{m}$	mass flow rate	kg/s
M	molecular weight	kg/mole
$\overline{M}$	average molecular weight of the air, that is 0.02894	kg/mole
p	pressure	Pa
Q	energy	J
R	gas constant, i.e. 8.31696	J/mole K
R_{xs}	Recirculation ratio in an air handling unit	–

T	absolute temperature	K
t	time	s
U	voltage	V
V	volume	m3
$\dot{V}$	volume flow rate	$m^3 s^{-1}$
Δp	pressure differential	Pa
Δt	time interval	s
Φ	power	W
ε	effectiveness	–
ϕ	angle	degree, radian
γ	ratio	–
η	efficiency, utilisation factor	–
ϕ	relative humidity	–
θ	temperature	°C
ρ	density	kg/m^3
τ	time constant	s
ω	specific humidity, i.e. mass of water vapour per kg of air	–
ξ	humidity ratio, i.e. the mass of water vapour per kg of dry air	–
COP	Coefficient of Performance	–
SNES	Specific Net Energy Saving	$Wh\,m^{-3}$

Subscripts

0	zero, reference	hs	heat storage
a	air	i	internal
c	cooling	inf	infiltration
da	dry air	l	loss
e	electricity	m	move, movement
exf	exfiltration	o	outdoor
f	fan	p	production
G	global	r	recovered
g	gain	s	solar
gl	gain/loss	V	ventilation
h	heating, heat	w	water vapour
HR	heat recovery	x	extract

Chapter 4

Modeling of ventilation airflow

James Axley and Peter V. Nielsen

4.1 Introduction

Over the past half century, a seemingly endless variety of methods have been developed to predict ventilation airflows in buildings. Nevertheless, these methods as well as related methods for modeling heat and other mass transfer phenomena in buildings fall within one of two broad categories – *microscopic methods* and *macroscopic methods*. *Macroscopic methods* are based on modeling buildings and their supporting heating, ventilating and air-conditioning (HVAC) systems as collections of finite-sized control volumes within which mass, momentum or energy transport is described in terms of straightforward algebraic and/or ordinary differential conservation equations. They provide the means to predict *bulk* (area integrated) airflows moving into, out of and within whole building/HVAC systems of practically arbitrary complexity and size. As macroscopic methods describe ventilation system behaviour in terms of relatively small systems of equations, they are commonly applied to predict, albeit only approximately, the time variation of these *bulk* airflows over extended seasonal or annual periods.

Microscopic methods, on the other hand, are based on continuum descriptions of mass, momentum and energy transport that are defined in terms of partial differential conservation equations. In principle, they provide the means to predict the details of both internal and external airflows in and around buildings, but because of the challenge of solving these demanding equations, the analyst must be satisfied with (approximate) analytical solutions for generic or *elemental* flow conditions within rooms, approximate numerical solutions for single or well-connected multiple rooms, and/or numerical solutions for airflows around buildings for approximate near-field boundary conditions. Furthermore, because of practical computational limits, microscopic methods are typically, but not exclusively, applied to the analysis of steady snapshots of these airflow details.

The strengths of each of these broad approaches are, fortuitously, complementary. As a result, together they have transformed the fields of ventilation research and are increasingly being used in concert in practice. Indeed, emerging methods of building ventilation analysis often seek to computationally, if not mathematically, couple these complementary methods to provide both researchers and practicing ventilation engineers with the ability to consider selected details of airflows in buildings while simultaneously accounting for the interaction of whole-building *bulk* airflows and their dynamic variation on these details.

This chapter will first consider available microscopic methods and introduce the reader to the range of analytical solutions that describe elemental flow conditions in rooms and then move on to consider computational methods of (continuum) fluid dynamics (CFD) from both a theoretical and a practical point of view. The final section will lay out the fundamental principles of macroscopic analysis and apply these principles to describe three general approaches to multi-zone ventilation analysis – the *nodal approach*, the *port plane approach* and the *loop approach*. The focus of this final section will be directed to the deterministic methods that have largely, but not completely, replaced the large variety of more empirical methods commonly presented in handbooks and building standards. Finally, an approximate macroscopic method to size components of ventilation systems will be introduced – the loop design method – and related methods to couple macroscopic ventilation analysis with thermal, contaminant dispersal and embedded microscopic models will be briefly reviewed.

4.2 Microscopic methods

Several models are available for the study of indoor climate at different stages in the design process. Simplified models that describe elemental flow conditions in rooms are easy to use at an early stage of design. These *room flow elements* often describe steady state and they may form the basis of zonal models.

Detailed models such as CFD models and Building Energy Performance models can be used both at the early stage of the design process and in the final design evaluation. The detailed models can be used in simplified geometry for the study of different scenarios and later in the actual geometry with boundary conditions for design evaluation. A CFD model can be either a steady state model or a dynamic model.

Scale-model and full-scale experiments are not considered to be mathematical models, but results from these experiments are used in the design process parallel with results from macroscopic and microscopic models.

4.2.1 Room flow elements

Elemental room flow models are available for volumes where the air movement can be fully described by upstream and local conditions. An isothermal free jet is a typical example that can be described by upstream conditions such as the supply opening geometry and supply velocity. A mathematical description of a flow that depends only on upstream conditions is often called parabolic because it can be expressed by parabolic equations (Anderson, 1995). The flow in a room ventilated by, for example, mixing ventilation is recirculating. Room flow elements may be influenced by this recirculating flow and to some extent is dependent on downstream conditions. In principle, it is much more difficult to describe the flow in this case. A mathematical description of this type of flow is called an elliptic flow because it is given by elliptic equations under steady-state conditions (Anderson, 1995). Room flow elements can also be used in this case because the downstream influence is restricted, and also because it is possible to make corrections as shown in later examples.

Air movement in a room may often be divided into a number of room flow elements that can be treated independently of the surrounding flow and geometry (Nielsen, 1998a). Sketches A–C in Figure 4.1 show three kinds of supply jets, whereas the sketches D and E show disturbance of the jets due to deflection at an end wall and flow around an obstacle. The flow around an exhaust opening is indicated in sketch F.

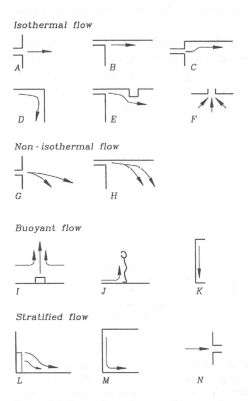

Figure 4.1 Room flow elements in room air movement.

A nonisothermal room flow element such as a free jet with deflection due to gravity is shown in sketch G and a wall jet with restricted penetration depth in sketch H. Buoyant flow is illustrated in sketches I–K. Sketch I shows a thermal plume above a heat source, J shows natural convection around a human body, and K shows cold downdraught at a wall. Stratified flow from a low-impulse opening and from a cold downdraught is illustrated in sketches L and M, respectively. The last sketch, N, represents an extract flow that removes air from an adjacent stratified horizontal layer in a room with a vertical temperature gradient (as for example in displacement ventilation). This flow is different from the isothermal situation in sketch F.

Turbulent free jets and wall jets

A free jet is a jet that in principle flows into an infinitely large space, and a wall jet is a jet that flows along a surface into a large space. The jets are not influenced by downstream conditions and can be described by an equation system based on boundary layer approximations that give a self-similar flow with universal flow profiles (Rajaratnam, 1976). These conditions are important because it is possible to describe the flow in the jets independently of the surrounding room dimensions.

Figure 4.2 shows that a wall jet is independent of the recirculating flow, even in cases with a narrow open space (1–3, Figure 4.2). The geometry of the surrounding room has only a significant influence on the velocity profile in the case where the height of the space is so small that the profile is measured outside the penetration depth of the jet, see Figure 4.2.

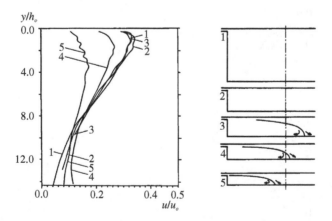

Figure 4.2 Velocity profile in a wall jet located at a given distance from a supply slot. The reciprocal height of the room h_0/H has the following values: $1 = 0.013$, $2 = 0.029$, $3 = 0.035$, $4 = 0.043$ and $5 = 0.057$. h_0 is slot height and H is room height.

For mixing ventilation, turbulent free jets and wall jets form the basis of practical design methods for air distribution systems.

The flow close to a supply opening

Figure 4.3 shows a circular opening with a diameter d_0 and a supply velocity u_0. There is a region immediately outside the opening called the constant velocity (potential) core and it has the velocity u_0. A turbulent mixing layer is formed between the core and the surrounding air. This mixing layer grows into the core at increasing distance from the opening, and the constant velocity core disappears at a distance of $\sim 6\,d_0$. The jet will continue to entrain air from the surroundings downstream, and the width of the jet increases proportional to the distance from a point called the virtual origin, x_0. This origin is in many situations located so close to the opening that the distance x_0 is ignored in comparison with the distance x. The angle γ is $\sim 24°$ for circular jets and $\sim 33°$ for plane jets, if the flow is fully turbulent.

The presence of a wall in the symmetry line – or symmetry plane – has only a small influence on the flow and the velocity profile. The boundary layer at the surface is thin compared with the turbulent mixing layer towards the entrainment side. The flow described in Figure 4.3 does not change much when a surface is located in the symmetry line. A wall jet has theoretically a length of constant velocity core which is $\sim 12\,h_0$, where h_0 is the height of the opening, and the angle of growth of wall jet is $\sim \gamma/2$.

It is the momentum flow from the opening which controls the jet and room air movement in case of mixing ventilation. The nozzle in Figure 4.3 gives a constant velocity u_0 everywhere in the opening (top hat profile), and the flow momentum to the room will be

$$I_0 = \rho_0 a_0 u_0^2 \tag{4.1}$$

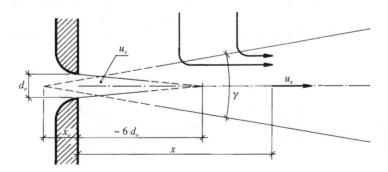

Figure 4.3 The airflow close to a circular opening.

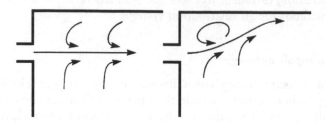

Figure 4.4 Airflow pattern in a plane jet close to a parallel surface.

where a_0 is the supply area and ρ_0 is the density. Many practical air terminal devices have a design with, for example, adjustable vanes which modify the velocity profile, and this will decrease the initial momentum flow (I_0) to the room. An example of change in momentum flow is shown in Figure 4.4. The flow from a slot is injected parallel to a surface as indicated in the left figure. The turbulent mixing layer on both sides of the plane jet entrains air from the surroundings, and a low pressure on the upper side deflects the jet upwards causing a wall jet to be established at some distance from the opening as shown in the right figure. The momentum flow in the opening I_0 is reduced and measurements by McRee and Moser (1967) show that the final momentum flow I has a value of $I/I_0 \sim 0.6-0.7$.

The effect that generates the deflection of the flow in Figure 4.4 is called the Coanda effect. This effect is present in the flow from a two-dimensional slot, but it also exists when the width of the slot is finite as shown by Jackman (1970).

Velocity decay in free jets and wall jets

The air movement in a free turbulent jet is a self-similar flow with a universal flow profile. The initial momentum of the flow at the air terminal device (Figure 4.5) is preserved in the jet because it is assumed that the direction of the flow is entrainment perpendicular to the main jet flow direction. It can be shown that the width of a circular jet increases linearly with the distance

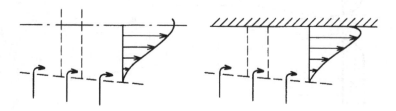

Figure 4.5 Section of the flow in free jets and wall jets.

$x + x_0$, which means that the area of the jet increases with the square root of the distance. The flow momentum is therefore preserved when the velocity is inversely proportional to the distance $x + x_0$.

$$\frac{u_x}{u_0} = \frac{K_a}{\sqrt{2}} \frac{\sqrt{a_0}}{x + x_0} \tag{4.2}$$

u_0 and u_x are the supply velocity and the velocity at distance $x + x_0$, respectively.

Equation (4.2) expresses that the dimensionless velocity u_x/u_0 in the symmetry line is inversely proportional to the dimensionless distance $(x + x_0)/\sqrt{a_0}$, where a_0 is the area of the supply opening. K_a is constant in case of high turbulent flow and its value depends on the type of diffuser. It has a maximum value of ~ 10 for a nozzle as the one shown in Figure 4.3.

The air movement in a wall jet can also be assumed to preserve the initial flow momentum from the supply opening. Figure 4.5 indicates an entrainment velocity perpendicular to the main flow, but it is also obvious from the figure that friction at the surface may decrease the flow momentum in the wall jet, although it is often of little practical importance.

The flow of a three-dimensional wall jet with a supply area a_0 is in practice identical to the flow of a free jet with supply area $2a_0$ (the opening and its mirror image). The velocity distribution in a three-dimensional wall jet is therefore given by

$$\frac{u_x}{u_0} = K_a \frac{\sqrt{a_0}}{x + x_0} \tag{4.3}$$

where u_x is the maximum velocity in the flow close to the surface.

The velocity distribution in a radial free jet is given by

$$\frac{u_r}{u_0} = \frac{K_r}{\sqrt{2}} \frac{\sqrt{a_0}}{r + r_0} \tag{4.4}$$

and the corresponding velocity distribution in a radial wall jet is given by (method of images)

$$\frac{u_r}{u_0} = K_r \frac{\sqrt{a_0}}{r + r_0} \tag{4.5}$$

where K_r is a constant whose value depends on the type of diffuser. r is the radial distance from the rim of the diffuser and r_0 is the distance to the virtual origin of the radial jet. The maximum value for K_r is approximately 1.2.

The width or the cross-sectional area of a plane (two-dimensional) jet increases linearly with the distance $x + x_0$. The flow momentum is therefore preserved when the velocity is inversely proportional to the square root of the distance $x + x_0$

$$\frac{u_x}{u_0} = \frac{K_p}{\sqrt{2}} \sqrt{\frac{h_0}{x + x_0}} \qquad (4.6)$$

where u_x/u_0 is a dimensionless velocity in the symmetry line and $(x + x_0)/h_0$ is the dimensionless distance from the virtual origin. h_0 is the equivalent height of the supply slot and K_p is a constant. The maximum value that can be obtained is $K_p \sim 4$.

The velocity distribution in a plane wall jet is given by (method of images)

$$\frac{u_x}{u_0} = K_p \sqrt{\frac{h_0}{x + x_0}} \qquad (4.7)$$

Equations (4.2) to (4.7) are only valid for x-values where $u_x/u_0 \leq 1.0$ (outside the constant velocity core).

Figure 4.6 shows some practical air terminal devices that create different room flow elements. Product A is a three-core nozzle that is mounted at the end of a circular duct. The jet leaves the opening as a flow with a radial component and high spread. The flow merges further downstream to form a circular jet with a $K_a/\sqrt{2}$-factor of 3.5 [see Equation (4.2)]. The vanes shown in the figure can be reversed (turned by 180°), in which case a higher value of K_a is obtained.

Product B is a traditional wall-mounted grille with two sets of vanes. The horizontal vanes are often adjusted upwards for a deflection of 20° and the vertical vanes are adjusted for high spread. Two symmetrical wall

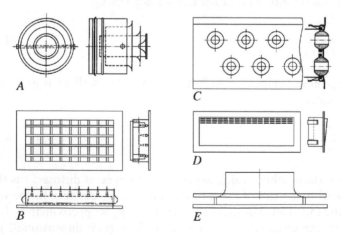

Figure 4.6 Examples of five different air terminal devices.

jets are generated at the ceiling because of the high spread and they merge to a single three-dimensional wall jet at some distance from the diffuser. This flow can be described by Equation (4.3) depending on the value of the Reynolds number. The K_a-value is 4.0 for $u_0 \sim 2\,\mathrm{ms}^{-1}$ and 5.5 for $u_0 \sim 5\,\mathrm{ms}^{-1}$ (Nielsen and Möller, 1985).

Product C consists of a number of nozzles that can be adjusted in different directions (Figure 4.6). The diffuser can either be ceiling-mounted or wall-mounted and it is made in different lengths. It is therefore possible to adjust the diffuser to produce two- or three-dimensional flow. The diffuser has been used in an international research work in a layout of four rows with eighty-four nozzles in all (IEA Annex 20). The diffuser was in this case wall-mounted and all the nozzles were adjusted with an upward direction of 40°. A three-dimensional wall jet is produced at the ceiling with $K_a = 4.8$ with some radial flow outside the symmetry plane (Nielsen, 1991).

Product D in Figure 4.6 is a wall diffuser that is often mounted 0.2 m below the ceiling. The airflow leaves the device with an upward direction of about 45° from the horizontal plane. The flow impinges on the ceiling and generates a jet that is partly three-dimensional jet and partly radial jet because it has different momentum flows in different directions. The jet in the symmetry plane can be described by Equation (4.3) and it has a K_a-value of 3.0, which is a very low value and typical of modern design (Nielsen and Möller, 1985). The K_a-value is independent of the Reynolds number for flow rates of practical interest.

The last diffuser shown in Figure 4.6 (Product E) is a ceiling-mounted diffuser. It has a rectangular design, but the flow from the diffuser is a radial wall jet with K_r-value of 0.9. The diffuser can be adjusted to form two separate wall jets in diametrically opposite directions. These jets represent sections of a radial wall jet with a high velocity in the symmetry plane with K_r equal 2.0.

Free convection flow

A number of room flow elements based on the buoyancy effect are discussed in the following section. As an example, Figure 4.7 shows a plume from a concentrated heat source.

In free convection, the buoyancy of the heated air in the room provides the driving force for air motion. Figure 4.7 shows the vertical flow above a heat source. The buoyancy generates a jet-like flow with a maximum velocity just above the source where the flow may expect to have a contraction. Air is entrained into the plume and the width and the volume of the flow increase with the height.

A concentrated heat source creates a circular flow that can be expressed by

$$q_y = 0.005\, \phi_K^{1/3} \left(y + y_0\right)^{5/3} \qquad (\mathrm{m^3 s^{-1}}) \tag{4.8}$$

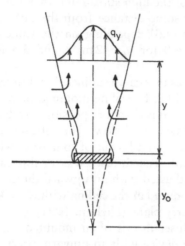

Figure 4.7 Free convection flow from a heat source.

where $y + y_0$ is height above a virtual origin of the flow. q_y and ϕ_K are volume flow at the height y and the convective heat emission from the source, respectively (Baturin, 1972).

Equation (4.8) is strictly valid for $y \gg d$, where d is the hydraulic diameter of the source, but practice shows that the equation can be used down to a height of $y \sim 2d$ (Kofoed and Nielsen, 1990). This is important because the size of a heat source can often be a large fraction of the height of the occupied zone.

The convective heat emission ϕ_K can be estimated from the energy consumption of the heat source ϕ

$$\phi_K = k\phi \quad \text{(W)} \tag{4.9}$$

The level of the coefficient k is 0.7–0.9 for pipes and channels, 0.4–0.6 for small components and 0.3–0.5 for large machines and components.

In practice, it can be difficult to judge the location of the virtual origin, but it is often assumed that $y_0 \sim 2d$ for a concentrated source. The surface temperature distribution close to lighting or heating equipment may extend the size of the heat sources due to radiation from the equipment to surfaces in the immediate surroundings.

A line source creates a two-dimensional flow that can be expressed by

$$q_y = 0.014 \left(\frac{\phi_K}{l}\right)^{1/3} (y + y_0)\, l \quad (\text{m}^3\text{s}^{-1}) \tag{4.10}$$

where l is the length of the source (Skåret, 1986). y_0 corresponds to one to two times the width of the heat source.

A heat source in the occupied zone may also be a contaminant source in the room. Typical examples are printers in offices, welding plumes in industrial areas and people in general. In such cases, the volume flow q_y is therefore the contaminant part of the flow in the room, and the concentration distribution across the flow is distributed in a manner similar to the velocity profiles indicated by Figure 4.7.

Displacement flow from a wall-mounted diffuser

The flow from a diffuser in a displacement ventilation system is addressed in the following section. Figure 4.8 shows how cold air from the diffuser drops into the floor region creating a radial flow in the occupied zone. Measurements show that the entrainment into this stratified flow is small because of the vertical temperature gradient, and they also show that the thickness δ of the cold layer is constant at high Archimedes numbers. The maximum air velocity u_x in the centreline of radial flow with constant thickness and zero entrainment is inversely proportional to the distance x from the diffuser. Measurements confirm this behaviour and Nielsen (1992a, 2000) has shown that the velocity u_x is given by

$$u_x = q_0 K_{Dr} \frac{1}{x+x_0} \qquad (\text{ms}^{-1}) \tag{4.11}$$

where q_0 $(\text{m}^3\text{s}^{-1})$ is the flow rate from the diffuser. K_{Dr} is a variable that is a function of the type of diffuser and of the Archimedes number involved. (A large Archimedes number gives a reduced thickness and therefore a high velocity in the layer). Typical values of K_{Dr} are from 4 to $10\,\text{m}^{-1}$. x_0 is the distance from the front of the diffuser to the virtual origin of the radial flow, which is often located slightly behind the diffuser.

A small temperature difference between room air T_{oc} (measured at a height of 1.10 m above floor) and supply air T_0 decreases the buoyancy effect, and the flow will, to some extent, behave like a three-dimensional wall jet or a radial wall jet. The preservation of momentum will in this case produce a flow in which the thickness δ is linearly proportional to the distance $x+x_0$.

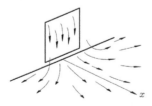

Figure 4.8 Air distribution in a room ventilated by displacement ventilation generated by a wall-mounted air terminal device.

It can be seen that the structure of the Equations (4.11) and, for example, (4.3) is similar, which implies that Equation (4.11) can be used as an empirical description of the flow from such diffusers over the whole temperature range.

Buoyant flow from an open window

The following measurements have been made on a side-hinged window. The area of the opening a_0 is defined as the smallest geometrical cross-section for the flow through the window.

Figure 4.9 shows that the first part of the flow from the window acts as a thermal jet. The distance x_0 to the point where the jet reaches the floor is dependent on the airflow rate (pressure difference) and the temperature difference. Measurements show that the flow along the floor is a radial flow, and it is obvious that it is necessary to take the virtual origin into consideration in this flow as indicated in Figure 4.9.

The flow along the floor is partly a stratified flow with a constant thickness δ, which is similar to the flow from a wall-mounted diffuser for displacement system. The maximum velocity u_x in the symmetry plane of the flow can therefore be described by an equation similar to Equation (4.11) (Nielsen *et al.*, 2000), as below.

$$u_x = q_0 K_w \frac{1}{x + x_0} \quad (\text{ms}^{-1}) \tag{4.12}$$

It is important to consider x_0 in this equation also.

K_w is a constant that is dependent on the geometry of the test setup as, for example, type of window and location of the window in the wall. It is also dependent on the degree of opening a_0 and on the flow rate and temperature difference (Archimedes number).

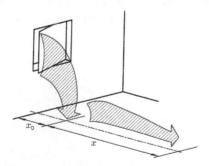

Figure 4.9 Buoyant cold flow from an open window in the case of single-sided natural ventilation.

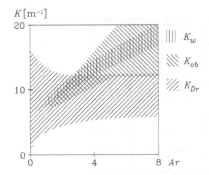

Figure 4.10 K-values for different types of radial flows. K_{Dr}: flow from wall-mounted diffuser; K_{ob}: flow from opening between obstacles in a room ventilated by displacement ventilation; K_w: flow from an open window.

It is surprising to see that the K-values for different types of stratified and semi-stratified flows are of similar value, i.e. 4–$20\,\mathrm{m}^{-1}$, as indicated in Figure 4.10, which shows K-values for different types of flows. The flow from an open window behaves like the flow from an opening between obstacles in a room ventilated by displacement ventilation K_{ob}, see Nielsen (2000). K_w has also the same value as K_{Dr} for the first generation of wall-mounted diffusers where the flow was directed into the occupied zone.

4.2.2 Computational fluid dynamics

This section describes computer-based model that can predict the air movement, temperature and contaminant distribution, as well as many other parameters of room air distribution. Such a model is called a CFD model. The aim of this section is to familiarize the reader with some fundamental statements, principles and abbreviations which are used in CFD application to indoor environment. Basic knowledge of CFD, biographical sources and the use of CFD in the design process are given here. The general development of CFD is discussed by Anderson (1995).

The development of CFD models for room air movement, as well as the progress in the general fluid dynamics research area, is strongly influenced by the increase in computer power that has been made available in the past few decades. This development is obvious in Figure 4.11, which shows the increase in computer speed over the last 30 years.

The development during the last few years has not only brought increasing computer speed but also decreasing costs. The latter is reduced by a factor of ten every eighth year. Chapman (1979) cites an impressive example of this trend in computing efficiency. He mentions that a numerical calculation of the flow over an airfoil would have taken 30 years if it had started in

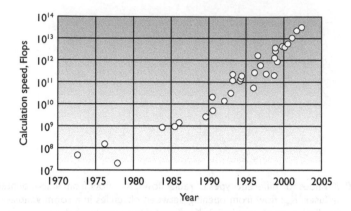

Figure 4.11 Development of computer speed with year of introduction (Kunugiyama, 2002).

1960, and it would have cost $10 million. Twenty years later, in 1980, the same calculation would have taken half an hour and have cost $1000. By 2000, the cost and effort of the same calculation had become insignificant.

There are more reasons for this development. First, the computer speed has increased more rapidly than its cost, and this tendency seems to be continuing. Secondly, different kinds of software show increasing flexibility such as preprocessor and postprocessor software, and furthermore, there is a continuous development of new software. Improvements in the fundamental routines such as the process of grid generation and the numerical methods also contribute to an increasing speed.

First predictions of room air movement were made in the 1970s. Nielsen (1973) showed an early CFD prediction of the flow in a ventilated room, whereas Jones and Whittle (1992) discussed the status and capabilities in the 1990s. Russel and Surendran (2000) have reviewed additional work on the subject.

The prediction of airflow is based on a solution of the fundamental flow equations. These consist of the equation of continuity, three momentum equations (one for each coordinate direction), the energy equation and, perhaps, a transport equation for contaminant distribution. All the equations are time-averaged, and the local turbulence is expressed as a variable diffusion coefficient called the turbulent viscosity. This viscosity is often calculated from two additional transport equations, namely the equation for turbulent kinetic energy and the equation for dissipation of turbulent kinetic energy. The equations are called the RANS (Reynolds-averaged Navier–Stokes) equations. Therefore, the total description of the flow consists of eight differential equations, which are coupled and nonlinear. Further details can be found in Chapter 1. Although it is not possible to solve these

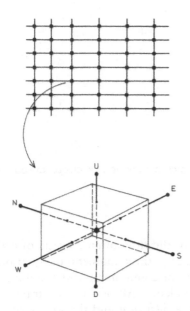

Figure 4.12 Grid point distribution and a control volume around a grid point P.

differential equations directly for the flow regime of a room, a numerical method can be applied. The room is divided into grid points (Figure 4.12). The differential equations are transformed into discretization equations formulated around each grid point; see the lower sketch in the figure.

If the room is divided into $90 \times 90 \times 90$ cells, the eight differential equations are replaced by eight algebraic equations at each point, which for the $90 \times 90 \times 90$ grid gives a total of 5.8 million equations with the same number of unknown variables. An iterative procedure is used in the numerical method. It typically involves up to 3,000 iterations, which means that a total of 17 billion grid point calculations are made for the prediction of a flow field.

Figure 4.13 shows the elements in the numerical prediction method as they are described in the following sections. *Equation systems* and *turbulence models* describe the flow equations inside the room where the indoor climate is studied. The equation system needs a description of the *boundary conditions* around the surfaces of the flow regime to be completely described. As mentioned, it is necessary to use an appropriate *numerical method* for the solution of the equation system. Some suitable methods are described in this section and the last section gives some examples and case studies involving predictions of the indoor climate and related parameters.

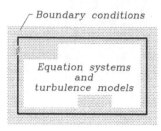

Figure 4.13 A room and the three main topics in numerical prediction of room air distribution.

Equation systems and turbulence models

The equation system for the prediction of the flow often consists of eight equations, namely the equation of continuity, three momentum equations, the energy equation, a transport equation for contaminant distribution and two equations for the description of turbulence. All of these are transport equations that express the convection, the diffusion and the source of the variables included. The equations are given in Chapter 1.

Figure 4.14 shows how the convection in the transport equation for the contaminant (mass fraction per unit of mass of mixture) can be considered as a change in the concentration of flow times velocity in the x-direction, $\rho u \frac{\partial c}{\partial x}$, as well as the change in the two other directions. Here, ρ, u and c are density, velocity and concentration, respectively, where $\partial c/\partial x$ is the concentration gradient in the x-direction.

Diffusion over the cell surfaces in Figure 4.14 is generated by a concentration gradient in the flow field. The diffusion through a surface in the x-direction is equal to $-\Gamma_c \frac{\partial c}{\partial x}$, and the total change in this direction is

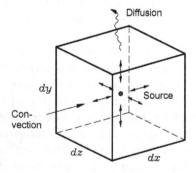

Figure 4.14 Structure of a transport equation around centre point P. Convection, diffusion and source processes are indicated in the figure.

thus $-\frac{\partial}{\partial x}\left(\Gamma_c\frac{\partial c}{\partial x}\right)$, where Γ_c is the diffusion coefficient. Similar expressions are valid for the two other directions.

The source term S_c in the contaminant transport equation is located inside the cell and it can be zero, positive or negative.

The time-averaged version of a transport equation is given as

Convection = diffusion + source

The steady three-dimensional transport equation for contaminant distribution is therefore given by the following equation

$$\rho u \frac{\partial c}{\partial x} + \rho v \frac{\partial c}{\partial y} + \rho w \frac{\partial c}{\partial z} = \frac{\partial}{\partial x}\left(\Gamma_c\frac{\partial c}{\partial x}\right) + \frac{\partial}{\partial y}\left(\Gamma_c\frac{\partial c}{\partial y}\right) + \frac{\partial}{\partial z}\left(\Gamma_c\frac{\partial c}{\partial z}\right) + Sc$$

$$(4.13)$$

Similar transport equations can be formulated for all the other variables in the equation system (see Chapter 1).

The air movement in a ventilated room is turbulent. Figure 4.15 shows the instantaneous velocity measured in a radiator-heated room (Olesen, 1979). Although it can be fully described by the Navier–Stokes equation, it is impossible to make a direct numerical simulation (DNS) of this flow, because it requires an extremely high number of cells to describe the turbulence and the dissipation of turbulence expressed by the fluctuations.

It is possible to obtain a practical number of cells if the numerical simulation is based on averaged variables. The flow is averaged with respect to time, as in Equation (4.13), and the flow in Figure 4.15 is considered as steady flow of high viscosity (fluctuations influence the flow in such a way that it exhibits an apparent increase in resistance to deformation).

The apparent increase in viscosity is described as a turbulent viscosity, or eddy viscosity μ_t added to the physical viscosity μ, which gives the effective viscosity.

$$\mu_{eff} = \mu_t + \mu \qquad\qquad (4.14)$$

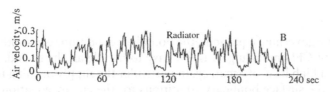

Figure 4.15 Recording of velocity in a radiator-heated room. The velocity is sampled at a frequency of 1.7 Hz during a period of 240 s.

The diffusion coefficient Γ_c is related to the turbulence by

$$\Gamma_c = \frac{\mu_{eff}}{\sigma_c} \qquad (4.15)$$

where σ_c is the turbulent Schmidt number.

A very simplified expression of the eddy viscosity is a zero-equation model like

$$\mu_t = const\,\mu \qquad (4.16)$$

but a more widely used method is to predict the eddy viscosity μ_t from a turbulence model that consists of two transport equations for turbulence kinetic energy k and dissipation of turbulence kinetic energy ε (Launder and Spalding, 1974). The k–ε turbulence model is valid only for a fully developed turbulent flow. Although the flow in a room is not always a high Reynolds number flow, i.e. fully developed everywhere in the room, good predictions can generally be obtained by the k–ε model. Nielsen (1998b) gives a general discussion of different turbulence models for room airflow.

The turbulence can also be predicted by more elaborate models such as the Reynolds stress model (RSM). This model closes the equation system with additional transport equations for the Reynolds stresses (Launder, 1989). The RSM is superior to the standard k–ε model because anisotropic effects of turbulence are taken into account. For example, the wall reflection terms dampen the turbulent fluctuations perpendicular to the wall and convert the energy to fluctuations parallel to the wall. This effect may be important for the prediction of a three-dimensional wall jet flow (Schälin and Nielsen, 2004).

In general, the RSMs give better results than the standard k–ε model, but the improvements are not always significant (Chen, 1996; Kato et al., 1994). Murakami et al. (1994) compared the k–ε model, the algebraic model (simplified RSM) and the RSM in the prediction of room air movement induced by a horizontal nonisothermal jet. The RSM prediction of mean velocity and temperature profiles in the jet shows slightly better agreement with the experiments than a prediction based on the k–ε model. Further treatment is given in Chapter 1.

Boundary conditions

Together with the governing equations, the boundary conditions are another important part of CFD modeling. It is not easy to handle all the boundary conditions with a high level of accuracy, in particular the supply opening for the airflow and the surface boundary conditions for the energy equation.

Usually the flow from a diffuser is influenced by small details in the design. This means that a numerical prediction method should be able to handle small details in geometry (sometimes of one tenth of a millimetre), as well as dimensions of several metres. This wide range of the geometrical detail necessitates a large number of cells in the numerical procedure, which increases the prediction cost and computing time to a rather high level. The very fine details in the flow domain that are generated by a diffuser are also very difficult to measure and specify numerically.

Various simplifications are suggested. The most obvious method is to replace the actual diffuser with one of less complicated geometry that supplies the same momentum of airflow to the room. This may be obtained from a single opening with a height-to-width ratio similar to that of the diffuser and with an area equivalent to the total supply area of the diffuser (simplified boundary conditions). The flow from the opening is specified to be in the same direction as that in the actual diffuser. It is necessary to compare the predicted flow in the jet with the measured flow in front of the actual diffuser when simplified boundary conditions are used.

The Box Method is based on a wall jet flow generated close to the diffuser (Nielsen, 1973). Figure 4.16 shows the location of the boundary conditions around the diffuser. The details of the flow in the immediate vicinity of the supply opening are ignored, and the jet supplied is described by values along the surfaces a and b. Two advantages are obtained by using these boundary conditions. First, it is not required to use a grid as fine as the one needed for full numerical prediction of the wall jet development. Secondly, it is possible to make two-dimensional predictions for supply openings that are three-dimensional, provided that the jet develops into a two-dimensional wall jet or a free jet at a given distance from the openings.

The 'Prescribed Velocity Method' has also been successfully used in the numerical prediction of room air movement. Figure 4.17 shows the details of the method. The inlet profiles are given as boundary conditions at the diffuser in the usual way (simplified boundary conditions), although they are represented by only a few grid points. All the variables, except the velocities u and w, are predicted in a volume close to the diffuser (x_a, y_b), as well as in the rest of the room. The velocities, u and w, are prescribed for the volume in front of the diffuser as the analytical values obtained for a wall jet starting from the diffuser, or they are given as measured values in front of the diffuser (Gosman et al., 1980 Nielsen, 1992b).

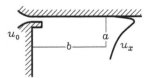

Figure 4.16 Location of boundary conditions around the diffuser, as used in the Box Method.

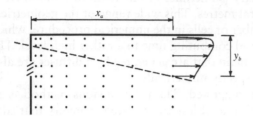

Figure 4.17 Prescribed velocity field close to the supply opening.

The data for velocity distribution in a wall jet (or a free jet) generated by available commercial diffusers can be obtained from diffuser catalogues or design guidebooks, such as ASHRAE Fundamentals (2005), or from textbooks, e.g. by Awbi (2003), Rajaratnam (1976), Etheridge and Sandberg (1996).

Continuous development of computer capacity and speed will undoubtedly make the direct methods with local grid refinements possible. Promising results have been shown by Bjerg *et al.* (2002).

The boundary conditions for the velocities on a solid surface are simply $u = 0$, $v = 0$ and $w = 0$. Although the airflow close to a surface is laminar, it changes from laminar to the actual (turbulent) flow at an increasing distance from the surface. The steep profile and the change in turbulence level necessitate a high number of grid points close to a surface, as indicated in Figure 4.18A. In a standard k–ε model, the flow close to the surface is ignored and is replaced by wall functions that describe the flow and heat transfer by analytical expressions, as shown in Figure 4.18B.

By this procedure, it is possible to obtain a large reduction in the number of grid points required for a solution. A grid of $50 \times 50 \times 50$, for example, could be replaced with wall functions using a grid of $34 \times 34 \times 34$, which

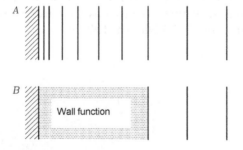

Figure 4.18 Grid line distribution close to a surface. A: without wall functions, and B: with wall functions.

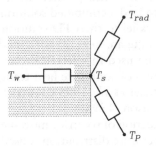

Figure 4.19 Wall surface temperature T_s. This temperature is influenced by conduction T_W, radiation T_{rad} and local air temperature T_P.

would reduce the total number of grid points to 30 per cent of the original value.

Surface temperature and heat transmission are often complicated variables of time and position. Figure 4.19 shows how the surface temperature T_s depends on the heat conduction to or from the surroundings (T_w), radiation to and from the surfaces (T_{rad}) in a room, and also on the air temperature close to the surface (T_p).

The boundary conditions for the temperature or the energy flux can be either found from measurements or obtained by a Building Energy Performance Simulation (BEPS) program. A detailed BEPS program predicts the energy flow in the building structure, internal and external radiation, detailed dynamic energy flow and energy use of the whole building for a given period (Figure 4.20). The method can be structured in different ways. A detailed BEPS program can be connected to a separate CFD program that

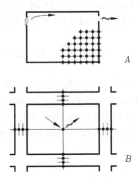

Figure 4.20 A combination of CFD and BEPS programs. The CFD program predicts the flow in the room based on the heat flux calculated from the BEPS program. There is an iterative interaction between the two programs.

predicts the energy flow in selected situations. It is also possible to work with a CFD program that includes an option to find a combined solution of radiation, conduction and thermal storage parallel to the CFD solution of the flow field. This model is often called a conjugate heat transfer model. Another possibility is to add a CFD code to be used in selected rooms as an extension of a large BEPS program. Examples of conjugate heat transfer models and combined models are given by Moser *et al.* (1995), Kato *et al.* (1995), Schild (1997) and Nielsen and Tryggvason (1998).

When CFD models are used for detailed prediction of room air movement, it is recommended that a prescribed wall temperature distribution is used as boundary conditions and that predictions are compared with experiments in which this temperature distribution is measured. A prediction where a CFD program is combined with a BEPS program can only give indications of the air movement, because of the uncertainties of the boundary conditions for the energy equation. It is also recommended that a turbulence model without wall functions is used when heat flow and mass flow from surfaces are the important parameters. This is because predictions of the actual flow at the surfaces (Figure 4.18A) are more accurate in a given situation than analytical values found from wall functions. Recommendations on gridding close to surfaces are given in Awbi (1998).

Numerical method

As mentioned in the introduction, it is not possible to make a direct analytical solution of the system of differential equations. Therefore, it is necessary to reformulate the differential equations into difference equations for which solutions can be found by a numerical method.

This section is based on a simple one-dimensional case to facilitate understanding. Although the case is one-dimensional, it can also be considered as a small part of a complicated flow, which is one-dimensional in certain areas, parallel with grid lines and steady, and therefore part of the general situation (Figure 4.21).

The flow domain is divided into grid points with cells of size $\Delta x \cdot \Delta y \cdot \Delta z$ around each grid point P (Figure 4.22).

The difference equation for a grid point P can either be developed from the transport Equation (4.13) or be established directly from the cell shown in the figure. The convective mass transport to the volume is the difference between the convective mass transport through the two surfaces e and w.

$$\rho(u_e c_e - u_w e_w)\Delta y \Delta z$$

The total diffusion over the two surfaces is equal to

$$\Gamma_c \left[\left(\frac{dc}{dx} \right)_e - \left(\frac{dc}{dx} \right)_w \right] \Delta y \Delta z$$

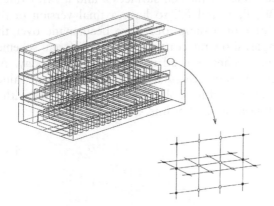

Figure 4.21 Four grid points, with neighbouring points, in a flow domain where the flow is one-dimensional, parallel with the grid lines and steady.

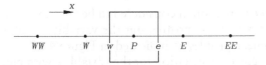

Figure 4.22 Five grid points, WW, W, P, E and EE, and a cell around P with the two surfaces w and e.

and the source term is equal to

$$(S_c + S_p \cdot c)\Delta x \Delta y \Delta z$$

The steady-state one-dimensional transport equation for contaminant distribution (mass fraction per unit mass of mixture) is therefore given by

$$\rho\,(u_e c_e - u_w c_w) = \Gamma_c \left[\left(\frac{dc}{dx} \right)_e - \left(\frac{dc}{dx} \right)_w \right] + (S_c + S_p c)\,\Delta x \qquad (4.17)$$

Equation (4.17) is called a control-volume formulation because it can be considered to be an integration of the transport equation [differential Equation (4.13)] over the length Δx.

An important feature is the integral conservation of quantities such as mass, momentum and energy. This feature is valid not only for each control volume but also for the total flow domain, and it is independent of the grid distribution. Even a coarse-grid solution exhibits exact integral balances (Patankar, 1980).

It is necessary to replace values at the cell surfaces e and w with values at the grid points WW, W, P, E and EE to have a final version of the discretization equation. Different assumptions have been made over the years. The following examples show the consequences of different schemes and the development of new schemes used in CFD software today. An original strategy was to let values on cell surfaces be given as mean values of the two neighbouring values, and to replace the gradient dc/dx with a piecewise linear profile between the grid points, as for example:

$$c_e = \frac{(c_E + c_P)}{2} \qquad (4.18)$$

and

$$\left(\frac{dc}{dx}\right)_e = \frac{c_E - c_P}{\Delta x} \qquad (4.19)$$

The two assumptions for convection and diffusion are both of second-order accuracy.

The value of c in the source term is often considered to be constant within Δx, and the profile is called a stepwise profile, but the variable c can also be composed of piecewise linear profiles as indicated in Figure 4.23.

The one-dimensional discretization equation for the variable c with piecewise linear profiles for convection and diffusion terms, and stepwise profile for the source term, is therefore

$$4\Gamma_c c_P = (2\Gamma_c + \rho \Delta x u)\, c_W + (2\Gamma_c - \rho \Delta x u)\, c_E + 2\, (S_c + S_P c_P)\, \Delta x^2 \qquad (4.20)$$

The equation shows the connection between the concentration c_P and the concentration in the neighbouring points c_W and c_E.

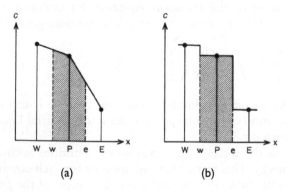

Figure 4.23 Two profile assumptions. Piecewise linear profile (A) and stepwise profile (B).

As an example, we consider the flow in a case for which the length x is equal to 4 (Figure 4.24), the velocity u, the density ρ_0 and the diffusion constant Γ_c are all equal to 1.0, S_c is equal to 0.0 and S_p is equal to -2.0. The temporary boundary values c_0 and c_4 are equal to 1.0 and 0.0, respectively.

Figure 4.25 shows the prediction of one-dimensional concentration distribution at low velocity $u = 1.0$. The prediction is shown for both stepwise and piecewise linear profiles in the source term. It is also possible to compare the numerical solution based on three internal grid points with the analytical solution that is given from the one-dimensional convective/diffusion equation.

Experience shows that unstable (oscillatory or wiggly) solutions are obtained when there is an increased velocity u or an increased grid point distance Δx. It is shown that the Peclet number

$$Pe = \frac{\rho \Delta x u}{\Gamma_c} \tag{4.21}$$

must be smaller than 2 to ensure convergence and stable solutions. This is a very disadvantageous situation because most engineering applications have a high Reynolds number or a high convective flux and a small diffusion.

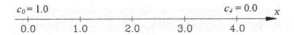

Figure 4.24 Grid point distribution with four internal grid points. Points 0 and 4 are temporary boundaries for the one-dimensional predictions.

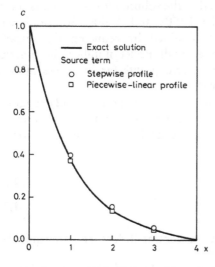

Figure 4.25 Concentration distribution c versus the distance x for an exact solution, and for two numerical schemes based on the control-volume method. Velocity $u = 1.0$ and $\Delta x = 1.0$.

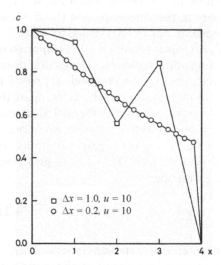

Figure 4.26 Solution of the discretization equation when the velocity is high ($u =$ 10.0).

Figure 4.26 shows the solution of Equation (4.20) for a velocity of $u = 10.0$. The numerically unstable solution is typical of a control-volume formulation with a central difference in the convection term and a large Peclet number ($Pe = 10.0$).

A decrease in the Peclet number can be obtained by decreasing the grid point distance Δx. Figure 4.26 shows that the solution for $\Delta x = 0.2$ is stable.

This situation was typical in the 1950s and 1960s. Solutions with increasing Reynolds numbers were obtained by reducing the distance between the grid points to obtain a small Peclet number. In practice, this remedy often led to a number of grid points far in excess of what computers of that time could deal with.

A large step forward was taken when Courant *et al.* (1952) suggested the upwind scheme, which has almost unconditional stability. This scheme defines the values, for example, on the control-volume surface e for the convection term by

$$c_e = c_P \qquad u \geq 0 \tag{4.22}$$

$$c_e = c_E \qquad u < 0 \tag{4.23}$$

instead of the mean value given in the central difference. An equivalent formulation is used on the surface w.

The main idea of the upwind scheme can be seen from the 'tank-and-tube' model suggested by Patankar (1980). Figure 4.27 shows a row of control volumes that are visualized by stirred tanks connected in series by

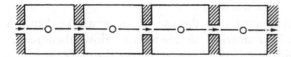

Figure 4.27 Tank-and-tube model.

short tubes. The flow through the tubes represents convection, whereas the conduction through the tank walls represents diffusion. The fluid flowing in each connecting tube has the temperature which prevails in the tank at the upstream side and it will not be influenced by the downstream temperature.

A solution of a transport equation with an upwind scheme for the convection term and high Peclet number ($Pe = 10.0$) is shown in Figure 4.28. The solution is physically correct with a continuously decreasing c-value as a function of the distance x, and it is close to the central difference solution with $\Delta x = 0.2$ and 19 internal grid points.

The hybrid scheme is an improved version of the upwind scheme suggested by Spalding (1972). The hybrid scheme is identical to the central-difference scheme at low velocities ($|Pe| \leq 2$), and it is identical to an upwind scheme in which the diffusion has been set equal to zero for large velocities ($|Pe| > 2$).

Predictions with the hybrid scheme are also shown in Figure 4.28. Comparisons can be made with the central-difference scheme in order to see the influence of the number of grid points. A reasonable solution is obtained

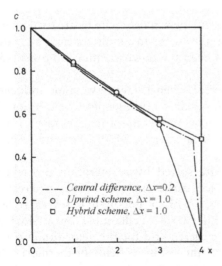

Figure 4.28 Concentration distribution c versus the distance x for a central-difference scheme, an upwind scheme and a hybrid scheme. Velocity $u = 10$.

with three internal grid points. (In this case the central-difference scheme for $\Delta x = 0.2$ is also a hybrid scheme.)

In the early 1970, it seemed that the use of an upwind scheme had opened the way to make numerical simulations of flow phenomena at indefinitely high Reynolds numbers. However, before the end of the decade it had become clear that there were errors in the predictions, although high stability was obtained. The error is connected with situations where the flow has an angle to the grid lines; it had a maximum at 45°. The result is a 'false diffusion' that is proportional to the velocity and to the distance between the grid points. Huang et al. (1985) conclude that many studies at the end of the 1970s had false diffusion that was larger than the actual physical diffusion.

The QUICK scheme by Leonard (1979) is an improved scheme for the convection term, which has a small false diffusion and a higher accuracy (second-order accuracy in two and three-dimensional flow) than the first-order accuracy of the upwind scheme. The scheme can be interpreted as a central-difference scheme with a stabilizing upstream-weighted curvature correction arising from the second-order polynomial fit. The value c_e on the control-volume surface has the following formulation:

$$c_e = \frac{1}{2}(c_E + c_P) - \frac{1}{8}(c_W - 2c_P + c_E) \qquad u \geq 0 \qquad (4.24)$$

$$c_e = \frac{1}{2}(c_E + c_P) - \frac{1}{8}(c_P - 2c_E + c_{EE}) \qquad u < 0 \qquad (4.25)$$

The second-order upwind scheme and the van Leer scheme are two others, which take into account the fact that the upstream conditions have a greater influence on the variable at point P than the downstream conditions (van Leer, 1974).

The selection of the right numerical scheme may have some influence on the results when it is difficult to obtain a grid-independent solution. It is always recommended to use a scheme of second-order accuracy, if it is available in the software (Casey and Wintergerste, 2000; Chen and Srebric, 2002; Sørensen and Nielsen, 2003).

The finite difference equations are solved by an iteration procedure. A Gauss–Seidel procedure was used earlier in the 1960s, whereas a tri-diagonal matrix is used together with a line-by-line solver today. The coupling between pressure and velocity is handled by a SIMPLE procedure. This procedure uses a staggered grid for the velocities in order to avoid nonphysical oscillations in the pressure field; furthermore, the continuity equation is rewritten as an equation for pressure correction. A detailed description is given by Pantankar (1980) and Versteeg and Malalasekera (1995).

Examples of CFD predictions

This section examines two examples of predictions. The first is a study that is conducted at an early stage of the design process for an exposition pavilion, whereas the second is an existing building where problems with indoor air quality are encountered.

The Danish Pavilion at the EXPO'92 World Exhibition was equipped with a special air distribution system that was examined using by two-dimensional CFD predictions (Figure 4.29). The exhibition pavilion consists of two main elements: a steel framed structure, the tower (facing west) with a floor area of 45.0 m × 2.5 m and a height of 24 m, and a fibreglass construction (facing east), which leans against the steel structure. The large room formed between the fibreglass surface and the steel tower is enclosed by glass walls to the north and to the south.

The design load for the occupied part of the exhibition hall is 48 kW, corresponding to 300 persons in the pavilion. The equipment for slides and video projection generates another 130 kW, which is expected to produce a convective flow upwards and high temperatures in the upper part of the pavilion.

The underlying idea of the ventilation system is to use an extract fan in the north top of the exhibition hall (smoke ventilation) and cooling elements in the south gable. Air is drawn through the cooling elements where it is cooled, and then falls to the floor. Because it is difficult to estimate the downdraught from the 12-m high diffuser in the gable, it was decided to conduct CFD

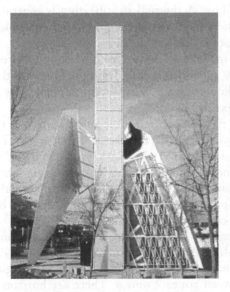

Figure 4.29 The Danish Pavilion, at the World Exhibition EXPO'92 in Seville, and the cooling elements in the south gable.

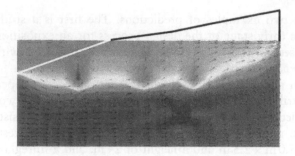

Figure 4.30 Velocity and temperature distribution in the Danish Pavilion. Velocity is given by vectors and the temperature is shown by the shaded area. The real dimensions of the room are indicated by the black and white line.

predictions the flow. Although the hall has a complicated geometry, it was decided to use a two-dimensional prediction to get preliminary results at the early stage. The main interest was the draught in the occupied zone, and the ability to obtain a stratified flow with a large vertical temperature gradient, which prevents the heated air in the upper part of the hall from reaching the occupied zone below.

Figure 4.30 shows the results of a simulation for a load in the occupied zone of $27\,\mathrm{Wm}^{-2}$ and a load in the upper part of the room of $300\,\mathrm{Wm}^{-2}$. The velocity in the occupied zone is restricted to $0.4–0.5\,\mathrm{ms}^{-1}$. Although this is a high velocity compared with the situation in an ordinary room, it is acceptable for an exhibition pavilion. A thermal stratification is generated in the room, and the heat emission from the slides and video equipment moves to the upper part of the room (Nielsen, 1995).

Detailed predictions should be made at a later stage in the design process, but scale-model experiments were used in this case. Figure 4.31 shows that the initial two-dimensional predictions were in sufficient agreement with the results from detailed model experiments.

The second example of the use of CFD predictions is an evaluation of the air distribution system in a library hall (Nielsen and Tryggvason, 1998). The work is a combined use of a CFD program and a BEPS program (Figure 4.20).

The Regional Library of Northern Jutland consists of 170 rooms ventilated partly by a constant air volume (CAV) system and partly by a variable air volume (VAV) system. Most of the rooms are heated by radiators, but the main hall is heated by an air-conditioning system.

The main library hall is connected by large openings to two other rooms, as shown in Figure 4.32. Figure 4.33 shows the library hall as zone 1 (Z1) and the two other rooms as zone 2 (Z2) and zone 3 (Z3). The figure shows a vertical connection between zones 2 and 3. There are horizontal connections between the zones 1 and 2, as well as between zones 1 and 3. All three zones are equipped with supply and return openings for individual air

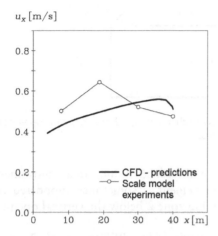

Figure 4.31 Maximum velocity in the occupied zone of the exhibition pavilion given by two-dimensional CFD predictions and detailed scale-model experiments.

Figure 4.32 Library hall and openings to two other lending departments.

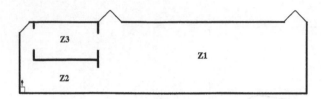

Figure 4.33 The library hall, Z1, and two other zones, Z2 and Z3, with open connections between the zones.

distribution. Pressure and temperature differences induce an air movement between the three zones. The air movement is sometimes supported by thermal plumes from radiators located in zone 2 below the vertical opening to zone 3.

The energy use of the building was calculated by a BEPS program. Because the program cannot predict the flow between the zones 1, 2 and 3 as a function of the pressure and temperature distribution in the zones, it is assumed, as an initial guess, that the energy flow between the zones is low and therefore unimportant.

Figure 4.34 shows the development of temperatures in the three zones for the a week in June (week 23), using the ambient climate conditions for the Danish reference year. The library hall (zone 1), which is heated by solar

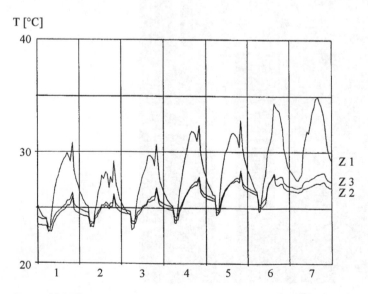

Figure 4.34 The temperature development in zones 1, 2 and 3 of the library building for a week in June (23). It is assumed that there is no air exchange between the zones.

radiation through skylights, sometimes reaches a temperature of 8° C above the temperature in the surrounding zones. It is obvious that this temperature difference induces energy flow between the zones. Consequently, it is necessary to extend the program to include information on the air movement between the zones.

The flow between zones 1, 2 and 3 is found from CFD predictions for all three zones. Different situations are selected from the reference year for this analysis. The boundary conditions for the energy flow are obtained by the BEPS program. They are based on given values of the heat transfer coefficient. The boundary conditions could also be given as a temperature distribution, but it is difficult to make an accurate prediction of the heat transfer coefficient using the wall functions in a CFD program (Chen and Jiang, 1992; Nielsen, 1998b).

It is necessary to obtain the final solution by iteration between the BEPS program and the CFD program, because a change in the airflow between the three zones changes their temperatures, which causes a change in the energy flow to the surrounding surfaces.

Figure 4.35 shows the stages in this iteration. The initial predictions are made without air exchange between the zones 1, 2 and 3. The corresponding CFD predictions provide an air exchange, which is introduced into the BEPS program. The new energy flow at the surfaces is read into the CFD program, which results in a new air exchange between the three zones.

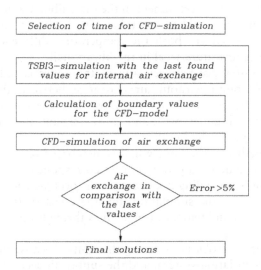

Figure 4.35 Flow chart of the iterations between the building energy performance simulation program and the CFD program.

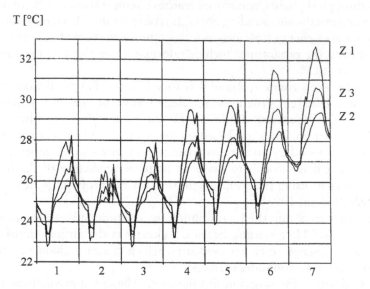

Figure 4.36 The temperature development in the three library zones 1, 2 and 3, in June, week 23. The air exchange between the zones is obtained by CFD predictions.

The iterations continue until the change in the airflows is below 5 per cent compared with the latest values.

Figure 4.36 shows the temperature development in the three library zones during the week in June, when the air exchange is taken into consideration. The airflow between the zones reduces both the temperature difference between them, as well as the temperature level in the library hall.

The greatest temperature difference between any of the zones is only 3 K, in contrast to the 8 K found without air exchange between the zones (Figure 4.34). It must be assumed that the most accurate energy consumption calculations are obtained by taking into account the air exchange between the open zones in the building.

An evaluation of the air quality in the library hall is also based on CFD predictions. The prediction is made for a geometry that corresponds to the layout of the library hall (zone 1) including bookcases in the occupied zone. Zones 2 and 3 are excluded from the simulation, but the effect of these zones is introduced by flow rates and temperature profiles through vertical surfaces in the hall.

Indoor air quality problems were observed during the winter because the library hall is heated by the ventilation system, and the supply air seems to bypass the occupied zone due to the buoyancy effect. Figure 4.37 shows the air distribution in the hall on 8th January at eleven o'clock in the morning. The heated jets ($4.56\,m^3s^{-1}$, $32.5°$ C) move upwards and generate a radial

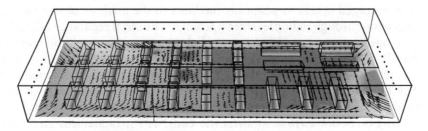

Figure 4.37 Air movement and concentration distribution at floor level (0.25 m) in the library hall, zone I. The arrows show the air movement, and the grey scale indicates concentration levels. The darkest area corresponds to the highest level.

flow below the ceiling. Air moves down into the corners of the hall because three of the corners do not have supply openings. The flow in the bottom left-hand corner of the figure comes from zone 2, which, in this situation, supplies air to the occupied zone ($2.07\,\mathrm{m^3s^{-1}}$, $18.6°$ C). The flow in the top left-hand corner of the figure is partly cold downdraught from a glazed surface in this corner of the hall, partly induced flow from the upper part of the building. The velocity is very low in most of the occupied zone, which corresponds to low ventilation efficiency in this part of the hall.

The concentration distribution was predicted for a situation where the lower surfaces in the hall are the contaminant sources. The maximum concentration in the occupied zone c_P/c_R has a level of 2.05, which corresponds to a local ventilation index ε_P of 0.49 (c_P is the concentration in the air and c_R is the mean concentration in the return openings). The average concentration in the occupied zone c_{oc}/c_R is 1.3, which corresponds to a mean ventilation efficiency of 0.77. This is not a high value, and it must be kept in mind that recirculation in the ventilation system would further decrease the level of outdoor air in the occupied zone.

4.3   Macroscopic methods

As buildings are invariably configured as collections of rooms, it was quite natural for building researchers to first turn to macroscopic methods and model buildings as single or multiple control volumes. Thus, Sir Napier Shaw, the prominent British meteorologist, effectively introduced macroscopic methods of ventilation analysis by idealizing buildings as control volumes linked to the outdoor environment via flow-limiting openings that he modeled using the orifice equation (Shaw, 1907). Although progress was slow and interest was limited to infiltration airflows, by mid-twentieth century Dick was able to lay out the key principles that were to shape the development of the field to the present day (Dick, 1949; Dick, 1950; Dick, 1951; Thomas and Dick, 1953).

The success of Dick's single control volume (*single-zone*) infiltration models encouraged others to adapt his methods to the analysis of wind-driven ventilation through multiple larger openings. The serial resistance offered by these openings were modeled using a linked series of control volumes that directly corresponded to the rooms along the cross-ventilation flow path (see, for example, (Aynsley *et al.*, 1977b)). Turning to digital computation, Dick's guiding principles were again adapted to develop methods to analyse buildings of more general configurations, first, for isothermal wind-driven airflows (*network airflow models*) (de Gids, 1978; Liddament and Allen, 1983; Walton, 1984) then later accounting for buoyancy effects (*multi-zone models*) (Walton, 1988; Walton, 1989; Feustel, 1990; Feustel and Raynor-Hoosen, 1990; Wray, 1990; Li, 1993). Immediately thereafter, these models were further modified to account for the interaction of mechanically ducted air distribution systems (*duct network models*) (Walton, 1994; Feustel and Smith, 1997; Pelletret and Keilholz, 1997; Walton, 1997; Dols, *et al.*, 2000; Dols and Walton, 2000) (see Figure 4.38).

This section will consider the application of the macroscopic principles set out by Dick, now a half-century ago, to the development of deterministic methods of ventilation analysis. These principles will, however, be presented with a level of rigour not considered by Dick to highlight their limitations and establish a more general basis upon which future developments may be made.

4.3.1 Motivation and purpose

Although improved air quality has been associated with purpose-provided ventilation since at least the twelfth century (i.e. when Eleanor of Aquitaine allegedly promoted the use of chimneys for smoke control in France and

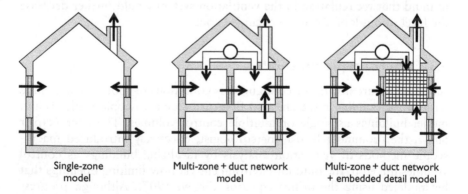

Single-zone model   Multi-zone + duct network model   Multi-zone + duct network + embedded detail model

Figure 4.38 Representative models of a given building of increasing detail corresponding to the evolution of building ventilation analysis methods of the past half century.

England), quantitative links between air quality and ventilation were not widely established until Yaglou, at the Harvard School of Public Health, published his seminal papers in the mid-1930s (Yaglou, *et al.*, 1936; Yaglou and Witheridge, 1937). With an interest in limiting odours, carbon dioxide and tobacco smoke generated by building occupants, Yaglou and his colleagues set about to quantitatively establish ventilation standards for air quality control. These standards were expressed in terms of minimal volumetric ventilation airflow rates Q (ls^{-1} or cfm) per occupant needed to assure acceptable indoor air quality. With these quantitative objectives in mind, Dick in the following decade turned to the development of mathematical models to predict these ventilation airflow rates. Furthermore, he recognized that with an estimate of the ventilation rate in hand, one could also directly estimate ventilation heat losses. Thus, early on, macroscopic ventilation analysis was directed to the dual and conflicting purposes of quantitative air quality control and energy accounting with the motivation to provide healthy environments for the former (i.e. achieved by fostering ventilation) and to conserve energy for the latter (i.e. achieved by limiting ventilation).

As computational methods of multi-zone building air quality and airflow analysis matured, however, researchers and practitioner in the building smoke control community found that these tools could be usefully applied to the analysis and design of building smoke control systems as well (Klote and Milke, 1992; Wray, 1993; Wray and Yuill, 1993). More recently, these multi-zone tools have found a new purpose to not only predict the consequences of chemical and biological terrorist (CBT) attacks on building ventilation systems but to provide for faster-than-real time control of these building ventilation systems in the event of a CBT attack.

4.3.2 Theoretical basis

The macroscopic principles that have become the basis of conventional semi-empirical and deterministic methods of single-zone and multi-zone building ventilation analysis will be presented in this section. The presentation of these principles will not, however, be conventional. Instead a subtly modified approach will be presented that allows consideration of (a) nonuniform pressure and velocity distributions in flow-limiting resistances, (b) mechanical energy conservation in zone airflows and its relation to simplifying *zone field assumptions* and (c) alternative envelope boundary conditions that may better account for wind-driven airflow especially in porous buildings.

The purpose of this section is not, however, intended to be revolutionary – the conventional theory will be presented fully. Rather, the presentation has been contrived to highlight the underlying and often tacit assumptions of the conventional approach to allow both a critical evaluation of its shortcomings and a deeper appreciation of its strengths. Furthermore, the theory will be presented so that simpler applications of it may readily be implemented using any one of a number of commercially available mathematical processing programs [e.g. MATLAB®, Mathematica© or MathCad (MathWorks, 2000; MathSoft, 2004; Wolfram_Research, 2004)].

Building idealization and analytic approach

In conventional single- or multi-zone airflow analysis, building systems are idealized as collections of zones and duct junctions linked by discrete, limiting-resistance airflow paths; external wind pressure boundary conditions acting on the building are defined in terms of an approach reference wind velocity; temperatures within the zones and duct junctions are specified (typically but not necessarily as uniform); and specific flow relations are assigned to each of the discrete flow paths or *flow elements* of the building idealization (Figure 4.39 on the left). Then, equations governing the behaviour of the system as a whole are formed by demanding that zone mass airflow rates be conserved. Finally, these equations are then complemented by assuming hydrostatic conditions exist in each of the modeled zones to achieve closure. The resulting nonlinear algebraic *system equations*, defined in terms of zone and duct junction *node pressures*, are then solved and the solution is back-substituted into the *flow element* equations to determine airflow rates within the system. This most common and thus conventional approach will be identified as the *nodal approach* to multi-zone airflow analysis.

Alternatively, building systems may be idealized as shown to the right in Figure 4.39 – here both flow paths and zones are treated as finite-size control volumes separated by distinct *port planes*. In this more general approach, airflow variables associated with each port plane include, as

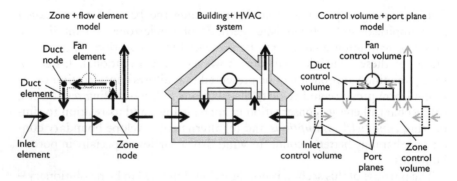

Figure 4.39 The conventional zones + flow element model (left) of a representative integrated building + HVAC system (centre) and an alternative model of control volumes and separating port planes with associated port plane airflows (right).

primary variables, pressures and airflow velocities, and, as related secondary variables, volumetric and mass airflow rates. Again, system equations may be formed, but now the conservation of both mass and mechanical energy may, optionally, be used to form the system equations that will, therefore, be expressed in terms of the primary pressure and velocity variables. Again, boundary conditions must be specified and zone field assumptions must be imposed to effect closure. The resulting nonlinear equations may then be solved (often with difficulty) to determine port plane pressures and airflows velocities (i.e. the airflow rates are determined directly).

Short of imposing full mechanical power balances – or, equivalently, the Bernoulli relation for two-port control volumes – one may instead use the flow element relations from the conventional nodal approach to (indirectly) account for dissipation within the multi-zone flow system. Thus the conventional approach to multi-zone analysis may be formulated in terms of port plane variables rather than zone node pressures. Consequently, a *port plane approach* to analysis will be understood here to be any approach defined in terms of the port plane variables while a port plane approach imposing both mass and mechanical energy conservation principles will be identified as a *power balance port plane approach*.

Finally, limiting consideration to conditions of steady flow, one may form system equations by demanding mass airflow conservation, as in the other approaches, and, in addition, demand that pressure changes encountered as one progresses around a continuous *flow loop* in a building system sum to zero. In forming the *pressure loop equations*, one must also impose boundary conditions and makes zone field assumptions as in the other approaches, although it will be seen that this is done in forming the loop equations. Although not immediately evident from this introduction, the resulting system equations share the same theoretical basis of the nodal approach but are defined in terms of flow element mass airflow rates instead of zone pressures. This less popular but nevertheless useful approach will be identified as the *loop approach*.

Although these approaches are seemingly straightforward in these simple descriptive terms, when limiting assumptions, physical and mathematical details and numerical and computational strategies needed to actually form and solve the system equations are considered, the strengths and weaknesses of each of the approaches may be understood in their full complexity.

Flow variable representation and notation

To introduce a level of rigour not commonly considered, it must be recognized that the detailed (time-averaged) velocity, $v_i(r, s)$, and pressure, $p_i(r, s)$, distributions associated with a port i will, in general, vary across the section (r, s) of the port, (Figure 4.40). In macroscopic analysis, the spatial average

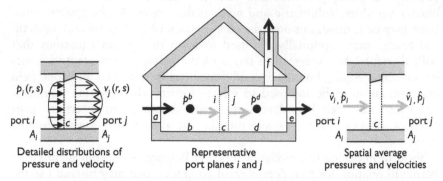

Figure 4.40 Representative port planes i and j of a building system illustrating detailed pressure and velocity distributions on the left and spatial averages of these on the right.

of these distributions $\hat{v}_i$ and $\hat{p}_i$ (i.e. taken over the port plane cross-sectional area A_i) are reasonably taken as the primary variables that characterize system behaviour. By definition, these averages are

$$\hat{v}_i = \int_{A_i} v_i dA \Big/ A_i \quad ; \qquad \hat{p}_i = \int_{A_i} p_i dA \Big/ A_i \qquad (4.26, 4.27)$$

With these primary port plane variables in hand, two secondary flow quantities may be directly defined – the volumetric flow rate $\dot{V}_i$ through port i is simply the product of the spatial averaged velocity and port plane cross-sectional area A_i, and the air mass flow rate $\dot{m}_i$ is this product multiplied by the density of the airflow ρ_i, assuming that this density is uniform across the section:

$$\dot{V}_i = \hat{v}_i A_i \qquad \text{and} \qquad \dot{m}_i = \rho_i \dot{V}_i = \rho_i \hat{v}_i A_i \qquad (4.28, 4.29)$$

In conventional analysis, system equations are formulated in terms of *node pressures* associated with zones or duct junctions, where it is tacitly assumed that these node pressures are time-averaged values. Thus, in control volume b of Figure 4.40, which is nominally a 'zone', pressure p^b is associated with a specific location (*node*) at a specified elevation z^b.

In addition, conventional analysis typically limits consideration to two-port flow elements of constant cross-section. Thus, for the conventional approach, one may associate the cross-sectional area with the flow element control volume instead of the port planes. For example, for control volume a of Figure 4.40, the cross-sectional area will be identified as A^a and will be equal to the inflow and outflow port plane areas of the two-port flow element. For control volume c on the other hand, a single control volume cross-section is not defined and the inflow and outflow port plane areas are not equal, $A_i \neq A_j$. As mass airflow must be conserved, one may however demand equality of the port plane mass airflow rates (i.e. for the usual case of steady flow) and, thereby, define a single value associated with all two-port control volumes, as for control volume c of Figure 4.40:

For steady flow in two-port control volumes

$$\dot{m}^c \equiv \dot{m}_i = \dot{m}_j \tag{4.30}$$

The use of subscript indices to associate variables with port planes and superscript indices to associate variables with control volumes will be used consistently within this section. These notational conventions, and the use of the hat embellishment (e.g. as for $\hat{p}$) to distinguish spatially averaged variables from others, allow a unified treatment of the commonly used *nodal* approach and the less commonly used *loop* and *power balance* approaches to macroscopic building airflow analysis.

Conservation and loop principles – the fundamental bases of airflow analysis

As outlined descriptively above, the theory of macroscopic airflow analysis is based on a number of well-established principles. Of these, three principles are truly fundamental – the conservation of mass, the conservation of mechanical energy and the consistency of variations of pressure changes around any *loop* through the flow system (e.g. see Figure 4.41). Other relations that establish driving wind and buoyancy pressures, the resistance of system components to flow, the ideal gas law, etc., although essential to ventilation analysis theory, are not truly fundamental as they have an empirical basis. Consequently, we will begin with these fundamental principles.

To ground the discussion in reality, these principles will be presented in relation to the representative but hypothetical building idealization shown in Figure 4.41. This system is idealized as an assemblage of nine control volumes lettered from a to i, plus the outdoor control volume, which will be identified by the letter o, with a total of twelve separating port planes numbered from 1 to 12. In the conventional approach, control volumes b and d would be identified as *zones*; control volumes f, g and i as duct control volumes; and control volume h as a fan control volume; control volumes a, c, and e would normally be treated as flow-limiting resistances and identified as *flow elements*. In this figure, the conventional flow variables are explicitly identified – e.g. element air mass flow rates $\dot{m}^a$, $\dot{m}^c$ and $\dot{m}^e$; zone node pressures p^b and p^d; and duct junction node pressures p^i and p^g. The flow variables are complemented by control volume air temperatures

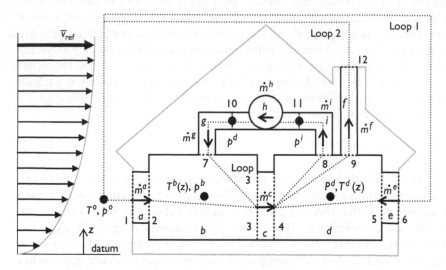

Figure 4.41 A representative but hypothetical building idealization consisting of an integrated building airflow system and duct network. Port planes are indexed by number, and control volumes and pressure nodes by letter, and flow loops are indicated by dotted lines.

(shown only for zones b and d in Figure 4.41), which may vary from zone to zone but are nevertheless specified. As indicated in Figure 4.41, these control volume temperatures may be specified to vary spatially but are typically assumed uniform. That is to say, the *well-mixed* control volume assumption, although commonly made in analysis, is not theoretically necessary.

Port planes are indicated by dashed lines but the variables associated with them are not explicitly represented in Figure 4.41. A line linking the centres of one port plane to the next along any possible airflow path and continuing back to the first port plane is a *loop*. For example, loop 1 shown in Figure 4.41 links port planes 1, 2, 3, 4, 5, 6 and returns to port plane 1.

Finally, a hypothetical approach wind profile characterized by a reference approach, time-smoothed average wind speed $\bar{v}_{ref}$; an outdoor air temperature T^o, which is commonly, but not necessarily, assumed to be spatially uniform yet vary with time $T^o(t)$; and an outdoor ambient air pressure p^o associated with an outdoor node (i.e. at elevation z^o) together define the environmental conditions governing the analysis.

System variables

With a building idealization in hand, the primary unknown variables – the *system variables* – may then be defined. Given the two alternative building idealizations presented in Figure 4.39 and the three fundamental principles, there are a number of alternative strategies for forming the system equations and, therefore, defining system variables. Here, we shall consider three.

Nodal approach: If system variables are defined in terms of the node pressures – here collected into a single vector $\{p\}$ – then system equations may be formed based on the imposition of mass conservation at each node of the system idealization. This is the basis of the conventional approach to multizone airflow analysis. Thus for the hypothetical building of Figure 4.41, the conventional system variables will include four unknown pressures as

$$\{\mathbf{p}\} = \{p^b, p^d, p^g, p^i\}^T \tag{4.31}$$

Port plane approach: Alternatively, for the power balance port plane approach, the system variables would be defined in terms of the port plane velocities $\hat{v}_i$ and pressures $\hat{p}_i$ and collected into a partitioned vector $\{\hat{\mathbf{v}}|\hat{\mathbf{p}}\}^T$, whereas for a port plane implementation of the conventional approach, the system variables would be defined in terms of only the port plane pressures $\hat{p}_i$. Thus for the hypothetical building of Figure 4.41, the power balance port plane system variables will include twenty-four unknown variables – twelve velocities and twelve pressures – whereas the conventional theory may be implemented using twelve unknown port plane pressures as

$$\{\hat{\mathbf{v}}|\hat{\mathbf{p}}\}^T = \{\hat{v}_1 \quad \hat{v}_2 \quad \cdots \quad \hat{v}_{12} \mid \hat{p}_1 \quad \hat{p}_2 \quad \cdots \quad \hat{p}_{12}\}^T \tag{4.32a}$$

$$\{\hat{\mathbf{p}}\} = \{\hat{p}_1 \quad \hat{p}_2 \quad \cdots \quad \hat{p}_{12}\}^T \tag{4.32b}$$

Loop approach: Finally, if system variables are defined in terms of flow-limiting mass airflow rates – here collected into a single vector $\{\dot{\mathbf{m}}\}$ – then system equations may be formed by demanding the consistency of pressure changes as one proceeds around a *loop* in the building airflow system. Thus for the hypothetical building of Figure 4.41, the *loop* system variables would include seven unknown mass airflow rates as

$$\{\dot{\mathbf{m}}\} = \{\dot{m}^a, \dot{m}^c, \dot{m}^e, \dot{m}^f, \dot{m}^g, \dot{m}^h, \dot{m}^i\}^T \tag{4.33}$$

From these definitions, it may be tempting to conclude the conventional nodal approach is to be preferred as it may be defined, in general, in terms of far fewer variables. While this is true, the port plane approach allows the imposition of both the conservation of mechanical energy and mass flow, thus although defined in terms of a larger set of system variables, it provides a more complete and rigorous approach to analysis. As the details of these methods are unfolded, it will become clear that each offers certain advantages. Furthermore, it will be seen that the conventional approach is, in fact, a special case of the port plane power balance approach (Axley and Chung, 2005a).

Conservation of mass

At the most fundamental level, air mass flowing into and out of each control volume must be conserved – more precisely, its difference must equal the rate at which the mass of air m^l within a control volume l is accumulated. For two-port control volumes, mass conservation may be determined directly by integration of the detailed air mass flow rates over the port sections. This leads to the familiar macroscopic relation, defined here for ports i and j with ρ_i being the air density distribution over port i (assumed, here, to be uniform):

Two-port mass balance:

$$\left(\rho_i A_i \hat{v}_i - \rho_j A_j \hat{v}_j\right) = \frac{dm^l}{dt} \tag{4.34}$$

For multi-port control volumes, mass flow rates are simply summed over the inflow and outflow ports, but now the conservation relation may be expressed in terms of the adjoining port plane velocities, Equation 4.35a, or if the building idealization includes zones linked by only two-port flow elements, the mass balance may be expressed in terms of the linking mass flow rates, Equation 4.35b.

Multi-port mass balance

Port-plane variables:

$$\underset{\substack{\text{inflow port planes}\\ i=i1,i2,\dots}}{\sum \rho_i A_i \hat{v}_i} - \underset{\substack{\text{outflow port planes}\\ j=j1,j2,\dots}}{\sum \rho_j A_j \hat{v}_j} = \frac{dm^l}{dt} \tag{4.35a}$$

Control volume variables:

$$\underset{\substack{\text{inflow control volumes}\\ i=i1,i2,\dots}}{\sum \dot{m}^i} - \underset{\substack{\text{outflow control volumes}\\ j=j1,j2,\dots}}{\sum \dot{m}^j} = \frac{dm^l}{dt} \tag{4.35b}$$

Commonly, airflow is modeled as a steady phenomenon so the accumulation term dm^l/dt is assumed to be zero. Two programs in the CONTAM family of programs, CONTAMW 2.0 and CONTAM97, however allow accumulation (e.g. for smoke generation and dispersal studies) (Walton, 1998; Dols, 2001a,b).

In forming the system equations, mass balances would be formed for each control volume l of the system idealization. To establish a compact notation for the resulting system equations, we will designate each as the *mass balance function* F_m^l. For a system of n control volumes, these may, then, be collected into a *system mass balance vector* as: $\{\mathbf{F}_m\} = \{F_m^1, F_m^2, \dots F_m^n\}^T$.

Conservation of mechanical energy and related dissipation relations

For flow systems, one may unambiguously account for flow dissipation through the application of the conservation of mechanical energy (Bird, *et al.*, 1960 and 2002), yet this conservation principle has been largely ignored by the building ventilation community. Instead, dissipation in flow-limiting paths has been indirectly modeled using semi-empirical relations derived using the Bernoulli equation. Here instead, the conservation of mechanical energy will be considered first, and then from it a more complete Bernoulli relation and a variety of flow element relations will be derived as special cases of this more general conservation principle.

The details of the macroscopic mechanical energy balance are rather subtle as they can only be completely derived by integrating the microscopic equations of motion (e.g. see Section 7.8 of Bird, *et al.*, 2002). However, the individual terms of both the detailed and macroscopic mechanical power balance have direct physical meanings. For example, consider the detailed unsteady conservation of mechanical energy for a two-port control volume, such as that illustrated in Figure 4.40 left, with unequal port plane cross-sectional areas, $A_i \neq A_j$, to add generality:

$$\overbrace{\left[\int_{A_i} (p_i + \rho_i g z_i) v_i dA - \int_{A_j} (p_j + \rho_j g z_j) v_j dA \right]}^{\text{pressure-work rate difference}} + \overbrace{\tfrac{1}{2} \left(\int_{A_i} \rho_i v_i^2 v_i dA - \int_{A_j} \rho_j v_j^2 v_j dA \right)}^{\text{kinetic energy rate difference}}$$

$$- \dot{E}_d^l = \frac{d(K_{\text{tot}}^l + \Pi_{\text{tot}}^l)}{dt} \tag{4.36}$$

As indicated, the first term accounts for the difference in the rate at which *pressure-work* is done on the flow – here accounting for density changes through the buoyancy term $\rho_i g z_i$, with z_i being the port plane elevation and g the acceleration of gravity. The second term accounts for kinetic energy rate contributions, where v_i is the airflow velocity resultant and v_i is the normal velocity at port plane i. The third term $\dot{E}_d^l$ accounts for the rate of viscous energy dissipation in the control volume l – an unambiguous quantitative measure of the power dissipated. The fourth term accounts for the rate of accumulation of total kinetic energy K_{tot}^l and geopotential energy Π_{tot}^l within the control volume l (i.e. evaluated, in principle, via integration of $0.5\rho v^2$ and $\rho g z$, respectively, over the volume of the control volume).

This detailed power balance may be recast in terms of the spatially averaged macroscopic variables illustrated in Figure 4.40 right, through the introduction of two terms, α_i and β_i, that account for the nonuniformity of the pressure and velocity distributions at the port sections assuming, again, that air density is uniform over port sections:

Two-port power balance:

$$\overbrace{\left[(\beta_i\hat{p}_i + \rho_i g\hat{z}_i)\hat{v}_i A_i - (\beta_j\hat{p}_j + \rho_j g\hat{z}_j)\hat{v}_j A_j\right]}^{\text{pressure-work rate difference}} + \overbrace{\tfrac{1}{2}\left(\alpha_i\rho_i\hat{v}_i^3 A_i - \alpha_j\rho_j\hat{v}_j^3 A_j\right)}^{\text{kinetic energy rate difference}}$$

$$-\dot{E}_d^l = \frac{d(K_{\text{tot}}^l + \Pi_{\text{tot}}^l)}{dt} \qquad (4.37)$$

where:

$$\alpha_i \equiv \int_{A_i} v_i^2 v_i \, dA \bigg/ (\hat{v}_i^3 A_i); \qquad \beta_i \equiv \int_{A_i} p_i v_i \, dA \bigg/ (\hat{p}_i \hat{v}_i A_i)$$

$$(4.38, 4.39)$$

For nearly uniform pressure and velocity distributions, $\beta_i \to 1.0$. If, in addition, the port velocity is normal to the port section, then $\alpha_i \to 1.0$. It should be noted that the velocity distribution correction factor, α_i, is sometimes included in the published literature (e.g. Bird et al., 2002; ASHRAE, 2005) but is most often assumed to be close to unity for practical applications, whereas the pressure-work correction factor β_i is not commonly presented. For flow-limiting resistances found in building ventilation systems, there appears to be little justification to assume these factors are close to unity. Indeed, some preliminary studies have indicated quite the opposite (Axley and Chung, 2005a,b; Axley and Chung, 2006).

For multi-port control volumes, flow contributions are simply summed:

Multi-port power balance:

$$\left[\overbrace{\underset{\substack{\text{inflow}\\ i=i1,i2,\dots}}{\sum}(\beta_i\hat{p}_i + \rho_i g\hat{z}_i)\hat{v}_i A_i - \underset{\substack{\text{outflow}\\ j=j1,j2,\dots}}{\sum}(\beta_j\hat{p}_j + \rho_j g\hat{z}_j)\hat{v}_j A_j}^{\text{pressure work rate difference}}\right]$$

$$+ \tfrac{1}{2}\left(\overbrace{\underset{\substack{\text{inflow}\\ i=i1,i2,\dots}}{\sum}\alpha_i\rho_i\hat{v}_i^3 A_i - \underset{\substack{\text{outflow}\\ j=j1,j2,\dots}}{\sum}\alpha_j\rho_j\hat{v}_j^3 A_j}^{\text{kinetic energy rate difference}}\right) \qquad (4.40)$$

$$-\dot{E}_d^l = \frac{d\left(K_{\text{tot}}^l + \Pi_{\text{tot}}^l\right)}{dt}$$

For any two-port control volume l, the viscous dissipation rate $\dot{E}_d^l$ may be related to a characteristic kinetic energy rate of the flow (that, for convenience here, will be taken as that at the inflow port) and a dimensionless *friction loss factor* ζ^l (Bird *et al.*, 2002) as

$$\dot{E}_d^l = \tfrac{1}{2}\rho_i \hat{v}_i^3 A_i \zeta^l \tag{4.41}$$

Friction loss factors have been published for a large variety of two-port flow components – most obviously for HVAC duct network components, but also for simple and operable openings similar in geometry to common building ventilation devices (Fried and Idelchik, 1989; Idelchik, 1994; Blevins, 2003; ASHRAE, 2005). Yet few have applied them to the analysis of these common devices. The recent attention directed to wind-driven airflow through porous buildings by efforts of European, American and Japanese researchers (Sandberg, 2002; True *et al.*, 2003; Kato, 2004; Sandberg, 2004; Seifert *et al.*, 2004; Axley and Chung, 2006; Kobayashi *et al.*, 2006) and to evaluating the efficacy of stack ventilation devices (Hunt and Syrios 2004) has, however, fostered new interest in investigating mechanical power balances in building airflows and, consequently, the use of friction loss factors to characterize dissipation.

Nevertheless, most building ventilation analysts have preferred instead to follow the tradition establish a century ago by Sir Napier Shaw. Most commonly, building openings are modeled with a simplified version of the orifice equation defined in terms of the familiar discharge coefficient C_d – a relation that is strictly valid only for isothermal, steady flow in, of course, orifice meters:

Simplified orifice model (isothermal):

$$(\hat{p}_i - \hat{p}_j) = \frac{1}{2}\rho\,\hat{v}_i^2 \left(\frac{1}{C_d^2}\right) \tag{4.42}$$

By combining the two-port mass and mechanical energy balances above, Equations 4.34 and 4.37, and simplifying them for isothermal steady flow conditions as well, a mathematically similar two-port flow model may be derived:

Complete model (isothermal):

$$(\hat{p}_i - \hat{p}_j) \approx \frac{1}{2}\rho\,\hat{v}_i^2 \frac{\left[\zeta^l + \left(\alpha_j \frac{A_i}{A_j} - \alpha_i\right)\right]}{\beta^l} \tag{4.43}$$

This relation is approximate because the individual port plane β factors have been assumed of similar magnitude so that they may be replaced by a single component value $\beta^l \approx \beta_i \approx \beta_j$.

Comparing this *complete model* for airflow in two-port building openings with that of the simplified orifice equation suggests that measured discharge coefficients may be related to the parameters of the *complete model* as

$$C_d \approx \sqrt{\frac{\beta^l}{\zeta^l + [\alpha_j (A_i/A_j) - \alpha_i]}} \tag{4.44}$$

This result establishes a direct relation between the mechanical power balance and the commonly used simplified orifice equation. Importantly, by lumping dissipation, kinetic energy differences and pressure-work differences into the single discharge coefficient, C_d, the simplified orifice equation obscures the impact of these phenomena on flow – the very phenomena that Bernoulli famously identified as central to fluid dynamics over 250 years ago! With this in mind, a number of special cases of the full mechanical power balances will be considered.

The mechanical power balance for two-port control volumes, Equation 4.37, simplifies to a modified form of the Bernoulli equation (i.e. augmented with the α and β terms) if one assumes steady flow conditions and the inflow, outflow and control volume densities are nearly equal in the kinetic energy terms, $\rho_i \approx \rho_j = \rho^l$, as, given the mass conservation Equation 4.34, $\hat{v}_i A_i = \hat{v}_j A_j$:

Two-port (Bernoulli) total pressure balance:

$$\alpha, \beta \neq 1.0 : \quad \begin{aligned} (\beta_i \hat{p}_i + \rho_i g \hat{z}_i + \tfrac{1}{2}\alpha_i \rho^l \hat{v}_i^2) - (\beta_j \hat{p}_j + \rho_j g \hat{z}_j + \tfrac{1}{2}\alpha_j \rho^l \hat{v}_j^2) \\ -\tfrac{1}{2}\rho^l \hat{v}_i^2 \zeta^l = 0 \end{aligned} \tag{4.45a}$$

$$\alpha, \beta = 1.0 : \quad \begin{aligned} (\hat{p}_i + \rho_i g \hat{z}_i + \tfrac{1}{2}\rho^l \hat{v}_i^2) - (\hat{p}_j + \rho_j g \hat{z}_j + \tfrac{1}{2}\rho^l \hat{v}_j^2) \\ -\tfrac{1}{2}\rho^l \hat{v}_i^2 \zeta^l = 0 \end{aligned} \tag{4.45b}$$

These relations, in more familiar terms, defined total pressure balances. Although seldom used in building ventilation analysis, total pressure balances are favoured by the piping and duct network analysis communities (Jeppson, 1976; Wood and Funk, 1993; Saleh, 2002). Note that here the actual inflow and outflow densities are retained for the geopotential terms (i.e. to properly account for buoyancy effects) whereas the mean zone density ρ^l is employed for the kinetic energy terms. This modeling strategy is a macroscopic form of the better known microscopic *Boussinesq assumption*.

The (isothermal, steady) Bernoulli relation defined by Equation 4.45 may be simplified further for two-port control volumes with equal inflow and outflow port areas $A_i = A_j$ and uniform port velocity and pressure distributions (i.e. so that $\alpha_i = \alpha_j = \beta_i = \beta_j = 1.0$):

Two-port (modified) pressure balance:

$$[(\hat{p}_i + \rho_i g \hat{z}_i) - (\hat{p}_j + \rho_j g \hat{z}_j)] - \zeta^l \tfrac{1}{2}\rho^l \hat{v}_i^2 = 0 \qquad (4.46)$$

This simplified relation, defined in terms of port static pressures modified by buoyancy terms, $(p_i + \rho_i g z_i)$, is, essentially, the basis of conventional multi-zone airflow analysis – not the Bernoulli equation as is often claimed. That is to say, by comparing this relation to any of the semi-empirical two-port *element* airflow models used in conventional analysis (i.e. that relate static pressure differences to the square of flow velocity $\hat{v}_i^2$, mass flow rate $\dot{m}^{l^2}$ or volumetric flow rate $\dot{V}^{l^2}$), one may directly relate the semi-empirical model parameters to the friction loss factor ζ^l.

At extremely low flow intensities, airflow velocities are linearly related to (modified) static pressure differences and may be modeled by the classic Hagen–Poiseuille equation (Bird *et al.*, 2002):

Two-port (Hagen–Poiseuille) linear static pressure balance:

$$[(\hat{p}_i + \rho_i g z_i) - (\hat{p}_j + \rho_j g z_j)] - \frac{32\mu L}{D_h^2}\hat{v}_i = 0 \qquad (4.47)$$

where μ is the viscosity of air, L the effective length from inlet to outlet and D_h the effective hydraulic diameter of the two-port control volume under consideration. Although this flow element model seldom appears in the literature, it, or a similar variant of it, is invariably used internally in multi-zone airflow analysis programs for low-flow conditions (e.g. $Re = \rho \hat{v}_i D_h / \mu < 100$) to provide greater accuracy and to assure convergence of iterative solvers. It also provides a useful relation to provide the initial estimate (iterate) of airflows in a multi-zone model needed for iterative solution of the otherwise nonlinear systems of equations that govern airflow.

In conventional multi-zone airflow analysis zone pressures, $p(z)$ are assumed to vary hydrostatically within zones – that is to say a *hydrostatic field assumption* is applied to modeled zones. Thus for each pair of port planes i and j of a given zone or, alternatively, for the zone node l and any given port plane i, the *hydrostatic field assumption* is imposed as

Conventional zone (inviscid, hydrostatic) control volume:

$$(\hat{p}_i + \rho^l g \hat{z}_i) - (\hat{p}_j + \rho^l g \hat{z}_j) = 0 \qquad (4.48a)$$

$$(p^l + \rho^l g z^l) - (\hat{p}_i + \rho^l g \hat{z}_i) = 0 \qquad (4.48b)$$

For two-port control volumes, this is yet another simplification – this time of the *modified static pressure model* assuming there is no dissipation – although now the zone density is used as conditions *within* the zone are assumed to be hydrostatic.

Library of dissipation functions

The macroscopic conservation of mechanical energy establishes the general principle for modeling dissipation in building ventilation and infiltration systems, yet for practical analysis a large and varied *library* of specific dissipation functions are needed to properly account for the variety of both ventilation system components and infiltration leakage paths that may be expected to be encountered in practice. Most of the dissipation functions that have been applied to macroscopic airflow analysis may be presented as special semi-empirical cases of the mechanical power balances presented above, Equations 4.37 and 4.40, yet other more empirical relations have also proven useful. A relatively complete selection of these dissipation functions is presented in Table 4.1. Detailed data needed to use these models and additional details relating to such matters as accuracy, application suitability and limiting assumptions are published in a number of handbooks and program users manuals (Liddament, 1986a,b; Pelletret and Keilholz, 1997; Orme *et al.*, 1998; Orme, 1999; Dols *et al.*, 2000; Dols, 2001a,b; Persily and Ivy, 2001; ASHRAE, 2005).

Three related aspects of this table should be noted. First, the vast majority of dissipation functions are limited to modeling two-port control volumes (the so-called flow *elements* or *components*), as until recently multi-port control volumes (e.g. building *zones* and duct plenums) have been modeled as being inviscid. Second, when applying the *loop approach* or the *power balance approach* to building airflow analysis, it is most convenient to express the dissipation functions for a given two-port control volume l in the so-called *inverse form* (e.g. as presented in Equations 4.42 and 4.43):

Inverse form:

$$\Delta P^l = G_d(\hat{v}^l, \vartheta^l) \quad or \quad G_d(\dot{m}^l, \vartheta^l) \tag{4.49}$$

where $\dot{m}^l$ is the air mass flow rate in the element or component (i.e. $\dot{m}^l = \rho^l \hat{v}^l A^l$) and ϑ^l is (one or more) model parameters that establish the size of the flow element or component. For the conventional approach to airflow analysis, on the other hand, the dissipation functions are expresed in the *forward form*:

Forward form:

$$\dot{m}^l = F_d(\Delta P^l, \vartheta^l) \tag{4.50}$$

Finally, the pressure difference across a two-port control volume with port planes i and j must, in general, be understood to be a *modified pressure* difference that accounts for hydrostatic effects:

Modified pressure difference:

$$\Delta P^l \equiv (\hat{p}_i + \rho_i g z_i) - (\hat{p}_j + \rho_j g z_j) \tag{4.51a}$$

Table 4.1 Library of dissipation models for steady building airflow analysis with both the inverse and forward form of most models given

Model	Model Relations	Comments
Multi-port models		
Multi-port power balance	pressure–work rate difference $$\overbrace{\sum_{i=i1,i2,\ldots}^{inflow} (\hat{p}_i + \rho_i g\hat{z}_i)\hat{v}_i A_i - \sum_{j=j1,j2,\ldots}^{outflow} (\hat{p}_j + \rho_j g\hat{z}_j)\hat{v}_j A_j}$$ kinetic energy rate difference $$+ \overbrace{\sum_{i=i1,i2,\ldots}^{inflow} \tfrac{1}{2}\rho_i \hat{v}_i^3 A_i - \sum_{j=j1,j2,\ldots}^{outflow} \tfrac{1}{2}\rho_j \hat{v}_j^3 A_j}$$ $$-\dot{E}_d^l = 0$$	Equation 4.40 simplified for cases where port plane velocity and pressure distributions may be assumed to be effectively uniform. $\dot{E}_d^l$ is the total power dissipated within the control volume – commonly assumed negligible for building zones.
Multi-port hydrostatic field assumption	For all ports of control volume $(\hat{p}_i + \rho^l g\hat{z}_i) - (\hat{p}_j + \rho^l g\hat{z}_j) = 0$	The conventional zone field assumption models control volumes as hydrostatic, thus modified pressures are uniform.
Linear form two-port models		
Darcy's Law	$\Delta P^l = R^l \dot{m}^l$ $\dot{m}^l = \frac{1}{R^l}\Delta P^l$	For 1D airflow in porous media. R^l is the measured porous media resistance. Here, $\vartheta^l \equiv 1/R^l$.
Hagen–Poiseuille Law	$\Delta P^l = \frac{128\mu^l}{\rho^l \pi D^{4l}}\dot{m}^l$ $\dot{m}^l = \frac{\rho^l \pi D^{4l}}{128\mu^l}\Delta P^l$	For low Reynolds number airflow (e.g. Re < 500) in circular ducts of diameter D^l. Here, $\vartheta^l \equiv D^l$.

Table 4.1 (Continued)

Model	Model Relations	Comments
Quadratic form two-port models		
'Complete' friction loss factor model	$$\Delta P^l = \frac{[\zeta^l + (\alpha_j A_j/A_i - \alpha_i)]}{2\rho^l A_i^2 \beta^l}\,\dot{m}^{l2}$$ $$\dot{m}^l = A_i \sqrt{\frac{2\rho^l \beta^l}{[\zeta^l + (\alpha_j A_j/A_i - \alpha_i)]}\,\Delta P^l}$$	Derived using both mass and mechanical power conservation, Equations 4.34 and 4.37 for steady flow conditions. See Duct Fitting model below for components of constant cross-section with uniform pressure and velocity distributions.
Classic orifice model	$$\Delta P^l = \frac{[1-(A^l/A)^2]}{2\rho^l (C_d^l A^l)^2}\,\dot{m}^{l2}$$ $$\dot{m}^l = C_d^l A^l \sqrt{\frac{2\rho^l \Delta P^l}{1-(A^l/A)^2}}$$	For airflow in a duct of cross-section A with a sharp edge orifice plate with opening A^e installed. $C_d^l \approx 0.61$ for turbulent flow (Re > 2000).
Simplified orifice model or 'discharge coefficient model'	$$\Delta P^l = \frac{\dot{m}^{l2}}{2\rho^l (C_d^l A^l)^2}$$ $$\dot{m}^l = C_d^l A^l \sqrt{2\rho^l\,\Delta P^l}$$	Semi-empirical model used for larger building openings. Typically, measured discharge coefficients fall in the range $0.2 \leq C_d^l \leq 1.2$. Here, $\vartheta^l \equiv A^l$.
Effective leakage area model	$$\Delta P^l = \frac{\dot{m}^{l2}}{2\rho^l A_L^{l2}}$$ $$\dot{m}^l = A_L^l \sqrt{\rho^l\,\Delta P^l}$$	Similar to the discharge coefficient model, used for distributed leakage openings in building construction. Measured values of the *effective leakage area* A_L^l are found in handbooks (e.g. ASHRAE, 2005). Here, $\vartheta^l \equiv A_L^l$.
Duct segment model	$$\Delta P^l = \frac{f^l l^l \dot{m}^{l2}}{\rho^l D_h^l A^{l2}}$$ $$\dot{m}^l = A^l \sqrt{\frac{\rho^l D_h^l \Delta P^l}{f^l l^l}}$$ $$f = 16/Re;\ Re < 2100$$ $$f \approx 0.0791/Re^{1/4};\ Re \geq 2100$$	A^l, D_h^l and l^l are the cross-sectional area, hydraulic diameter and length of the duct segment. The Fanning friction factor may be expected to fall in range $0.01 \leq f \leq 0.05$ for natural or hybrid ventilation systems. Here, the sizing parameter may be taken as $\vartheta^l \equiv D_h^l A^{l2}$ or $\vartheta^l \equiv D_h^l A^{l2}/l^l$.
Duct fitting or simplified friction loss factor model	$$\Delta P^l = \zeta^l \frac{\dot{m}^{l2}}{2\rho^l A^{l2}}$$ $$\dot{m}^l = A^l \sqrt{\frac{2\rho^l \Delta P^l}{\zeta^l}}$$	Duct fitting friction loss coefficients ζ^l can be found in a number of handbooks (e.g. ASHRAE, 2005). This model is identical to the 'Complete' model for uniform cross-sections with uniform velocity and pressure distributions. Here $\vartheta^l \equiv A^l$.

Polynomial form two-port models

Quadratic polynomial model

$$\Delta P^l = C_1^l \frac{\dot{m}^l}{\rho^l} + C_2^l \frac{\dot{m}^{l2}}{\rho^l}$$

Theoretically consistent model for extended cracks and joints in construction that are likely to experience laminar or transitional flow. Model parameters may be measured or estimated from crack geometry (Etheridge, 1998; Chiu and Etheridge, 2002).

Fan curve model

$$\Delta P^l = a_0 \left(\frac{D^l}{D_0}\right)^2 \left(\frac{N^l}{N_0}\right)^2 +$$
$$a_1 \frac{\dot{m}^l}{\frac{D^l}{D_0}}\left(\frac{N^l}{N_0}\right) + a_2 \frac{\dot{m}^{l2}}{\left(\frac{D^l}{D_0}\right)^4}$$

Coefficients a_0, a_1 and a_2 are empirical constants fitted to the performance of a base fan of diameter D_0 and rotational speed N_0. Here $\vartheta_1^l = D^l$ is the fan diameter and $\vartheta_1^l = N^l$ is the rotational speed of a geometrically similar fan.

Fully empirical two-port models

Power-law model

$$\Delta P^l = \frac{(\dot{m}^l/\rho^l)^{1/n}}{(C^l)^{1/n}}$$
$$\dot{m}^l = \rho^l C^l \Delta p^n$$

Fully empirical model used typically for cracks and joints in construction with measured exponents in range $0.5 \leq n \leq 1.0$, but typically ≈ 0.60. Measured values for $\vartheta^l = C^l$ can be found in handbooks (Persily, 1994; Orme et al., 1998; Orme, 1999; Persily and Ivy, 2001).

Logarithmic model for self-regulating vents

$$\Delta P^l = -\Delta P_0^l \ln\left(1 - \frac{\dot{m}^l}{\dot{m}_0^l}\right)$$

Fully empirical but intuitively satisfying model that faithfully captures full range of vent behaviour. The threshold of vent regulation is modeled by parameter ΔP_0^l and the nominal regulated flow is modeled by $\vartheta^l = \dot{m}_0^l$.

Additional relations

Component reynolds number

$$Re = \frac{D^l \sqrt{A^l \vartheta^l}}{\mu} = \frac{\sqrt{A^l}\dot{m}^l}{\mu A^l}$$

To provide an order of magnitude estimate of flow intensity and, thus, flow regime. A^l is a characteristic cross-sectional area of the component and μ is the viscosity of air.

Ideal gas law for dry air

$$\rho = \frac{352.6K}{(T_{^\circ C}+273.15)\,K} \quad (\text{kg m}^{-3})$$

For dry air. For preliminary calculations, this ideal gas law may be acceptable except in those instances where significant moisture changes occur.

Note, for the usual (but not theoretically necessary) assumption that temperatures within a control volume are uniform (*well-mixed*), the modified pressure difference may be expressed as the sum of a static pressure difference and a hydrostatic increment:

Well-mixed control volume:

$$\Delta P^l = \Delta \hat{p}^l + \rho^l g \Delta z^l \tag{4.51b}$$

where:

$$\Delta \hat{p}^l = (\hat{p}_i - \hat{p}_j); \quad \Delta z^l = (z_i - z_j) \tag{4.52, 4.53}$$

Most of the functions presented in Table 4.1 are likely to be familiar to the reader as they are well established in the ventilation literature. The *logarithmic model for self-regulating vents* is likely to be an exception. It models the characteristics of self-regulating vents that have proven essential to passive natural ventilation design strategies (Axley, 2000a and 2001).

The dissipation functions presented in Table 4.1 are valid for unidirectional flow. In larger building openings, bi- and multi-directional airflows commonly occur. A number of approaches have been proposed to model these *large-opening* conditions – most based on applying an infinitesimal form of the discharge coefficient model to infinitesimal lamina over the plane of the opening without theoretical justification (van der Maas, 1992; van der Maas *et al.*, 1994; Dascalaki and Santamouris, 1996; Etheridge and Sandberg, 1996; Feustel and Smith, 1997; Etheridge, 2004).

In forming the system equations, power, total pressure, modified pressure or hydrostatic balance relations (i.e. Equations 4.37, 4.40, 4.42, 4.43, 4.45, 4.46, 4.47, 4.48 or those from Table 4.1) would be formed for each control volume l of the system idealization. To establish a compact notation for the resulting system equations, we will designate each as the control volume *dissipation function* of the corresponding dissipation relation – F_d^l if formed using the forward form and G_d^l if formed using the inverse form. The dissipation functions for a system of n control volumes may, then, be collected into a *system dissipation vector* as $\{F_d\} = \{F_d^1, F_d^2, \ldots F_d^n\}^T$ or $\{G_d\} = \{G_d^1, G_d^2, \ldots G_d^n\}^T$ for the forward or inverse forms, respectively.

Boundary conditions

Equations governing the airflow in building systems may be formed using any one of the three approaches introduced above – i.e. the conventional *nodal approach*, the *port plane approach* or the *loop approach*. Alone, however, these equations would be indeterminate and, thus, must be complemented by boundary conditions and, in general, zone field assumptions to establish a determinate (*closed*) system of equations that may then be solved.

In macroscopic analysis, 'boundary' conditions may be established in terms of the defined system variables of the analytical approach taken. Although pressure conditions acting at envelope ports of a building system often define these boundary conditions, one may also specify internal zone or port plane pressures (e.g. to investigate building pressurization strategies for smoke control) or component or port plane air mass flow rates or velocities (e.g. to model well-controlled mechanically induced airflows). Thus, boundary conditions need not be actually associated with the physical boundary of the building as a whole.

Port plane approach: For the port plane approach, one may directly specify static pressure values $\hat{p}^*$, total pressure values $\hat{p}^*_{tot}$ or velocity values $\hat{v}^*$ as boundary conditions at any of the port planes of the system. The algebraic specification of these boundary conditions may be achieved in a number of ways. Here, they will be represented as additional *boundary condition functions* F_{bi} for each port plane i involved, as follows:

Port plane pressure BC $\qquad \hat{p}_i - \hat{p}_i^* = 0$ $\qquad\qquad\qquad$ (4.54)

Port plane total pressure BC $\qquad \hat{p}_i + \frac{1}{2}\rho^k \hat{v}_i^2 - \hat{p}_{itot}^* = 0$ $\qquad$ (4.55)

Port plane velocity BC $\qquad \hat{v}_i - \hat{v}_i^* = 0$ $\qquad\qquad\qquad$ (4.56)

where, $\hat{p}_i^*$, $\hat{p}_{toti}^*$ and $\hat{v}_i^*$ are the corresponding specified numerical values and ρ^k is the air density of the control volume associated with the boundary condition specification.

Nodal approach: For the nodal approach, one may specify node pressures directly:

Node pressure BC $\qquad\qquad p^l - p^{*l} = 0$ $\qquad\qquad\qquad$ (4.57)

One may also specify the airflow rate for a given limiting flow resistance (element) l during the formation of the nodal system equations when applying the mass conservation relation, Equation 4.35b, simply by summing its specified numerical value m^{*l}.

When using the nodal approach, the specification of envelope and other port plane pressures, however, demands definition of a relation between port plane and nodal pressures. For building zones, the hydrostatic field assumption, Equation 4.48, could be used. More typically, wind-induced envelope port plane pressures are commonly assumed to be linked to the ambient outdoor zone node pressure p^o via *sealed building* wind pressure coefficients $C_{p,i}$ for each envelope port plane i (where here a hydrostatic increment relative to the ambient pressure node datum has been included):

Envelope pressure BC $\qquad \hat{p}_i - C_{p,i}\frac{1}{2}\rho^o \bar{v}_{ref}^2 + \rho^o g(z_i - z^o) - p^o = 0$ (4.58)

(The hydrostatic increment and ambient outdoor node pressure p^o of this relation are often not included in the literature yet are necessarily included in multi-zone airflow analysis programs.)

Wind pressure coefficients are available in a number of handbooks (Dascalaki and Santamouris, 1996; Allard, 1998; Orme, *et al.*, 1998; Santamouris and Dascalaki, 1998; Persily and Ivy, 2001; ASHRAE, 2005). Increasingly, CFD is also used to predict wind pressure coefficients (Murakami, 1993; Holmes and McGowan, 1997; Kurabuchi *et al.*, 2000; Jensen *et al.*, 2002a,b). Wind pressure coefficients in general vary with wind direction. A number of empirical correlations and computational tools have been developed that account for this variation (Walker and Wilson, 1994; Dascalaki and Santamouris, 1996; Knoll and Phaff, 1996; Knoll *et al.*, 1997; Orme *et al.*, 1998). Finally, the determination of the reference wind velocity often presents a challenge as wind data is normally available only from local airport weather records where wind conditions may vary significantly from those at the building site. The handbooks listed above provide guidance for adjusting airport wind data for building height, location, surrounding site topography and shielding effects of nearby buildings.

The uncertainty of wind pressure coefficients combined with that associated with estimating the reference wind conditions are thought to introduce the greater part of the uncertainty of computed results in multi-zone ventilation analysis (Bassett, 1990; Fürbringer *et al.*, 1993; Fürbringer *et al.*, 1996a,b; Roulet *et al.*, 1996). Nevertheless, two additional sources of error may, in some instance, be equally important: (a) sealed building wind pressure coefficients may not properly represent driving forces on porous buildings and (b) wind-induced turbulence may significantly contribute to ventilation flow rates (Etheridge and Sandberg, 1996; Sirén, 1997; Girault and Spennato, 1999; Saraiva and Marques da Silva, 1999; Haghighat *et al.*, 2000; Etheridge, 2002).

As airflow through envelope openings alters wind pressure distributions, significant inaccuracies result when opening areas exceed approximately 20 per cent of any given building surface (Aynsley *et al.*, 1977; Aynsley, 1999). Recent research directed to better understand wind-driven airflow in *porous buildings* suggests that windward envelope pressures may be better modeled through the specification of total pressure rather than static pressure (Aynsley *et al.*, 1977; Seifert *et al.*, 2000; Sandberg, 2002; True *et al.*, 2003; Karava *et al.*, 2004; Ohba *et al.*, 2004; Sandberg, 2004; Seifert *et al.*, 2004), yet the issue remains unresolved.

Loop approach: As will be shown below, wind- and buoyancy-induced boundary conditions are directly included in the formation of the loop system equations; thus these system equations, when formed, will be determinate. In addition, given the system variables are defined in terms of control volume mass airflow rates $\dot{m}^l$, one may specify selected component flow rates as well as (e.g. to implement a first-order approximation of a fan's contribution to the system equations):

$$\text{Loop mass flow BC} \qquad \dot{m}^l - \dot{m}^{*l} = 0 \qquad\qquad (4.59)$$

Zone field assumptions

The zone field assumptions introduced above, Equation 4.48a and 4.48b, for both node and port plane variables, are specific cases of the assumption of hydrostatic conditions within a flow regime:

$$\text{Hydrostatic field:} \qquad p + \rho g z = \text{constant} \qquad\qquad (4.60)$$

With this field assumption in mind two additional possibilities may be identified – a uniform pressure field and a uniform total pressure field. The latter is a condition that is associated with *irrotational* flow in (nearly) inviscid fluids (Chorin and Marsden, 1993):

$$\text{Uniform pressure field:} \qquad p = \text{constant} \qquad\qquad (4.61a)$$

$$\text{Irrotational field:} \qquad p + \rho g z + \frac{1}{2}\rho v^2 = \text{constant} \qquad\qquad (4.61b)$$

These two alternative field assumptions may be employed to establish approximate relations between port plane pressures (i.e. following the example of Equation 4.48a) or between node and port plane pressures (i.e. after the example of Equation 4.48b).

The uniform pressure field assumption is commonly applied to duct junctions whereas the hydrostatic field assumption is preferred for zones in conventional multi-zone analysis – both quite reasonable approximations, but nevertheless approximations. Use of the irrotational field assumption has been investigated for ventilation airflows characterized by a dominant stream tube both as a strategy to correct for violations of mechanical energy conservation in the (conventional) nodal approach and as a means to account for the near-irrotational character of these dominant stream tubes in the port plane power balance approach to multi-zone airflow analysis (Axley *et al.*, 2002b; Axley and Chung, 2005a). In the few cases investigated, however, the hydrostatic field assumption appears to provide a better approximation than the irrotational field assumption and, importantly, avoids the complexity of the added nonlinearity introduced.

4.3.3 Multi-zone system equations

With fundamental mass and mechanical energy conservation relations, semi-empirical and fully-empirical dissipation relations, boundary conditions and zone field assumptions in hand, one may then *assemble* these equations to form *system equations* governing the airflows in multi-zone building systems. Here, the assembly of these system equations will be presented both in simple algebraic notation and in concise matrix notation for each of the three general approaches to multi-zone building ventilation analysis – the conventional *nodal approach*, the more general *port plane approach* and the often overlooked *loop approach*. The concise matrix formulation of these approaches is most appropriate for computational implementation whereas the simple algebraic formulations are more directly useful for manual analysis. Any one of the popular mathematical processing programs such as MATLAB®, Mathematica© or MathCad (MathWorks, 2000; MathSoft, 2004; Wolfram_Research, 2004) greatly facilitate 'manual' analysis whereas the higher level programming languages of these same tools are useful for prototyping computational tool development.

Nodal approach

Given a nodal building idealization (e.g. Figure 4.39 left), one implements the nodal approach by simply demanding zone mass airflow rates are conserved (i.e. Equation 4.35b applied to zone control volumes) for each zone z:

Airflow mass conservation function:

$$F_m^z \equiv \underbrace{\sum \dot{m}^i}_{\substack{\text{inflow} \\ i=i1,i2,\ldots}} - \underbrace{\sum \dot{m}^j}_{\substack{\text{outflow} \\ j=j1,j2,\ldots}} = \frac{dm^z}{dt} \qquad (4.62)$$

where, most commonly, zone mass accumulation is assumed to be negligible: $dm^z/dt \approx 0$. Element mass airflow relations are then applied for each of the limiting flow resistances i by selecting appropriate semi- or fully-empirical dissipation relations (e.g. from Table 4.1) expressed in forward form, Equation 4.50:

Element airflow dissipation function: $\dot{m}^i = F_d^i(\Delta P^i, \vartheta^i)$ (4.63)

As each of the element flow relations is defined in terms of the modified port plane pressures, r and s, of the adjacent zones they link, $\Delta P^i = (\hat{p}_r - \hat{p}_s) + \rho^i g(\hat{z}_r - \hat{z}_s)$, zone field assumptions are then imposed for interior port planes s and, typically, wind pressure boundary conditions are imposed for envelope port planes r to establish the dependency on the zone node pressures p^z:

Hydrostatic field assumption: $\quad (p^z + \rho^z g\, z^z) - (\hat{p}_s + \rho^z g\, \hat{z}_s) = 0$

$$(4.64)$$

Envelope pressure BC: $\quad \hat{p}_r - C_{p,r}\frac{1}{2}\rho^o \bar{v}_{\text{ref}}^2 + \rho^o g(z_r - z^o) - p^o = 0$

$$(4.65)$$

When these equations are then combined for the k zones of the building idealization, one will have k equations defined in terms of the k zone node pressures.

In manual analysis, numerical values would be substituted into each of these equations for element model parameters, port plane elevations, wind pressure coefficients, and zone elevations and densities. Then the series of substitutions outlined above would be made. That is to say, the actual assembly of the flow element, field assumption and boundary condition relations depends on the specific topology and details of the building idealization (model). Although the details of this assembly may be presented in concise matrix notation (Axley, 2006), the procedures needed to do so are complex and involved and, therefore, beyond the scope of this chapter.

One may, however, resort to functional notation to gain an essential understanding of the computational procedures used to implement the nodal method. The substitution of specific model parameters into the envelope pressure boundary condition relations, the hydrostatic field assumptions and the definition of modified pressure difference and the subsequent substitution of these relations into the element airflow dissipation relations have the net effect of mapping these dissipation relations from functions that depend on modified pressure differences to ones that depend on the system zone pressures $\{\mathbf{p}\}$ as $F_d^i(\Delta P^i) \rightarrow F_d^i(\{\mathbf{p}\})$. Consequently, the mass conservation function for a zone z will be defined as a (nonlinear) function of the same system zone pressures as, by Equation 4.62, it is simply an algebraic sum of these dissipation relations (i.e. with the convention that mass inflows will be positive and outflows negative):

Airflow mass conservation function:

$$F_m^z(\{\mathbf{p}\}) = \sum_{\substack{\text{linked elements} \\ i = i1,\, i2,\, \ldots}} F_d^i(\{\mathbf{p}\}) = \frac{dm^z}{dt} \quad (4.66)$$

The equations governing ventilation airflows in the system as a whole – the *system equations* – are defined by these mass conservation equations for all zones. Here they will be collected into a single vector of functions as

Nodal system equations: $\{\mathbf{F}_m(\{\mathbf{p}\})\} = \{F_m^1(\{\mathbf{p}\})\ F_m^2(\{\mathbf{p}\})\cdots\}^T = \dfrac{d}{dt}\{\mathbf{m}\}$

(4.67)

where $\{\mathbf{m}\} = \{m^1\ m^2\cdots\}^T$ is a vector of zone air masses (i.e. the instantaneous value of the mass of air within each zone). In as much as each of these functions is coupled to the others through shared dependencies on select zone node pressures and, furthermore, these couplings are nonlinear, the solution of these system equations is generally difficult.

Most computational tools use the Newton–Raphson method, or one of several variants of it, to solve the system equations. The Newton–Raphson method is a well-known method based on using a truncated Taylor series approximation of the system equations in an iterative manner where given an initial iterate for the zone pressure vector $\{\mathbf{p}\}_{[0]}$, an improved estimate is formed $\{\mathbf{p}\}_{[1]} = \{\mathbf{p}\}_{[0]} + \{\Delta\mathbf{p}\}_{[1]}$ by solving the truncated Taylor series approximation (Press *et al.*, 1992; Kelley, 1995; Lorenzetti and Sohn, 2000; Kelley, 2003). Using this improved estimate, one can repeat the process until the solution converges to sufficient accuracy.

More specifically for the typical case of steady flow $d\{\mathbf{m}\}/dt = \{0\}$, at iterate k of the iterative process one

1   forms the *system Jacobian matrix*: $[\mathbf{J}]_{[k]} = \dfrac{\partial\{\mathbf{F}_m\}}{\partial\{\mathbf{p}\}}\bigg|_{\{\mathbf{p}\}_{[k]}}$ (4.68)

2   solves the truncated Taylor series approximation:

$$[\mathbf{J}]_{[k]}\{\Delta\mathbf{p}\}_{[k+1]} \approx -\{\mathbf{F}_m(\{\mathbf{p}\}_{[k]})\} \tag{4.69}$$

3   updates the solution estimate: $\{\mathbf{p}\}_{[k+1]} = \{\mathbf{p}\}_{[k]} + \{\Delta\mathbf{p}\}_{[k+1]}$ (4.70)

4   forms the current *system residual* (i.e. by Equation 4.66):

$$\left\{\mathbf{F}_m\left(\{\mathbf{p}\}_{[k+1]}\right)\right\} \tag{4.71}$$

5   evaluates convergence:

$$\left\|\left\{\mathbf{F}_m\left(\{\mathbf{p}\}_{[k+1]}\right)\right\}\right\| \leq \tau_{\text{rel}}\left\|\left\{\mathbf{F}_m\left(\{\mathbf{p}\}_{[k]}\right)\right\}\right\| + \tau_{\text{abs}} \tag{4.72}$$

where both relative, τ_{rel}, and an absolute, τ_{abs}, error tolerances may be considered. If the convergence criterion is met, the approximate solution is deemed satisfactory.

Port plane approach

The port plane approach is superficially similar to the nodal approach in that one assembles airflow conservation and element dissipation relations and imposes boundary conditions and zone field assumptions to form the system equations. The relations used are, however, expressed in terms of the port plane variables of velocity $\hat{v}_i$ (or port plane mass airflow rate $\dot{m}_i$) and pressure $\hat{p}_i$ of each port plane i of the building model. Alternatively, for conventional building system models involving zones linked by constant cross-section two-port flow elements l, one may formulate the port plane system equations in terms of element mass airflow rates $\dot{m}^l$ and port plane pressures $\hat{p}_i$.

Limiting consideration to the most general alternative and steady conditions of airflow for simplicity, we get the constituent relations of the port plane approach:

Mass conservation function (for control volume l):

$$F_m^l = \sum_{\substack{\text{inflow port planes}\\ i=i1,i2,\dots}} \rho_i A_i \hat{v}_i - \sum_{\substack{\text{outflow port planes}\\ j=j1,j2,\dots}} \rho_j A_j \hat{v}_j = 0 \tag{4.73}$$

Two-port dissipation function alternatives (for control volume l):

Bernoulli model:
$$F_d^l = (\hat{p}_i + \rho_i g \hat{z}_i \tfrac{1}{2} \rho^l \hat{v}_i^2) - (\hat{p}_j + \rho_j g \hat{z}_j + \tfrac{1}{2} \rho^l \hat{v}_j^2)$$
$$- \tfrac{1}{2} \rho^l \hat{v}_i^2 \zeta^l = 0 \tag{4.74a}$$

Table 4.1 models: $\quad F_d^l = (\hat{p}_i + \rho^l g \hat{z}_i) - (\hat{p}_j + \rho^l g \hat{z}_j) - G_d^l(\hat{v}^l, \vartheta^l) = 0$
$$\tag{4.74b}$$

Multi-port dissipation or power balance function (for control volume l):

$$F_d^l = \left[\overbrace{\sum_{\substack{\text{inflow}\\ i=i1,i2,\dots}} (\hat{p}_i + \rho_i g \hat{z}_i)\, \hat{v}_i A_i - \sum_{\substack{\text{outflow}\\ j=j1,j2,\dots}} (\hat{p}_j + \rho_j g \hat{z}_j)\, \hat{v}_j A_j}^{\text{pressure-work rate difference}} \right]$$

$$+ \tfrac{1}{2} \left(\overbrace{\sum_{\substack{\text{inflow}\\ i=i1,i2,\dots}} \rho_i \hat{v}_i^3 A_i - \sum_{\substack{\text{outflow}\\ j=j1,j2,\dots}} \rho_j \hat{v}_j^3 A_j}^{\text{kinetic energy rate difference}} \right) - \dot{E}_d^l = 0 \tag{4.74c}$$

Boundary condition functions (for port plane i adjacent to control volume k):

Pressure BC: $\quad F_{b,i} = \hat{p}_i - \hat{p}_i^* = 0 \tag{4.75a}$

Total pressure BC: $F_{b,i} = \hat{p}_i + \frac{1}{2}\rho^k \hat{v}_i^2 - \hat{p}_{itot}^* = 0$ (4.75b)

Velocity BC: $F_{b,i} = \hat{v}_i - \hat{v}_i^* = 0$ (4.75c)

Zone field assumption functions (for control volume l):

Uniform pressure: $F_f^l = \hat{p}_i - \hat{p}_j = 0$ (4.76a)

Hydrostatic: $F_f^l = (\hat{p}_i + \rho^l g \hat{z}_i) - (\hat{p}_j + \rho^l g \hat{z}_j) = 0$ (4.76b)

Irrotational: $F_f^l = (\hat{p}_i + \rho^l g \hat{z}_i + \frac{1}{2}\rho^l \hat{v}_i^2) - (\hat{p}_j + \rho^l g \hat{z}_j + \frac{1}{2}\rho^l \hat{v}_j^2) = 0$

(4.76c)

where, for the Bernoulli and multi-port power balance, the velocity and pressure distribution terms have been assumed to be unity (i.e. $\alpha = \beta = 1.0$) to simplify the presentation.

Given a specific building idealization, the mass balance functions would be formed for each control volume and collected into a vector of functions $\{F_m\} = \{F_m^1 \ F_m^2 \cdots\}^T$; the dissipation conservation functions would be formed for each control volume and collected into a vector of functions $\{F_d\} = \{F_d^1 \ F_d^2 \cdots\}^T$; the boundary conditions functions would be formed for each specified port plane m, n, ... (e.g. envelope port planes) and collected into a vector of functions $\{F_b\} = \{F_{b,m} \ F_{b,n} \cdots\}^T$; and, finally, zone field assumptions would be applied to each pair of port planes m, n, ...in excess of two and collected into a vector of functions $\{F_f\} = \{F_{f,m} \ F_{f,n} \cdots\}^T$. The system equations $\{F\}$ would then be defined by the collection of these functions, again collected into a vector: $\{F\} = \{F_m | F_d | F_b | F_f\}^T$. This system of equations may then be solved using the Newton–Raphson method, or a variant of this popular method, given an initial iterate for the port plane system variables $\{\hat{v}| \hat{p}\}_{[0]}$ by following the schematic algorithm presented above, Equations 4.68–4.72 – where, now, the Jacobian matrix would be formed in terms of the port plane system variables $\{\hat{v}| \hat{p}\}$.

Although the port plane approach may seem rather formidable, it can be applied manually to simple systems as demonstrated in Section 4.3.4 below. More importantly, it is a general method that reveals the limiting assumptions of other approaches used to model flow networks. If, for a given building model, one models dissipation using power balance dissipation functions for all control volumes (or, equivalently, Bernoulli dissipation for two-port control volumes), then the building model will be theoretically most complete – especially if one is able to include the velocity and pressure distribution correction factors α and β. If such a model includes only two-port control volumes (i.e. is a serial assembly), then zone field assumptions will not be required for closure and, therefore, the system equations will be 'exact' in the macroscopic sense. Furthermore, the imposition of total pressure boundary conditions (i.e. that appear more appropriate for upwind envelope ports (Sawachi, 2002; Kato, 2004; Kurabuchi *et al.*, 2004; Ohba *et al.*, 2004; Axley and Chung, 2005a; Axley, 2006) is inherently direct in the port plane approach whereas it is not so for the nodal and loop approaches.

If, on the other hand, one utilizes Bernoulli dissipation function for all two-port control volumes (e.g. the flow-limiting resistances) and applies either the hydrostatic or the uniform field assumptions to multi-port control volumes (e.g. zones), then the analytical model will correspond to those preferred by the piping network analysis community (Jeppson, 1976; Wood and Funk, 1993; Saleh, 2002). Finally, if one models dissipation using any one of the semi-empirical or fully empirical modified pressure difference models presented in Table 4.1 and applies either the hydrostatic (i.e. for zones) or the uniform pressure field (i.e. for duct junctions) assumptions to multi-port control volumes (e.g. zones), then the analytical model will correspond to that used conventionally within the multi-zone building ventilation analysis community. Consequently, the conventional approach is theoretically incomplete in that mechanical energy conservation is ignored and field assumption approximations are required. Although research is under way to evaluate the significance of these shortcomings, no general conclusion yet can be made (Guffey and Fraser, 1989; Murakami *et al.*, 1991; Kato, 2004; Axley and Chung, 2005a).

Loop approach

Equations governing *steady* airflow in building system may be formed by imposing mass balances for each (assumed) hydrostatic zone using Equation 4.35b and demanding that the port plane pressure changes due to the airflow encountered while traversing a flow *loop* in the airflow system simply sum to zero. Formally, the imposition of the *loop consistency* requirement may be stated mathematically as

$$\text{Loop consistency:} \quad F_l = \sum_{\text{loop} l} \Delta \hat{p}^k = 0 \qquad (4.77)$$

where k is permuted through linked control volumes (including the outdoor 'zone') as one proceeds around a given loop l. To properly account for wind-induced effects, one must include an outdoor ambient pressure node in the loop for loops that traverse the outdoor zone. The individual loop equations (functions) may then be collected into a vector of loop functions $\{F_l\} = \{F_{l,1} \ F_{l,2} \cdots\}^T$, although one must be sure that the collection of these loop functions are an independent set of all the loop equations that could conceivably be formed. This last condition has been the principle barrier to the use of the loop method. Recently, however, graph theoretic algorithms have been employed to automatically identify the independent loops needed to achieve this critical objective (Jensen, 2005).

The loop consistency functions are then complemented by mass conservation for each zone , the inverse form of appropriate element airflow relations that account for dissipation (e.g. from Table 4.1), the hydrostatic relation for pressure changes encountered within zones and the wind pressure relation that accounts for pressure increments at envelope ports:

Mass conservation: $$F_m^z = \sum_{\substack{\text{inflow} \\ i=i1,i2,\dots}} \dot{m}^i - \sum_{\substack{\text{outflow} \\ j=j1,j2,\dots}} \dot{m}^j = 0 \qquad (4.78)$$

Element dissipation: $$\Delta \hat{p}^k = G_d^k(\dot{m}^k, \vartheta^k) - \rho^k g \Delta z^k \qquad (4.79a)$$

Hydrostatic zone assumption: $$\Delta \hat{p}^k = -\rho^k g \Delta z^k \qquad (4.79b)$$

Wind pressure increments: $$\Delta \hat{p}^k = C_{p,i} \tfrac{1}{2} \rho^o \bar{v}_{\text{ref}}^2 \qquad (4.79c)$$

Equations of the form of 4.79a–c would simply contribute to the sum of the loop consistency relation, as appropriate, as one traverses a given loop. Consequently, for steady flow conditions through well-mixed control volumes, the loop consistency relation may be simplified to the intuitively satisfying result that control volume (*element*) pressure losses simply balance the sum of wind-induced (Δp_w) and buoyancy-induced (Δp_b) pressure changes:

$$F_l = \overbrace{\sum_{\substack{\text{loop l} \\ k=1,2,\dots;}} G_d^k(\dot{m}^k, \vartheta^k)}^{\text{element losses}}$$

$$- \left(\overbrace{\frac{1}{2}\rho^o \bar{v}_{\text{ref}}^2 \underset{\substack{\text{envelope} \\ i}}{\sum} C_{p,i}}^{\text{wind pressure } \Delta p_w} + \overbrace{\sum_{\substack{\text{loop l} \\ k=0,1,2,\dots}} \left(\rho^k g \Delta z^k \right)}^{\text{bouyancy pressure } \Delta p_b} \right) = 0 \qquad (4.80)$$

where the envelope wind pressure coefficients $C_{p,i}$ are summed as positive contributions for pressure increases as one moves along the loop in the forward direction. Note that zone node pressures need not be considered at all in forming the loop equations.

Finally, recognizing the mass conservation and loop consistency relations are defined in terms of the two-port control volume (i.e. flow-limiting resistances) mass airflow rates, one may take these airflow rates as the system variables, $\{\dot{m}\} = \{\dot{m}^1 \ \dot{m}^2 \cdots\}^T$, collect the mass conservation relations into a single vector $\{F_m\} = \{F_m^1 \ F_m^2 \cdots\}^T$ and, therefore, define the loop system equations as the set of the mass conservation and loop equations $\{F\} = \{F_m | F_l\}^T$. As before, this system of equations may then be solved using the Newton–Raphson method, given an initial iterate for the loop system variables $\{\dot{m}\}_{[0]}$ by following the schematic algorithm presented above, Equation 4.68–4.72 – where now the Jacobian matrix would be formed in terms of the loop system variables $\{\dot{m}\}$.

The loop approach, although overlooked by all but a few members of the building simulation community (Wray and Yuill, 1993; Nitta, 1994), provides an approach for sizing ventilation system components that will be discussed below (Axley, 1998, 1999a, 2000a,b and 2001; Axley *et al.*, 2002a; Ghiaus *et al.*, 2003) and has recently been employed for modeling the coupled thermal/airflow interactions in buildings (Jensen, 2005).

Determinacy of the system equations

One may examine the *determinacy* of these *system equations* without actually forming the equations by simply comparing the number of equations available with the number of unknown system variables to be determined.

Consider, for example, the two buildings illustrated in Figure 4.42 and their corresponding nodal and port plane idealizations. The upper is a *serial assembly* of two zones linked by three flow-limiting resistances – i.e. a serial

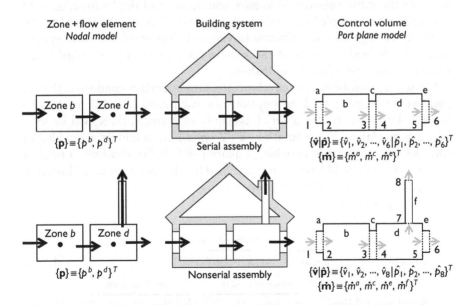

Figure 4.42 Representative buildings and building idealizations illustrating *serial* and *nonserial assemblies* of control volumes defining the ventilation systems. Port planes are indexed by number and control volumes by letter. The corresponding system variables are shown for the *nodal*, *port plane* and *loop* approaches to multi-zone airflow analysis.

assembly of five control volumes identified as a, b, c, d, and e, respectfully, separated by six port planes with IDs 1, 2, 3, 4, 5 and 6 as shown. The lower building presents an example of a nonserial assembly as an alternate flow path through the stack is provided in zone d (control volume d). For each of the idealizations, unknown system variable vectors are also shown for the three modeling approaches – i.e. $\{p\}$ for the nodal approach, $\{\hat{v}|\hat{p}\}$ for the port plane approach and $\{\dot{m}\}$ for the loop approach.

For the conventional nodal approach, in general, the number of system variables (node pressures) is equal to the number of zones, k. As a singe zone mass balance may be formed for each of these zones (i.e. by Equation 4.35b), one may conclude that the resulting system equations are determinate. However, zone mass balances are formed by summing the mass flow relations associated with each flow element linked to a given zone – relations that are defined in terms of the m port plane pressures. Consequently, one must add additional equations to establish the relationships between the port plane pressures and the node pressures; specifically one must establish b boundary conditions (typically establishing envelope pressures) and f field assumptions that link zone node pressures to each zone port plane pressure. Thus for the conventional nodal approach, $k+b+f$ equations must be assembled to solve for the $k+m$ variables, Equation 4.81a. In as much as the static pressure boundary conditions and the hydrostatic field assumptions are simple linear relations, however, they may be assembled by simple direct algebraic substitution leaving only k coupled nonlinear system equations defined in terms of the k zone node pressures that need to be solved by iterative solution algorithms.

For serial assemblies, two envelope pressure boundary conditions ($b = 2$) and two hydrostatic field assumptions for each zone ($f = 2k$) need to be formed. Furthermore, for a serial assembly of k zones, there will be $m = 2k+2$ port planes; thus a total of $3k+2$ equations would be assembled to solve for $3k+2$ pressure variables, Equation 4.81b. For example, a total of eight equations would have to be assembled for the two-zone case illustrated in Figure 4.42, upper left, Equation 4.81c.

Nodal approach:

General:
$$\overbrace{k+m}^{\text{unknown variables}} = \overbrace{k+b+f}^{\text{equations needed}} \tag{4.81a}$$

Serial assemblies:
$$\overbrace{k+(m=2k+2)}^{\text{unknown variables}} = \overbrace{k+(b=2)+(f=2k)}^{\text{equations needed}} \tag{4.81b}$$

Figure 4.42 upper left:
$$\overbrace{2+6}^{\text{unknown variables}} = \overbrace{2+(b=2)+(f=2\cdot2)}^{\text{equations needed}} \tag{4.81c}$$

Port plane power balance approach:

General:
$$\underset{\text{unknown variables}}{\overbrace{2m}} = \underset{\text{equations needed}}{\overbrace{2n+b+f}} \qquad (4.82\text{a})$$

Serial assemblies:
$$\underset{\text{unknown variables}}{\overbrace{2m}} = \overbrace{2n+(b=2)+(f=0)}^{\text{equations needed}};$$
$$as: m = n+1 \qquad (4.82\text{b})$$

Figure 4.42 upper right:
$$\underset{\text{unknown variables}}{\overbrace{2 \cdot 6}} = \underset{\text{equations needed}}{\overbrace{2 \cdot 5 + 2}} \qquad (4.82\text{c})$$

where

n = number of control volumes,
m = number of port planes,
l = number of (independent) flow loops,
k = number of zones,
b = number of boundary condition relations,
f = number of field assumption relations.

The analysis for the port plane approach is particularly straightforward if both mass and mechanical power balances are applied to each of n control volumes. Thus $2n$ conservation relations may be formed that would, in general, be augmented by b boundary conditions and f field assumptions. These equations may then be used to solve for the $2m$ unknown port plane pressure and velocity variables, Equation 4.82a.

For serial assemblies, the number of port planes is equal to the number of control volumes plus one, $m = n+1$, so that only two additional envelope pressure boundary conditions are needed to define a determinate system of equations, Equation 4.82b. That is to say, again for serial assemblies, the port plane power balance approach is an exact approach as zone field assumptions are not needed.

Applying the loop approach directly to either serial or nonserial assemblies, k independent mass conservation relations may be formed for the k modeled (hydrostatic) zones that would be augmented by l (independent) loop equations to form a system of equations to solve for the $(n-k)$ flow-limiting control volume mass flow rates, Equation 4.83a. Recall that the envelope pressure boundary conditions and hydrostatic field assumptions are included directly in forming the loop equations and thus need not be accounted for independently. For serial assemblies of alternating zone and flow-limiting control volumes, the number of unknown (flow limiting) mass airflow rates is simply $k+1$; consequently determinacy is defined by Equation 4.83b. Thus we conclude the rather obvious result that only one loop equation is needed for serial assemblies.

Loop approach (steady airflow):

$$\text{General}: \quad \overbrace{(n-k)}^{\text{unknown variables}} = \overbrace{(k+l)}^{\text{equations used}} \tag{4.83a}$$

$$\text{Serial assemblies}: \quad \overbrace{(k+1)}^{\text{unknown variables}} = \overbrace{(k+l)}^{\text{equations used}} \tag{4.83b}$$

However, if we apply the mass conservation relation a priori to a serial assembly, we conclude that all unknown (flow limiting) mass airflow rates are equal and, thus, may be replaced by a single *system* mass flow rate $\dot{m}$ (i.e. for Figure 4.42 upper right: $\dot{m} \equiv \dot{m}^a = \dot{m}^c = \dot{m}^e$). Thus, for the loop method, system equations for serial assemblies may be defined in terms of a single loop equation defined in terms of the single system mass flow rate $\dot{m}$!

For nonserial assemblies, the number of loops and flow-limiting control volumes that may be identified may become impractically large especially if all infiltration airflow paths are included. Thus, infiltration airflow analysis using the loop method is not likely to be competitive in comparison with the two alternative approaches to airflow analysis. On the other hand, for simpler nonserial assemblies where infiltration airflows may be assumed to be negligible (e.g. for typical natural and hybrid ventilation systems), the loop approach may well lead to small system equations as independent loops may readily be identified and are likely to be few in number. Furthermore, serial subassemblies of unknown flow-limiting mass airflow rates may be replaced by single subassembly mass airflow rate variables.

In summary, while the utility of each of the three approaches must be considered on a case-by-case basis, the loop method is likely to lead to relatively small system equations for simple systems, the nodal method is likely to lead to relatively small system equations for infiltration airflow analysis, and the port plane power balance method will, inevitably, lead to larger system equations than the nodal approach in all cases. However, the port plane power balance approach, by including consideration of the conservation of mechanical energy, provides a physically more complete and consistent approach to airflow analysis that, for serial assemblies, is exact (i.e. in the macroscopic sense).

4.3.4 Simple models and methods suitable for manual analysis

During the twentieth century, ventilation research and practice was largely directed to the problems associated with building infiltration airflows – specifically, the prediction of these airflows and the energy and indoor air quality consequences of them. Because of the general complexity of infiltration airflow paths, the indeterminacy and uncertainty of the characteristics of these airflow paths, and the uncertainty of the buoyancy and wind forces that drive these airflows, building ventilation researchers did not believe they could reasonably turn to fully deterministic ventilation models. Consequently, simplified models of building systems were formulated in terms of lumped parameters that then were calibrated to field measurements in an effort to establish semi-empirical models that could be used in practice.

As deterministic approaches to even infiltration airflow analysis have systematically displaced these semi-empirical models in the past two decades, the details of these semi-empirical models will not be considered here. Instead, the application of the fundamental theory to a selection of simplified building idealizations will be considered to (a) demonstrate that deterministic analysis may, in fact, be completed manually and (b) provide the more detailed understanding required to properly implement the computational tools needed for more complicated building systems.

For specific details of the semi-empirical models developed over the years, the reader may turn to a number of reviews. Awbi's early and useful text, *Ventilation of Buildings*, includes discussions of these models classified as either *empirical* or *simplified theoretical* models (Awbi, 1991); the second edition of this text provides additional detail and a discussion of *network* models (Awbi, 2003). Santamouris and Dascalaki present more specialized reviews of methods for predicting natural ventilation in buildings and sizing components of natural ventilation systems – first in 1996 (Dascalaki and Santamouris, 1996) then in more detail in 1998 (Santamouris and Dascalaki, 1998). The earlier publication provides practically essential details on modeling wind pressures (e.g. with Equation 4.60), with supporting data and a review of simplified and empirical methods to predict ventilation airflow rates. The later publication adds details on semi-empirical models to predict airflow velocities within naturally ventilated buildings. Liddament provides a broader review of empirical and simplified models for ventilation analysis (Liddament, 1996), whereas Irving and Uys present a comprehensive approach to practical natural ventilation analysis directed to the practitioner and offer an approach to the sizing of components of natural ventilation systems that may be considered to be a precursor to the *loop design method* that will be considered below (Irving and Uys, 1997). Finally, it should be noted that a number of these semi-empirical and simplified methods, and a large number of variants associated with them, have been included in building ventilation handbooks, codes and standards (e.g. CEN, 1999; ASHRAE, 2005).

Simple deterministic models

For preliminary analysis, entire smaller building systems or isolated portions of larger building systems, such as a single floor of a multi-storey building, often may be idealized using one-, two- or three-zone models (Figure 4.43),

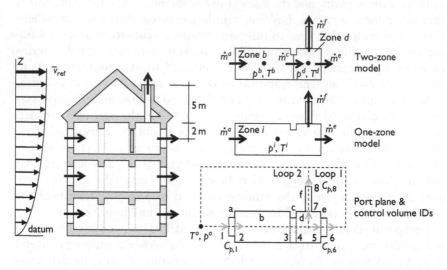

Figure 4.43 Single- and two-zone models of the upper floor of a multi-storey building with detailed control volume, port plane, and loop IDs and variables indicated.

as these simpler deterministic models may be solved using manual algebraic methods – albeit facilitated with commonly used mathematical processing tools (MathWorks, 2000; MathSoft, 2004; Wolfram_Research, 2004). Single-zone models may be acceptable when the interior volume being considered is isolated from other building volumes, is served by a single mechanical system, lacks significant resistance to airflow in relation to limiting resistance envelope airflow paths and/or has interior temperatures that may be expected to be more or less uniform. When these conditions are not met, then one may still be able to capture the essential behaviour of the building ventilation system with a small number of zones.

For our purposes here, we will consider the analysis of the upper floor of a multi-storey building with a single windward opening, a leeward opening, and a ventilation stack – control volumes a, e and f, respectively (Figure 4.43). Given a door that may be opened or closed (e.g. a bathroom door), this single floor could be modeled using either a one- or two-zone model. For the one-zone model, the zone indoor temperature T^i and flow characteristics of the limiting resistances would be specified.

For the two-zone model, both zone temperatures, T^b and T^d, and the flow characteristics of now four limiting resistances (i.e. control volumes a, c, e and f) would be specified. For both modeling options, the outdoor ambient temperature T^o and approach wind speed $\bar{v}_{ref}$ would also be specified – e.g. for critical design conditions. (In general, these model parameters could be specified as functions of time or of the system response, although this additional complexity would normally demand full computational rather than manual analysis.)

To simplify the presentation, we will follow conventional practice and assume all limiting resistances' control volumes have constant cross-sections and uniform air densities within. Thus the airflow through these flow-limiting resistances may be modeled in terms of single mass airflow rates rather than inflow and outflow port plane velocities. With this in mind, the objective of analysis will be to determine the steady mass airflow rates through these limiting resistances – i.e. $\dot{m}^a$, $\dot{m}^e$ and $\dot{m}^f$ for both single- and two-zone models and, in addition, $\dot{m}^c$ for the two-zone model.

Nodal approach – single-zone case: The conventional *nodal approach* appears deceptively simple at first as one simply demands mass airflows are conserved (i.e. by Equation 4.35b). Thus for the one-zone model we have

$$\dot{m}^a - (\dot{m}^e + \dot{m}^f) = 0 \tag{4.84}$$

Element flow relations: These limiting resistance mass airflow rates are, however, defined in terms of the port plane modified pressure differences ΔP^a, ΔP^e and ΔP^f, which, in turn, are defined in terms of spatially averaged port plane pressures $\hat{p}$ and hydrostatic increments $\rho g \Delta z$. If, for example, the inlet and outlets are modeled using the discharge coefficient model and the stack using the duct segment model (Table 4.1), we obtain the following *element* flow relations:

$$\dot{m}^a = C_d^a A^a \sqrt{2 \rho^a \, \Delta P^a} = C_d^a A^a \sqrt{2 \rho^a \left(\hat{p}_1 - \hat{p}_2 + \rho^a g \Delta \hat{z}^a \right)} \tag{4.85a}$$

$$\dot{m}^e = C_d^e A^e \sqrt{2 \rho^e \, \Delta P^e} = C_d^e A^e \sqrt{2 \rho^e \left(\hat{p}_5 - \hat{p}_6 + \rho^e g \Delta \hat{z}^e \right)} \tag{4.85b}$$

$$\dot{m}^f = A^f \sqrt{\frac{\rho^f D_b^f \Delta P f}{f \, l^f}} = A^f \sqrt{\frac{\rho^f D_b^f (\hat{p}_7 - \hat{p}_8 + \rho^f g \Delta \hat{z}^f)}{f \, l^f}} \tag{4.85c}$$

Boundary conditions: To these element flow relations, we must impose wind pressure boundary conditions on port planes 1, 6 and 8. As is commonly done in conventional analysis, the spatial average static pressures at these port planes will be modeled to be equal to that of a sealed building of similar geometry using the corresponding wind pressure coefficients and port plane elevations, Equation 4.58:

$$\hat{p}_1 = C_{p,1} \frac{1}{2} \rho^o \bar{v}_{\text{ref}}^2 - \rho^o g (\hat{z}_1 - z^o) + p^o \tag{4.86a}$$

$$\hat{p}_6 = C_{p,6} \frac{1}{2} \rho^o \bar{v}_{\text{ref}}^2 - \rho^o g (\hat{z}_6 - z^o) + p^o \tag{4.86b}$$

$$\hat{p}_8 = C_{p,8} \frac{1}{2} \rho^o \bar{v}_{\text{ref}}^2 - \rho^o g (\hat{z}_8 - z^o) + p^o \tag{4.86c}$$

Zone field assumption: Finally, the hydrostatic field assumption is applied to link the interior port plane pressures to the zone node pressure using Equation 4.48:

$$\hat{p}_2 = p^i + \rho^i g \left(z^i - \hat{z}_2\right) \tag{4.87a}$$

$$\hat{p}_5 = p^i + \rho^i g \left(z^i - \hat{z}_5\right) \tag{4.87b}$$

$$\hat{p}_7 = p^i + \rho^i g \left(z^i - \hat{z}_7\right) \tag{4.87c}$$

System equations: In general, the system equation may be formed by substituting the element flow relations, the boundary conditions and the zone field assumptions into the governing mass conservation relation. In the single-zone case, this will lead to a single equation defined in terms of the interior zone node pressure p^i. However, if done formally, the resulting equation becomes impractically long and complicated. In practice, whether done computationally or manually, specific values for each of the several known parameters are first substituted into the constituent equations before *assembling* these equations using the mass conservation relation.

For the purposes of this example, the approach wind velocity will be assumed to be $\bar{v}_{ref} = 4\,\text{ms}^{-1}$; the sealed building wind pressure coefficients will be assumed to be $C_{p,1} = 0.8$, $C_{p,6} = -0.5$ and $C_{p,8} = -0.4$; the atmospheric static pressure will be set to $p^0 = 0\,\text{Pa}$; the various elevations will be assumed to be $z^o = \hat{z}_1 = \hat{z}_2 = z^i = \hat{z}_5 = \hat{z}_6 = 0\,\text{m}$, $\hat{z}_7 = 2\,\text{m}$ and $\hat{z}_8 = 7\,\text{m}$ (thus $l^f = 5\,\text{m}$); the element discharge coefficients will be assumed to be $C_d^a = C_d^e = 0.6$ and the duct friction factor will be presumed to be $f = 0.008$ as an initial estimate; the stack diameter will be assumed to be $D^f = 0.40\,\text{m}$ (thus $D_h^f = D^f/2 = 0.20\,\text{m}$) and the element cross-sectional areas $A^a = 0.5\,\text{m}^2$, $A^e = 0.4\,\text{m}^2$ and $A^f = \pi D^{f2}/4 = 0.126\,\text{m}^2$. Finally, the indoor and outdoor air temperatures will be assumed to be $T^i = 25°\text{C}$ and $T^o = 10°\text{C}$; thus, using the ideal gas law relation given in Table 4.1, $\rho^o = \rho^a = 1.245\,\text{kgm}^{-3}$ and $\rho^i = \rho^e = \rho^f = 1.183\,\text{kgm}^{-3}$.

Substituting these parameters into the element, boundary condition and zone field assumption equations and then the results into the mass conservation relation yields a single equation expressed in terms of the zone node pressure p^i:

$$\dot{m}^a - (\dot{m}^e + \dot{m}^f) = 0.30\sqrt{-2.49p^i + 19.84} - 0.24\sqrt{2.37p^i + 11.78}$$

$$- 0.126\sqrt{5.92p^i + 48.75} = 0 \tag{4.88}$$

Even in this simple single-zone example, the resulting nonlinear equation is sufficiently complex that an analytical solution is practically out of reach. Nevertheless, the single-zone mass balance may be plotted to estimate the zone pressure p^i that will lead to mass conservation, as illustrated in Figure 4.44. Alternatively, any one of a number of commercially available

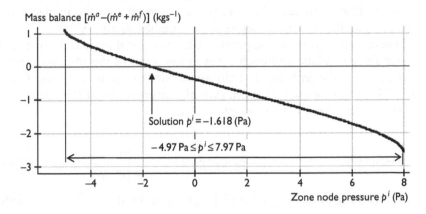

Figure 4.44 Plot of the single-zone mass balance, Equation 4.88, illustrating both the zone pressure for which mass conservation is realized and the real-valued limits of the mass balance.

math processing programs may be used to, more accurately, determine this zone pressure. For Equation 4.88, this was done to determine the value shown in Figure 4.44 – i.e. $p^i = -1.618\,\text{Pa}$.

Two features of Equation 4.88 should be noted as they may be expected to be encountered more generally. First, it is important to note that this mass balance is real-valued only for $-4.97\,\text{Pa} \le p^i \le 7.97\,\text{Pa}$ because of the sum within the first and second radicals of Equation 4.88. Thus for manual analysis, one should examine each of the terms summed in the mass balance to determine the range within which the mass balance evaluates to real values. Second, within this range, the mass balance function is likely to be quasi-linear; consequently the identification of a numerically approximate solution is likely to be revealed by simply plotting the function as in Figure 4.44.

Element mass flow rates: To obtain the individual mass airflow rates using the nodal approach, one must *back substitute* the zone pressure solution into the individual terms of the mass balance to determine the airflow rates in each of the limiting resistances (i.e. for $p^i = -1.914\,\text{Pa}$, $\dot{m}^a = 0.30\sqrt{-2.49p^i + 19.84} = 1.466\,\text{kgs}^{-1}$, $\dot{m}^e = 0.24\sqrt{2.37p^i + 11.78} = 0.677\,\text{kgs}^{-1}$ and $\dot{m}^f = 0.126\sqrt{5.92p^i + 48.75} = 0.789\,\text{kgs}^{-1}$).

Port plane approach – conventional assumptions – single-zone case: The *port plane approach* may be used to implement the conventional macroscopic theory or to implement a power balance analysis of the single-zone problem. For the former, the dissipation relations would include *inverse* forms of the element equations used above (i.e. the discharge coefficient model for the inlet and outlet and the duct model for the stack) augmented with the hydrostatic field assumption that link pairs of zone port plane pressures. That is to say, the port plane approach may be implemented without using the zone node pressures as variables at all.

However, to provide a comparison with the nodal approach, one may develop the port plane approach equations in terms of the zone node pressures by using the hydrostatic field assumptions employed above, Equation 4.71. For the one-zone model, then, the inverse element flow equations and these hydrostatic field assumptions are.

Element a: $\Delta P^a = (\hat{p}_1 - \hat{p}_2 + \rho^a g \Delta \hat{z}^a) = \dfrac{\dot{m}^{a^2}}{2 \rho^a \left(C_d^a A^a\right)^2}$ (4.89a)

Hydrostatic link: $(\hat{p}_2 + \rho^i g \hat{z}_2) - (p^i + \rho^i g z^i) = 0$ (4.89b)

Element e: $\Delta P^e = (\hat{p}_5 - \hat{p}_6 + \rho^e g \Delta \hat{z}^e) = \dfrac{\dot{m}^{e^2}}{2 \rho^e \left(C_d^e A^e\right)^2}$ (4.89c)

Hydrostatic link: $(p^i + \rho^i g z^i) - (\hat{p}_5 + \rho^i g \hat{z}_5) = 0$ (4.89d)

Hydrostatic link: $(p^i + \rho^i g z^i) - (\hat{p}_7 + \rho^i g \hat{z}_7) = 0$ (4.89e)

Element f: $\Delta P^f = (\hat{p}_7 - \hat{p}_8 + \rho^f g \Delta \hat{z}^f) = \dfrac{f \, l^f \dot{m}^{f^2}}{\rho^f D_h^f A^{f^2}}$ (4.89f)

where, in lieu of expressing the inverse element equations in terms of port plane velocities, control volume mass flow rates have been used (i.e. $\dot{m}^a = \rho^a A^a \hat{v}^a$, $\dot{m}^e = \rho^e A^e \hat{v}^e$ and $\dot{m}^f = \rho^f A^f \hat{v}^f$).

An examination of Equation 4.89a–f reveals that a simple sum of each successive pair of equations results in the algebraic cancellation of interior port plane pressures. If this is done and, in addition, the static envelope pressure boundary conditions, Equation 4.86a–c, are imposed, one obtains

$$\underbrace{\frac{\dot{m}^{a^2}}{2 \rho^a \left(C_d^a A^a\right)^2}}_{} = -p^i + \overbrace{C_{p,1} \frac{1}{2} \rho^o \bar{v}_{\text{ref}}^2}^{\text{wind-induced pressure}} + p^o + \overbrace{\left[\rho^a g \Delta \hat{z}^a + \rho^i g(\hat{z}_2 - z^i) - \rho^o g(\hat{z}_1 - z^o)\right]}^{\text{buoyancy-induced pressure difference}}$$
$$= -p^i + 7.97 \text{ Pa} + 0.00 \text{ Pa; for the specified parameters}$$
(4.90a)

$$\frac{\dot{m}^{e^2}}{2 \rho^e \left(C_d^e A^e\right)^2} = p^i - \overbrace{C_{p,6} \frac{1}{2} \rho^o \bar{v}_{\text{ref}}^2}^{\text{wind-induced pressure}} - p^o + \overbrace{\left[\rho^e g \Delta \hat{z}^e + \rho^i g(z^i - \hat{z}_5) + \rho^o g(\hat{z}_6 - z^o)\right]}^{\text{buoyancy-induced pressure difference}}$$
$$= p^i + 4.98 \text{ Pa} + 0.00 \text{ Pa; for the specified parameters}$$
(4.90b)

$$\overbrace{}^{\text{wind-induced pressure}} \qquad \overbrace{}^{\text{buoyancy-induced pressure difference}}$$

$$\frac{f\, l^f\, \dot{m}^{f2}}{\rho^f D_h^f A^{f2}} = p^i - C_{p,8}\frac{1}{2}\rho^o \bar{v}_{\text{ref}}^2 - p^o + \left[\rho^f g \Delta \hat{z}^f + \rho^i g(z^i - \hat{z}_7) + \rho^o g(\hat{z}_8 - z^o)\right]$$

$$= p^i + 3.98 \text{ Pa} + 4.26 \text{ Pa}; \text{ for the specified parameters}$$

$$(4.90c)$$

If one then solves these equations for the respective mass airflow rates, $\dot{m}^a$, $\dot{m}^e$ and $\dot{m}^f$, imposes the mass conservation relation for the zone, Equation 4.84, and substitutes the specified parameters enumerated above, then one obtains the same equation developed using the nodal approach.

Although this port plane implementation of the conventional nodal approach yields the same result, the form of Equations 4.90a–c are more intuitively direct than the nodal approach equations as they reveal the element losses (i.e. on the left hand side of these equations) are simply equal to the differences between adjacent node pressures offset by wind- and buoyancy-induced increments. Indeed, with practice, one may form the specific element equations by considering these physical contributions directly. Thus, for the assumed one-zone model, the buoyancy contribution to flow elements a and e are zero, given $z^o = \hat{z}_1 = \hat{z}_2 = \hat{z}_5 = \hat{z}_6$, and the buoyancy contribution to the stack element f is simply equal to the familiar *stack pressure* $(\rho^o - \rho^i)g(\hat{z}_8 - z^o)$, which, for the specified parameters, is 4.26 Pa.

Furthermore, and practically important, the right-hand sides of these inverse element equations are independent of element size parameters as they depend only on assumed wind conditions and zone temperatures – the so-called *design conditions*. If these equations are thus used for design, an analyst may systematically search for appropriate element sizes to achieve a specific ventilation objective without needing to alter the right-hand sides of the equations.

Loop approach – single-zone case: The astute reader will note that the loop approach may be implemented by simply summing Equations 4.90a and 4.90b to form the equation for loop 1 and Equations 4.90a and 4.90c to form the equation of loop 2 of the single-zone model shown in Figure 4.43. In summing these two sets of equations, the interior and exterior node pressure variables, p^i and p^o, cancel out leaving two equations defined in terms of the three individual element mass flow rates. These two loop equations would then be combined with the conservation of mass equation for the zone, Equation 4.84, to define the loop system equations:

Loop 1:

$$\frac{\dot{m}^{a2}}{2\,\rho^a\,(C_d^a A^a)^2} + \frac{\dot{m}^{e2}}{2\,\rho^e\,(C_d^e A^e)^2} = \overbrace{(C_{p,1} - C_{p,6})\frac{1}{2}\rho^o \bar{v}_{\text{ref}}^2}^{\Delta p_{w,1}} + \overbrace{g\left[\rho^i(\hat{z}_2 - \hat{z}_6) + \rho^o(\hat{z}_6 - \hat{z}_2)\right]}^{\Delta p_{b,1}}$$

$$\underbrace{\frac{\dot{m}^{a2}}{0.2241}}_{} + \underbrace{\frac{\dot{m}^{e2}}{0.1363}}_{} = \overbrace{12.95 \text{ Pa}}^{\Delta p_{w,1}} + \overbrace{0.00 \text{ Pa}}^{\Delta p_{b,1}}; \text{ for the specified parameters}$$

$$(4.91a)$$

Loop 2:

$$\frac{\dot{m}^{a^2}}{2\,\rho^a\,(C_d^a A^a)^2} + \frac{2f\,l^f\,\dot{m}^{f^2}}{\rho^f D_h^f A^{f^2}} = \overbrace{(C_{p,1}-C_{p,8})\frac{1}{2}\rho^o\bar{v}_{\rm ref}^2}^{\Delta p_{w,2}} + \overbrace{g\left[\rho^o(\hat{z}_8-\hat{z}_2)+\rho^i(\hat{z}_2-\hat{z}_8)\right]}^{\Delta p_{b,2}}$$

$$\frac{\dot{m}^{a^2}}{0.2241} + \frac{\dot{m}^{f^2}}{0.0939} = \overbrace{11.95\ {\rm Pa}}^{\Delta p_{w,2}} + \overbrace{4.26\ {\rm Pa}}^{\Delta p_{b,2}};\ \text{for the specified parameters}$$

$$(4.91b)$$

Mass conservation:

$$\dot{m}^a - (\dot{m}^e + \dot{m}^f) = 0 \tag{4.91c}$$

As in the port plane approach, the right-hand side of each pressure loop equation includes two terms – a wind-induced and a buoyancy-induced pressure difference for each loop – i.e. $\Delta p_{w,1}$ and $\Delta p_{b,1}$ for loop 1 and $\Delta p_{w,2}$ and $\Delta p_{b,2}$ for loop 2. Again, it should be noted that these driving pressure differences are independent of element size parameters as they depend only on the given *design conditions* of the posed problem.

With three equations defined in terms of the three unknown mass flow rates, this system of nonlinear equations is apparently determinate, although clearly more challenging to solve than the single nonlinear equation resulting from either the nodal or port plane approaches presented above. Inasmuch as these loop approach equations are simply linear combinations of the equations solved earlier, their solution is again $\dot{m}^a = 1.466\,{\rm kgs}^{-1}$, $\dot{m}^e = 0.677\,{\rm kgs}^{-1}$, and $\dot{m}^f = 0.789\,{\rm kgs}^{-1}$ (i.e. for the specified parameters).

Port plane approach – conventional assumptions – two-zone case: Following the same procedures used for each of the analytic approaches, one may directly form system equations for the two-zone model (i.e. without resorting to computational tools). Here this will be done using only the port plane approach as there is little to be gained by considering all approaches in detail again.

A single door (control volume c), that is now assumed to provide a limiting resistance, subdivides the original single zone into two adjacent zones b and d with their respective zone node pressures p^b and p^d. Consequently, the governing mass conservation condition yields two equations:

$$\dot{m}^a - \dot{m}^c = 0 \tag{4.92a}$$

$$\dot{m}^c - (\dot{m}^e + \dot{m}^f) = 0 \tag{4.92b}$$

For the two-zone case, the set of inverse flow element and hydrostatic field assumption relations used in the one-zone case (i.e. Equations 4.73a–f) must be augmented by an additional element relation for control volume c, two associated hydrostatic links and the node pressure variable p^i must be replaced by p^b or p^d as appropriate. For our purposes here, control volume

c will be modeled with the discharge coefficient model (Table 4.1) with $A^c = 0.25\,\text{m}^2$, a value characteristic of a slightly opened door, the discharge coefficients will be assumed to be $C_d^c = 0.6$, and air temperatures in both zone b and zone d will be assumed to be $25°\text{C}$ (i.e. $\rho^b = \rho^d = 1.183\,\text{kgm}^{-3}$).

If these element equations and hydrostatic links are combined and the static envelope pressure boundary conditions, Equations 4.70a–c, are imposed, one obtains

$$\frac{\dot{m}^{a^2}}{2\rho^a \left(C_d^a A^a\right)^2} = -p^b \overbrace{+C_{p,1}\frac{1}{2}\rho^o \bar{v}_{\text{ref}}^2 + p^o}^{\text{wind-induced pressure}} + \overbrace{\left[\rho^b g(\hat{z}_2 - z^i) - \rho^o g(\hat{z}_1 - z^o + \Delta\hat{z}^a)\right]}^{\text{buoyancy-induced pressure difference}}$$

$$\frac{\dot{m}^{a^2}}{0.2241} = -p^b + 7.97\,\text{Pa} + 0.00\,\text{Pa}; \text{ for the specified parameters}$$

(4.93a)

$$\frac{\dot{m}^{c^2}}{2\rho^c \left(C_d^c A_L^c\right)^2} = (p^b - p^d) + \overbrace{\left[\rho^b g(\hat{z}_3 - z^b + \Delta\hat{z}^c) - \rho^d g(\hat{z}_4 - z^d)\right]}^{\text{buoyancy-induced pressure difference}}$$

$$\frac{\dot{m}^{c^2}}{0.05323} = (p^b - p^d) + 0.00\,\text{Pa}; \text{ for the specified parameters}$$

(4.93b)

$$\frac{\dot{m}^{e^2}}{2\rho^e \left(C_d^e A^e\right)^2} = p^d \overbrace{-C_{p,6}\frac{1}{2}\rho^o \bar{v}_{\text{ref}}^2 - p^o}^{\text{wind-induced pressure}} + \overbrace{\left[\rho^d g(z^d - \hat{z}_5 + \Delta\hat{z}^e) + \rho^o g(\hat{z}_6 - z^o)\right]}^{\text{buoyancy-induced pressure difference}}$$

$$\frac{\dot{m}^{e^2}}{0.1363} = p^d + 4.98\,\text{Pa} + 0.00\,\text{Pa}; \text{ for the specified parameters}$$

(4.93c)

$$\frac{f\,l^f \dot{m}^{f^2}}{\rho^f D_h^f A^{f^2}} = p^d \overbrace{-C_{p,8}\frac{1}{2}\rho^o \bar{v}_{\text{ref}}^2 - p^o}^{\text{wind-induced pressure}} + \overbrace{\left[\rho^d g(z^d - \hat{z}_8) + \rho^o g(\hat{z}_8 - z^o)\right]}^{\text{buoyancy-induced pressure difference}}$$

$$\frac{\dot{m}^{f^2}}{0.09391} = p^d + 3.98\,\text{Pa} + 4.26\,\text{Pa}; \text{ for the specified parameters}$$

(4.93d)

Finally, by substituting Equations 4.93a–d into the mass conservation relations, Equations 4.92a–b, one obtains two equations expressed in terms of the two unknown zone pressures, p^b and p^d, as

$$\dot{m}^a - \dot{m}^c = 0.4734\sqrt{7.97 - p^b} - 0.2307\sqrt{p^b - p^d} = 0$$

(4.94a)

$$\dot{m}^c - (\dot{m}^e + \dot{m}^f) = 0.2307\sqrt{p^b - p^d}$$

$$- \left(0.3692\sqrt{p^d + 4.98} + 0.3064\sqrt{p^d + 8.24}\right) = 0$$

(4.94b)

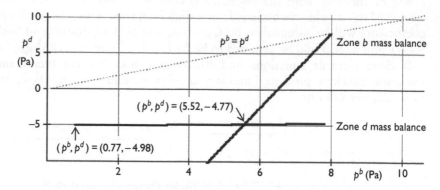

Figure 4.45 Plot of the two-zone mass balances, Equation 4.94a–b, illustrating both the zone pressures for which mass conservation is realized and the real-valued limits of the mass balance.

The nonlinear equations defined by Equations 4.94a–b may be plotted in $\langle p^b, p^d \rangle$ 'space' to estimate the solution (i.e. found at the intersection of the two curves), as shown in Figure 4.45. Again, an inspection of the terms within the radicals establishes the range over which these equations give real values – i.e. $p^b \leq 7.97$, $p^b \geq p^d$ and $p^d \geq -4.98$.

Finally, using the zone pressure solution, $(p^b, p^d) = (5.52, -4.77)\,\text{Pa}$, the individual element mass airflow rates may be determined by *back-substitution* to obtain $\dot{m}^a = 0.740\,\text{kgs}^{-1}$, $\dot{m}^c = 0.740\,\text{kgs}^{-1}$, $\dot{m}^e = 0.169\,\text{kgs}^{-1}$ and $\dot{m}^f = 0.571\,\text{kgs}^{-1}$. Comparing these results with the one-zone model results obtained above (i.e. $\dot{m}^a = 1.466\,\text{kgs}^{-1}$, $\dot{m}^e = 0.677\,\text{kgs}^{-1}$ and $\dot{m}^f = 0.789\,\text{kgs}^{-1}$), it is seen that the partially opened door reduces flow in the windward opening by 50 per cent, in the leeward opening by 75 per cent, and in the stack by 28 per cent.

These disproportionate changes result from the nonlinearity of the airflow system equations reflecting the rather subtle interaction of wind, buoyancy effects and limiting resistances. Indeed, if the door opening (control volume c) is constricted more, the airflow in the leeward opening will even reverse directions.

In general, the nonlinearity of the airflow system equations combined with the confounding uncertainty of element airflow direction limit the practical utility of manual solutions – solutions that inevitably depend on a priori knowledge of flow direction. Consequently, few analysts attempt manual solutions preferring instead to turn to one of the several computational tools now available [e.g. CONTAM, COMIS or BREEZE, (Solomons, 1990; Feustel and Smith, 1997; Dols and Walton, 2000)].

Simplified analysis based on circuit analogies

It has long been recognized that building ventilation systems share some similarities with electrical resistance networks. Most obviously, pressure differences that induce airflow are analogous to voltage differences that induce current. In addition, the principle that current flow at an electrical junction must be conserved is analogous the principle that mass airflow rates be conserved in building zones (i.e. for steady conditions). However, two key differences limit the utility of this analogy: (a) while current is linearly related to voltage drop for typical electrical resistors, the analogous relation between airflow and pressure differences of building ventilation components is typically nonlinear and (b) pressures within building zones, unlike voltages in electrical network junctions, must be expected to vary spatially (e.g. hydrostatically).

Nevertheless, as early as 1949, Dick verbally described strategies that could be used to simplify serial and parallel assemblies of flow 'resistances' for isothermal, wind-driven ventilation analysis (Dick, 1949). By the mid-1970s, equations governing airflow through series assemblies of discharge coefficient elements were well established (Aynsley *et al.*, 1977), and soon thereafter, an approach to simplify parallel assemblies was published (Gerhart and Gross, 1985). Most recently, Marques da Silva and Saraiva have combined and generalized these methods to provide an approximate procedure for the manual analysis of natural ventilation in simple dwelling layouts (Marques da Silva and Saraiva, 2005). Here, the approach employed by Marques da Silva will be used, although it will be generalized to include hydrostatic contributions more completely, parallel flows through wall, floor and branch paths will be distinguished, and their approximate approach to parallel circuits will not be used.

If building ventilation flow directions are evident a priori, one may isolate individual *serial* or *parallel* subassemblies of flow elements, (Figure 4.46) and replace them with single composite resistances to systematically transform complex ventilation *networks* into equivalent simple lumped parameter networks that may then be solved using relatively simple algebraic procedures.

Limiting consideration to flow elements of constant cross-section, we can express the forward and inverse element flow relations for all commonly used quadratic form elements (Table 4.1) in a general form as

$$\text{Inverse:}\quad \Delta P^l = K^l \dot{m}^{l^2} \qquad \text{Forward:}\quad \dot{m}^l = \sqrt{\frac{\Delta P^l}{K^l}} \tag{4.95}$$

where K^l is a composite model parameter – the element flow *resistance* – that is dependent on the specific flow element model used – e.g. for the discharge coefficient model $K^l = 1/[2\rho^l (C_d^l A^l)^2]$; for the simplified friction loss factor model $K^l = \zeta^l / (2\rho^l A^{l2})$; and for the duct segment model $K^l = (f\, l^e) / (\rho^l D_h^l A^{l2})$.

Serial assemblies: For serial subassembly (e.g. Figure 4.46 bottom right), one may apply the inverse form to each of the n limiting resistances and the (conventional) hydrostatic field assumption to the zones in between to obtain the following set of equations:

Element a: $\Delta P^a = (\hat{p}_1 - \hat{p}_2 + \rho^a g \Delta \hat{z}^a) = K^a \dot{m}^{a2}$ (4.96a)

Zone b: $-(\hat{p}_2 + \rho^b g \hat{z}_2) + (\hat{p}_3 + \rho^b g \hat{z}_3) = 0$ (4.96b)

Element c: $\Delta P^c = (\hat{p}_2 - \hat{p}_3 + \rho^c g \Delta \hat{z}^c) = K^c \dot{m}^{c2}$ (4.96c)

Zone d: $-(\hat{p}_3 + \rho^d g \hat{z}_3) + (\hat{p}_4 + \rho^d g \hat{z}_4) = 0$ (4.96d)

. . .

Element n: $\Delta P^n = (\hat{p}_m - \hat{p}_{m-1} + \rho^m g \Delta \hat{z}^m) = K^m \dot{m}^{m2}$ (4.96e)

Recognizing the element mass flow rates must be equal, $\dot{m}^S \equiv \dot{m}^a = \dot{m}^c = \ldots = \dot{m}^n$, we can sum these dissipation relations to obtain a pressure flow relation for the serial assembly 'S' as a whole:

$$\Delta P^S \equiv \left[\hat{p}_1 - \hat{p}_m + \sum_{i=a,b,c\ldots n} (\rho^i g \Delta z^i] \right] = \sum_{j=a,c,e\ldots n} K^j \dot{m}^{S2} \qquad (4.97a)$$

or $\Delta P^S = K^S \dot{m}^{S2}$ where $K^S = \displaystyle\sum_{j=a,c,e\ldots n} K^j$ (4.97b)

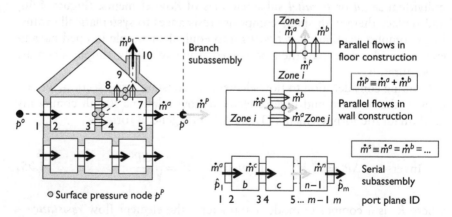

Figure 4.46 Parallel and serial subassemblies within a larger multi-zone model. To simplify the analysis of parallel subassemblies, we introduce surface pressure nodes.

Alternatively, one may apply the two-port Bernoulli relation, Equation 4.45b, to each of the zones instead – e.g. for zone b:

$$-\left[\hat{p}_2+\rho^b g\hat{z}_2+\tfrac{1}{2}\left(\frac{\dot{m}^{a^2}}{\rho^b A^{a^2}}\right)\right]$$

$$+\left[\hat{p}_3+\rho^b g\hat{z}_3+\frac{1}{2}\left(\dot{m}^{c^2}\rho^b A^{c^2}\right)\right]+\frac{1}{2}\left(\frac{\dot{m}^{a^2}}{\rho^b A^{a^2}}\right)\zeta^b=0 \qquad (4.98)$$

where the port plane velocities used in Equation 4.45b have been replaced with the identity $\hat{v}_2=\dot{m}^a/(\rho^a A^a)$ and ζ^b is a friction loss coefficient associated with the zone (e.g. typically assumed to be zero). Again if mass conservation is imposed, $\dot{m}^S\equiv\dot{m}^a=\dot{m}^c=\ldots=\dot{m}^n$, and the element and zone dissipation relations are summed, one now obtains

$$\Delta P^S\equiv\left[p_1-p_m+\sum_{i=a,b,c\ldots}\left(\rho^i g\Delta z^i\right)\right]$$

$$=\left[\sum_{k=a,c,e\ldots}K^k+\sum_{\substack{j=b,d,f\ldots \\ k=a,c,e,\ldots \\ l=c,e,g\ldots}}\frac{1}{2\rho^j}\left(\frac{1-\zeta^j}{A^{k^2}}-\frac{1}{A^{l^2}}\right)\right]\dot{m}^S \qquad (4.99a)$$

or $\quad\Delta P^S=K^S\dot{m}^S$

where $\quad K^S=\sum_{k=a,c,e\ldots}K^k+\sum_{\substack{j=b,d,f\ldots \\ k=a,c,e,\ldots \\ l=c,e,g\ldots}}\frac{1}{2\rho^j}\left(\frac{1-\zeta^j}{A^{k^2}}-\frac{1}{A^{l^2}}\right) \qquad (4.99b)$

It is important to note that Equation 4.99 was derived without imposing a zone field assumption. Consequently, this relation is exact in the macroscopic sense whereas Equation 4.97 is an approximate expression due to the assumption made that hydrostatic conditions prevail within each zone. For zones with negligible dissipation (i.e. $\zeta^j\approx 0$) and equal inflow and outflow port areas, the second sum of Equation 4.99b becomes zero and thus the resulting serial resistance becomes equal to that of Equation 4.97. The contribution of this second sum, which accounts for kinetic energy differences from zone inflow to outflow, has been ignored in the literature, yet it is likely to be significant in practical cases.

Parallel assemblies: Parallel flow paths may occur locally within wall or floor construction or more globally when a ventilation system offers parallel branches in a flow network (Figure 4.46). All three cases may be treated in the same manner if one introduces node pressures p^{Pi} and p^{Pj} at arbitrary positions (nodes), z^{Pi} and z^{Pj}, on the opposing surfaces of walls or floor construction, or, for the parallel flow branches at the commonly shared nodes that delimit the branches (e.g. the surface node at port plane 4 and the ambient outdoor node to the right in Figure 4.46).

For airflow in a series of parallel paths a, b, c, ... that share common nodes i and j, modified pressure changes following any one of the branch paths from i to j must equal the losses in the associated flow resistance:

$$\Delta P^{Pa} \equiv \left(p^{Pi} - p^{Pj} + \sum_{\text{path } a} \rho g \Delta z \right) = K^a \dot{m}^{a2} \qquad (4.100\text{a})$$

$$\Delta P^{Pb} \equiv \left(p^{Pi} - p^{Pj} + \sum_{\text{path } b} \rho g \Delta z \right) = K^b \dot{m}^{b2} \qquad (4.100\text{b})$$

$$\Delta P^{Pc} \equiv \left(p^{Pi} - p^{Pj} + \sum_{\text{path } c} \rho g \Delta z \right) = K^c \dot{m}^{c2} \qquad (4.100\text{c})$$

etc. $\qquad (4.100\text{d})$

where the $\Sigma \rho g \Delta z$ terms are the sum of hydrostatic pressure changes encountered as one follows each of the parallel paths.

In general, the modified pressures on the left-hand side of Equation 4.100a–d will not be equal. Nevertheless, four special cases may be identified when these pressures will be equal (i.e. when $\Delta P^P \equiv \Delta P^{Pa} = \Delta P^{Pb} = \Delta P^{Pc} = \ldots$):

• for parallel flows through a common floor construction, all $\Sigma \rho g \Delta z$ terms will be identical;
• for branch flows on a horizontal plane, all elevation changes Δz will be zero, thus the $\Sigma \rho g \Delta z$ terms will vanish;
• for isothermal, wind-driven flow conditions, all $\Sigma \rho g \Delta z$ terms will be identical;
• for parallel flows through interior walls, the $\Sigma \rho g \Delta z$ terms for each path will vanish if temperatures in the linked rooms are identical.

In fact, only nonisothermal branch flows with vertically offset branches like that illustrated in Figure 4.46 and nonisothermal parallel airflow through walls (e.g. envelope walls) constitute the primary exceptions to the rule.

As the airflow passing through all paths combined, $\dot{m}^P$, is simply the sum of the individual branch flows, then for these special, but often typical, cases, we may derive a simplified flow relation:

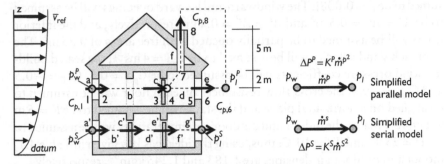

Figure 4.47 Examples of a serial flow (lower floor) and a mixed serial plus parallel flow (upper floor) 'circuits'. Control volumes are identified by lower case letter and port planes, for the upper floor only, by number IDs.

$$\dot{m}^P = \dot{m}^a + \dot{m}^b + \dot{m}^c + \cdots = \left(\sqrt{\frac{1}{K^a}} + \sqrt{\frac{1}{K^b}} + \sqrt{\frac{1}{K^c}} + \cdots\right)\sqrt{\Delta P^P}$$

(4.101a)

$$\text{or}\quad \dot{m}^P \approx \sqrt{\frac{\Delta P^P}{K^P}}\quad \text{where}\quad K^P = \left(1/\sum_{e=a,b,\dots}\frac{1}{\sqrt{K^e}}\right)^2\quad (4.101b)$$

To provide a worked example of the application of these circuit analogy simplifications, consider each floor of the two-storey building illustrated in Figure 4.47. The lower floor provides an example of a simple serial assembly ('circuit'), where we seek to develop a system equation expressed in terms of the pressure difference from a windward surface node pressure p_w^S and a leeward surface node pressure p_l^S. The upper floor provides an example of a mixed circuit – i.e. a serial assembly of resistances a and c that are connected to parallel branch resistances e and f. For the upper floor, mechanical energy conservation of the zone control volumes b and d will be ignored whereas it will be considered in the zones of the lower floor (b', d' and f'), thereby providing one illustration of the error introduced in conventional analysis due to violations of mechanical energy conservation. Finally, for the upper floor wind pressure boundary conditions on the leeward side, add a new complication that will be considered.

System model parameters will be identical to those used in the analyses presented above so that the results of the circuit simplifications may be compared with the results obtained earlier. Specifically, all resistances will be modeled using the simple discharge coefficient model [with all discharge coefficients assumed to be $C_d = 0.6$, except the stack that will be assumed to be a circular duct with a diameter of $D^f = 0.40\,\text{m}$ (i.e. $D_h^f = 0.20\,\text{m}$) and a cross-sectional area of $A^f = \pi D^{f^2}/4 = 0.126\,\text{m}^2$, and the duct friction factor will again be presumed to be $f = 0.008$]. The windward and leeward openings will be assumed to be $A^a = A^{a'} = 0.5\,\text{m}^2$ and $A^e = A^{e'} = 0.4\,\text{m}^2$, respectively, and the interior doors will be assumed to be partially opened with free areas of $0.25\,\text{m}^2$. The approach wind velocity will be assumed to be $\bar{v}_{\text{ref}} = 4\,\text{ms}^{-1}$; the sealed building wind pressure coefficients will be assumed to be $C_{p,1} = 0.8$, $C_{p,6} = -0.5$ and $C_{p,8} = -0.4$; the cross-flow resistances on each floor will be assumed to be aligned on a horizontal plane and the stack will be assumed to be $l^f = 5\,\text{m}$ as before. Finally, the indoor and outdoor air temperatures will be assumed to be $T^i = 25°\text{C}$ and $T^o = 10°\text{C}$; thus, using the ideal gas law given in Table 4.1, indoor and outdoor air densities are 1.183 and 1.245 kgm^{-3}, respectively.

Given these model parameters, the element resistances may be directly calculated:

$$K^a = K^{a'} = \frac{1}{2\rho^a(C_d^a A^a)^2} = \frac{1}{2 \cdot 1.245\,\text{kgm}^{-3}(0.6 \cdot 0.5\,\text{m}^2)^2} = 4.46\,\text{kg}^{-1}\text{m}^{-1}$$
(4.102a)

$$K^e = K^{g'} = \frac{1}{2\rho^e(C_d^e A^e)^2} = \frac{1}{2 \cdot 1.183\,\text{kgm}^{-3}(0.6 \cdot 0.4\,\text{m}^2)^2} = 7.34\,\text{kg}^{-1}\text{m}^{-1}$$
(4.102b)

$$K^c = K^{c'} = K^{e'} = \frac{1}{2\rho^c(C_d^c A^c)^2} = \frac{1}{2 \cdot 1.183\,\text{kgm}^{-3}(0.6 \cdot 0.25\,\text{m}^2)^2} = 18.78\,\text{kg}^{-1}\text{m}^{-1}$$
(4.102c)

$$K^f = \frac{f\, l^f}{\rho^f D_h^f A^{f^2}} = \frac{(0.008 \cdot 5\,\text{m})}{1.183\,\text{kgm}^{-3} \cdot 0.20\,\text{m} \cdot (0.126\,\text{m}^2)^2} = 10.65\,\text{kg}^{-1}\text{m}^{-1}$$
(4.102d)

Serial assembly: The lumped resistance of the lower floor serial assembly is simply the sum of the element resistances – i.e. if the contribution of kinetic energy differences is ignored:

$$K^S = K^{a'} + K^{c'} + K^{e'} + K^{g'} = 49.37\,\text{kg}^{-1}\text{m}^{-1}$$
(4.103)

If on the other hand, one includes the contribution due to kinetic energy differences yet, here, ignores the dissipation within the three consecutive zones (i.e. $\zeta^{b'} = \zeta^{d'} = \zeta^{f'} = 0$), one obtains

$$K^S = \sum_{k=a',c',e',g'} K^k + \sum_{\substack{j=b',d',f' \\ k=a',c',e' \\ l=c',e',g'}} \frac{1}{2\rho^j}\left(\frac{1-\zeta^j}{A^{k2}} - \frac{1}{A^{l2}}\right)$$

$$= 49.37 + \overbrace{\left(\frac{1}{0.5^2} - \frac{1}{0.25^2}\right)}^{-12.0} + \overbrace{\left(\frac{1}{0.25^2} - \frac{1}{0.25^2}\right)}^{0.0} \tag{4.104}$$

$$\underbrace{}_{9.75}$$

$$+ \overbrace{\left(\frac{1}{0.25^2} - \frac{1}{0.4^2}\right)} = 47.12\,\mathrm{kg^{-1}m^{-1}}$$

Thus in this case, the kinetic energy differences have the effect of reducing the overall resistance of the serial assembly by only 5 per cent due to the compensating differences of zone $b'(-12.0\,\mathrm{kg^{-1}m^{-1}})$ and zone $f'(+9.75\,\mathrm{kg^{-1}m^{-1}})$.

To complete the analysis, one may then compute the driving pressure difference ΔP^S, which, in this case, is due solely to the envelope wind pressure difference $\Delta C_P \frac{1}{2}\rho^\circ v_{\mathrm{ref}}^2 = 12.95\,\mathrm{Pa}$ as

$$\dot{m}^S = \sqrt{\frac{\Delta P^S}{K^S}}$$

$$= \begin{cases} \sqrt{12.95/49.37} = 0.512\,\mathrm{kgs^{-1}}\,(\text{ignoring kinetic energy effects}) \\ \sqrt{12.95/47.12} = 0.524\,\mathrm{kgs^{-1}}\,(\text{including kinetic energy effects}) \end{cases}$$
$$\tag{4.105}$$

Although, in this case, the violation of mechanical energy proved to have a marginal impact, from this example it should be clear that when serial assemblies have serially increasing or serially decreasing port plane areas, ignoring mechanical energy conservation may be expected to be significant.

Mixed serial and parallel assembly: The simplified circuit analysis of the upper floor involves both serial and parallel circuits. The latter is complicated by the fact that the pair of parallel branches terminate in the outdoor zone thus involving both hydrostatic and envelope wind pressure differences. Consequently, the total modified pressure drop (including pressure differences due to envelope wind pressures) along each branch will not be equal. This particular example, therefore, presents an unusual challenge for simplified circuit analysis.

Nevertheless, if the branch pressure differences are nearly equal, it may be reasonable to use a mean value of the driving pressures along each path to effect an approximate solution. For the problem at hand, one begins by writing the equations for each of the branches being careful to include the wind pressure changes as one follows each branch from the shared internal node to the shared outdoor node p_l^P as

$$\Delta P^{Pe} \equiv \left[p_4^P - p_l^P + \sum_{pathe} \rho g \Delta z - C_{P,6} \tfrac{1}{2} \rho^\circ \bar{v}_{ref}^2 \right] = K^e \dot{m}^{e2} \qquad (4.106a)$$

$$\Delta P^{Pf} \equiv \left[p_4^P - p_l^P + \sum_{pathf} \rho g \Delta z - C_{P,8} \tfrac{1}{2} \rho^\circ \bar{v}_{ref}^2 \right] = K^f \dot{m}^{f2} \qquad (4.106b)$$

For the given indoor and outdoor temperatures, node elevations and envelope wind pressure coefficients, the modified pressure differences for the path e is $\Delta P^{Pe} = (p_4^P - p_l^P + 4.98 \, \text{Pa})$ whereas that of path f is $\Delta P^{Pf} = (p_4^P - p_l^P + 8.24 \, \text{Pa})$. With little justification, then, both pressure differences will be approximated using a mean value of the driving pressures $\Delta P^{Pe} \approx \Delta P^{Pf} \approx (p_4^P - p_l^P + 6.61 \, \text{Pa})$ so that the lumped resistance of the parallel branches may be combined by Equation 4.101b as

$$K^{Pef} = \left(1 \bigg/ \sum_{i=e,f} \frac{1}{\sqrt{K^i}} \right)^2$$

$$= \left(\frac{1}{\sqrt{7.34 \, \text{kg}^{-1} \text{m}^{-1}} + \dfrac{1}{\sqrt{10.65 \, \text{kg}^{-1} \text{m}^{-1}}}} \right)^2 = 0.46 \, \text{kg}^{-1} \text{m}^{-1} \qquad (4.107)$$

As this lumped resistance is in series with airflow elements a and c (i.e. of the upper floor airflow system), it may then be combined with the element resistances of a and c using the serial circuit relation, Equation 4.97b, as

$$K^S = \sum_{j=a,c,Pef} K^j$$

$$= 4.46 \, \text{kg}^{-1} \text{m}^{-1} + 18.78 \, \text{kg}^{-1} \text{m}^{-1} + 0.46 \, \text{kg}^{-1} \text{m}^{-1} = 23.70 \, \text{kg}^{-1} \text{m}^{-1} \qquad (4.108)$$

To complete the analysis, one may then compute the driving pressure difference, which, in this case, is the sum of the windward envelope wind pressure $C_P \tfrac{1}{2} \rho^\circ v_{ref}^2 = 7.97 \, \text{Pa}$ (i.e. for $C_{P,1} = 0.8$) and the assumed effective driving pressure difference for the parallel branches of 6.61 Pa, and apply the governing simplified equation:

$$\dot{m}^S = \sqrt{\frac{\Delta P^S}{K^S}} = \sqrt{\frac{(7.97 + 6.61) \, \text{Pa}}{23.70 \, \text{kg}^{-1} \text{m}^{-1}}} = 0.78 \, \text{kgs}^{-1} \qquad (4.109)$$

This result compares well with the airflow rates computed for elements a and c by the nodal method presented above – i.e. $\dot{m}^a = \dot{m}^c = 0.745\,\mathrm{kgs}^{-1}$ from the exact nodal approach solution.

Given the topology of the flow 'circuit' this airflow rate would be equal to the sum of the airflow rate through the lumped branch as well – i.e. $\dot{m}^{Pef} = (\dot{m}^e + \dot{m}^f) = 0.80\,\mathrm{kgs}^{-1}$. Thus, using this value, we can compute the branch pressure difference:

$$\Delta P^{Pef} = K^{Pef}\dot{m}^{Pef^2} = 0.46\,\mathrm{kg}^{-1}m^{-1}(0.78\,\mathrm{kgs}^{-1})^2 = 0.28\,\mathrm{Pa} \qquad (4.110)$$

Finally, with this estimate of the branch pressure difference in hand, the flow through each branch may, in principle, be calculated as

$$\dot{m}^e = \sqrt{\frac{\Delta P^e}{K^e}} = \sqrt{\frac{0.29\,\mathrm{Pa}}{7.34\,\mathrm{kg}^{-1}\mathrm{m}^{-1}}} = 0.20\,\mathrm{kgs}^{-1}$$

and

$$\dot{m}^f = \sqrt{\frac{\Delta P^f}{K^f}} = \sqrt{\frac{0.29\,\mathrm{Pa}}{10.65\,\mathrm{kg}^{-1}\mathrm{m}^{-1}}} = 0.16\,\mathrm{kgs.}^{-1} \qquad (4.111a,b)$$

These values, however, not only vary considerably from the exact values computed using the nodal method (i.e. $\dot{m}^e = 0.092\,\mathrm{kgs}^{-1}$ and $\dot{m}^f = 0.653\,\mathrm{kgs}^{-1}$), they violate the continuity condition that $\dot{m}^{Pef} = (\dot{m}^e + \dot{m}^f) = 0.80\,\mathrm{kgs}^{-1}$! Thus, as asserted theoretically above, parallel branches may be combined using the branch circuit relation, Equation 4.101b, only if the parallel-branch-modified pressure differences, including wind pressure differences, are equal.

Although this case illustrates one limitation of simplified circuit analogies, not all is lost. Instead of attempting to simplify the ventilation network to a single quadratic form equation using the serial and parallel assembly relations, one may instead simplify the nominal two-zone problem to a single-zone problem. Specifically, by combining the mass conservation relation $\dot{m}^a = \dot{m}^c = (\dot{m}^e + \dot{m}^f)$ with the serial path a–c airflow relation $\dot{m}^{Sac} = \sqrt{\Delta P^{Sac}/(K^a + K^c)}$ and the branch path relations $\dot{m}^e = \sqrt{\Delta P^e/K^e}$ and $\dot{m}^f = \sqrt{\Delta P^f/K^f}$, we can form a single algebraic relation in terms of the single unknown node pressure identified above as p_4^P (or, alternatively, in terms of a zone d node pressure p^d). This single, albeit complicated, equation may then be solved to determine the 'correct' nodal values.

The general conclusion that may be drawn from these examples is simple. Under the special conditions outlined above – including, most importantly, a priori knowledge of airflow directions – the circuit analogy relations may be used to simplify serial, parallel and mixed ventilation networks to an equivalent single equation of simple quadratic form. When these conditions are not met, one may still use the circuit analogies to simplify the topology of many ventilation networks (i.e. from nominal multi-zone networks to effectively fewer zones).

Loop design method

In many cases, macroscopic ventilation *simulation analysis* is used to iteratively size ventilation system components; yet using the loop approach one may approach the sizing of ventilation system components *directly*, thereby avoiding the complexities of *simulation analysis*. In the *loop design method* – one of the very few methods of *design analysis* available for building component sizing in general – the design analyst establishes *design conditions* relating to prevailing wind, outdoor and indoor temperature conditions; defines *design criteria* that establish the needed ventilation airflow rates (e.g. to meet indoor air quality ventilation requirements or to provide adequate ventilation cooling given internal heat loads); and forms loop equations for each of the intended ('designed') ventilation airflow paths in the building. It will be shown that the resulting loop equations, now defined in terms of element or component size parameters, establish feasible combinations of ventilation system component sizes that will satisfy the *design criteria* given the specified *design conditions*.

The loop design method has been documented in detail in a number of publications (Axley, 1998, 1999a,b, 2000a,b, 2001; Axley *et al.*, 2002a; Dols and Emmerich, 2003; Ghiaus *et al.*, 2003; Mansouri, 2003; Mansouri and Allard, 2003; Santamouris, 2003). Here, the method will be introduced by way of a single worked example.

Consider then, the two-zone problem used in the examples above now from the point of view of a designer seeking to size windward and leeward openings *a* and *e*, an undercut at door *c* and the diameter of a 5-m-long ventilation stack *f* to provide a fresh air ventilation airflow of 1 air change per hour (ACH) to living spaces and 2 ACH to the kitchen of an apartment (Figure 4.48). For the purposes of this example, the living spaces will be assumed to have a floor area of $130\,\text{m}^2$ and the kitchen $20\,\text{m}^2$, both 4-m-high spaces – thus having volumes of $520\,\text{m}^3$ and $80\,\text{m}^3$, respectively. The design conditions will be those assumed in the earlier examples (i.e. an approach wind speed of $\bar{v}_{\text{ref}} = 4.0\,\text{m}^{-1}\text{s}$; outdoor and indoor temperatures of $T^o = 10°\text{C}$ and $T^b = T^d = 25°\text{C}$, thus corresponding air densities of 1.245 and $1.183\,\text{kg}^{-3}\text{m})$ On the other hand, envelope wind pressure coefficients will be assumed to be $C_{p,1} = +0.8$, $C_{p,6} = -0.4$ and $C_{p,8} = -0.5$ – a change from the earlier examples made to achieve the desired flow directions illustrated in Figure 4.48. From a practical point of view, this change could be realized by selecting a more optimal stack terminal device and intentionally shielding the leeward opening – both changes reflect the need to consider building

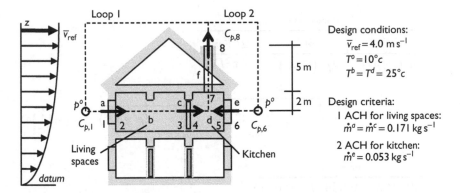

Figure 4.48 Hypothetical example used to introduce the loop design method.

form and envelope detail in preliminary design considerations of natural ventilation systems.

Given the design air densities and room volumes, the design criteria of 1 ACH for the living spaces is equivalent to mass airflow rates of $\dot{m}^a = \dot{m}^c = 0.171$ kg/s and the 2 ACH for the kitchen is equivalent to a mass airflow rate of $\dot{m}^e = 0.053$ kgs^{-1}. Furthermore, by mass conservation the stack airflow rate must equal the sum of these two contributions or $\dot{m}^e = 0.224$ kgs^{-1}. (Note that in the design case the airflow direction is not a matter of speculation – it is set as a design objective – i.e. here to provide fresh air directly to the individual spaces.)

To proceed, one simply forms the loop equations for each of the 'designed' ventilation airflow paths in the building. It will be recalled that the loop equations assume the intuitively satisfying general form given by Equation 4.80 or

$$
\overbrace{\sum_{\substack{loopl \\ k=1,2,\ldots;}} G_d^k(\dot{m}^k, \vartheta^k)}^{\text{element losses}} = \overbrace{\frac{1}{2}\rho^o \bar{v}_{\text{ref}}^2 \sum_{\substack{\text{envelope} \\ i}} C_{p,i}}^{\text{wind pressure } \Delta p_w} + \overbrace{\sum_{\substack{loopl \\ k=0,1,2,\ldots}} \left(\rho^k g \Delta z^k\right)}^{\text{bouyancy pressure } \Delta p_b} \quad (4.112)
$$

Thus modeling the windward opening, leeward opening and door undercut by the discharge coefficient model, with $C_d = 0.6$ in all cases, and the stack as a circular duct of diameter D^f (i.e. a hydraulic diameter of $D_h^f = D^f/2$ and cross-sectional area of $A^f = \pi D^{f^2}/4$), one may directly form loop equations for each of the loops illustrated in Figure 4.48 as

Loop 1:

$$
\overbrace{\frac{\dot{m}^{a2}}{2\rho^a\left(C_d^a A^a\right)^2} + \frac{\dot{m}^{c2}}{2\rho^c\left(C_d^c A^c\right)^2} + \frac{32f\,l^f\,\dot{m}^{f2}}{\rho^f\pi^2 D^{f5}}}^{\text{element losses}}
$$

$$
= \overbrace{\frac{1}{2}\rho^o\bar{v}_{\text{ref}}^2 \underset{\substack{\text{envelope}\\ i}}{\sum} C_{p,i}}^{\Delta p_w} + \overbrace{\underset{\substack{\text{loop 1}\\ k=0,1,2,\ldots}}{\sum}\left(\rho^k g\Delta z^k\right)}^{\Delta p_b} \qquad (4.113\text{a})
$$

$$
\overbrace{\frac{0.03262}{A^{a2}} + \frac{0.03433}{A^{c2}} + \frac{0.0055}{D^{f5}}}^{\text{element losses}}
$$

$$
= \overbrace{12.95\ \text{Pa}}^{\Delta p_w} + \overbrace{4.26\ \text{Pa}}^{\Delta p_b} \ \text{(for the specified parameters)}
$$
$$
(4.113\text{b})
$$

Loop 2:

$$
\overbrace{\frac{\dot{m}^{e2}}{2\rho^e\left(C_d^e A^e\right)^2} + \frac{32f\,l^f\,\dot{m}^{f2}}{\rho^f\pi^2 D^{f5}}}^{\text{element losses}}
$$

$$
= \overbrace{\frac{11}{2}\rho^o\bar{v}_{\text{ref}}^2 \underset{\substack{\text{envelope}\\ i}}{\sum} C_{p,i}}^{\Delta p_w} + \overbrace{\underset{\substack{\text{loop 1}\\ k=0,1,2,\ldots}}{\sum}\left(\rho^k g\Delta z^k\right)}^{\Delta p_b} \qquad (4.114\text{a})
$$

$$
\overbrace{\frac{0.003298}{A^{e2}} + \frac{0.0055}{D^{f5}}}^{\text{element losses}} = \overbrace{1.00\ \text{Pa}}^{\Delta p_w} + \overbrace{4.26\ \text{Pa}}^{\Delta p_b} \ \text{(for the specified parameters)}
$$
$$
(4.114\text{b})
$$

where, for preliminary analysis, the duct friction factor has been assumed to be $f = 0.008$ (i.e. appropriate for transitional flow with $\text{Re} = 2000$).

Upon substitution of the specific design conditions, design criteria and individual model parameters, one obtains two loop equations, Equations 4.113b and 4.114b, defined in terms of, in this case, four size parameters – i.e. the free areas of the windward opening, door undercut and leeward opening, A^a, A^c and A^e and the diameter of the stack, D^f. Given two equations and four size parameters, the 'design solution' remains under-determined. This will, in general, be the case when applying the loop design method as unique 'design solutions' seldom exist. Thus to determine a 'design solution', the design analyst must impose additional design constraints. As building components are invariably sold in discrete, finite size increments and a host of other architectural constraints must be considered, this apparent difficulty becomes one of the advantages of the loop design method.

On closer examination of the individual loop equations, it will be noticed that the element losses are inversely related to their size parameters. Thus by considering the limit of each loop equation as all but one of the size parameters is allowed to assume large values, one may determine the minimum feasible size of each parameter. Furthermore, by disaggregating the driving wind and buoyancy pressures for each loop, one may, in addition, size system components for two operating conditions – *with-wind* or *without-wind*. Thus for example, the minimum feasible sizes for the windward opening A^a would be determined by considering the limit of the loop 1 equation as both A^c and D^f approach infinity as

$$
\lim_{\substack{A^c \to \infty \\ D^f_b \to \infty}} \left(\overbrace{\frac{0.03262}{A^{a2}} + \frac{0.03433}{A^{c2}} + \frac{0.0055}{D^{f5}}}^{\text{element losses}} \right)
$$

$$
= \overbrace{12.95 \text{ Pa}}^{\Delta p_w} + \overbrace{4.26 \text{ Pa}}^{\Delta p_b} \text{ or } \overbrace{4.26 \text{ Pa}}^{\Delta p_b} \tag{4.115a}
$$

or $A^a \geq 0.044 \text{ m}^2$ *with-wind* or $A^a \geq 0.088 \text{ m}^2$ *without-wind*

$$\tag{4.115b}$$

Here, the *with-wind* driving pressure is the sum of the wind and buoyancy pressures whereas the *without-wind* driving pressure is that due to buoyancy alone.

In general, the limiting asymptotes for each of the size parameters would be computed for each operating condition to establish lower bounds for all component sizes. Table 4.2 presents the results of this exercise for the problem at hand.

Surveying these limiting asymptotes one design strategy becomes apparent. In as much as loop 2 experiences a low wind driving pressure, the limiting size of the stack is nearly identical for the *with-wind* and *without-wind* cases. Thus, recognizing a 0.30-m duct is a standard off-the-shelf size, one may elect to fix the duct size at $D^f = 0.30$ m and control airflow rates by varying the windward and leeward opening areas.

To determine the areas needed to implement this strategy, one simply substitutes the design decision of $D^f = 0.30$ m into the two loop equations to define a new system of equations that define feasible combinations of, now,

Table 4.2 Limiting asymptotes of loop I and loop 2 equations

Operating condition	A^a (m²)	A^c (m²)	A^e (m²)	D^f (m)
Loop I				
with-wind	0.044	0.045		0.200
without-wind	0.088	0.090		0.264
Loop 2				
with-wind			0.025	0.254
without-wind			0.028	0.264

A^a, A^c and A^e that satisfy the design conditions and criteria. Completing this exercise, we obtain

Loop 1:

$$\overbrace{\frac{0.03262}{A^{a2}} + \frac{0.03433}{A^{c2}}}^{\text{element losses}} + 2.264 = \overbrace{12.95 \text{ Pa}}^{\Delta p_w} + \overbrace{4.26 \text{ Pa}}^{\Delta p_b}$$

(4.116)

Loop 2:

$$\overbrace{\frac{0.00330}{A^{e2}}}^{\text{element losses}} + 2.264 = \overbrace{1.00 \text{ Pa}}^{\Delta p_w} + \overbrace{4.26 \text{ Pa}}^{\Delta p_b}$$

(4.117)

In this case, loop 2 relation now completely defines the needed opening areas for the leeward (kitchen) opening – i.e. solving Equation 4.117, we find that A^e must be $0.033 \, \text{m}^2$ for the *with-wind* condition and $0.041 \, \text{m}^2$ for the *without-wind* condition.

Given the door undercut would not normally be adjustable, we may set it at its limiting (*without-wind*) asymptote. From Equation 4.116, this limiting asymptote would be $A^c = 0.13 \, \text{m}^2$. As this would require an unusually large door undercut, it would be more reasonable to provide this opening using a purpose-provided grill in the door. In any event, one would then substitute this design decision into the loop 1 equation to obtain an updated version of the loop 1 equation which will now completely determine the windward openings needed:

Loop 1:

$$\overbrace{\frac{0.03262}{A^{a2}} + 1.994}^{\text{element losses}} + 2.264 = \overbrace{12.95 \text{ Pa}}^{\Delta p_w} + \overbrace{4.26 \text{ Pa}}^{\Delta p_b}$$

(4.118)

Finally, by solving Equation 4.118, we find that A^a must be $0.050 \, \text{m}^2$ for the *with-wind* condition and $0.061 \, \text{m}^2$ for the *without-wind* condition.

Although a number of refinements to the loop design method are considered in the publications cited above, this introductory example illustrates the basic approach of the loop design method. With a sectional diagram of the proposed ventilation system and airflow models identified for each of the ventilation system components, one defines design conditions and criteria and forms the loop equations for each of the 'designed' ventilation airflow loops. The resulting equations will define feasible combinations of component sizes for the specified design conditions that will achieve the design criteria. These design loop equations will, in general, be under-determined. Consequently, the designer would first examine the limiting asymptotes of these equations that establish minimum component sizes for the assumed design conditions (e.g. *with-wind* or *without wind* conditions) and then impose operational or practical design constraints to systematically devise an acceptable design solution. As each design decision is made (i.e. specific component sizes are set), the designer updates the loop equations and determines revised values for the limiting sizes of the remaining components. This iterative, *ad hoc* procedure is repeated until all components are sized for all considered operating conditions.

Of the number of refinements that could be considered, one is especially important. The driving wind (Δp_w) and buoyancy (Δp_b) pressure terms of the loop equations are independent of the ventilation system component sizes. Therefore, annual time variations of Δp_w and Δp_b may be computed a priori (i.e. given an annual record of outdoor air temperature, wind speed and direction and a schedule of desire indoor air temperatures) to determine statistically significant design conditions (e.g. probable low wind conditions and likely windy conditions) as a more reliable approach to system component sizing.

4.3.5 Related macroscopic methods

The development of the three broad approaches to multi-zone building ventilation analysis presented above – the *nodal approach*, the *port plane approach* and the *loop approach* – have effectively transformed the science of macroscopic building ventilation analysis from a discipline that relied on *ad hoc* and often fully empirical manual methods to one that is now based more rigorously on deterministic general computational methods of analysis. Building on these developments while, at the same time, recognizing their shortcomings, a number of related macroscopic methods are currently under development to extend the analytical scope of multi-zone building ventilation analysis.

At one end of the spectrum, some researchers have attempted to adapt the fundamental macroscopic principles introduced above to develop approximate macroscopic methods to model the details of airflows within selected zones, (Figure 4.49) (Chen, 1988; Bouia and Dalicieux, 1991; Inard and Buty 1991a,b; Rodriguez *et al.*, 1993). Most recently, macroscopic *sub-zone* (or *zonal*) models of individual rooms have been coupled to multi-zone models of the building system in which these rooms are embedded (Mora

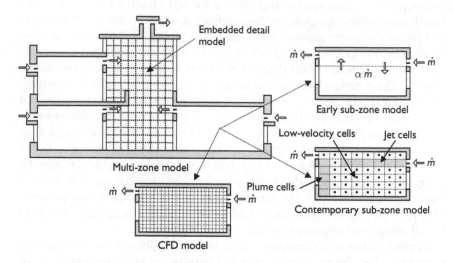

Figure 4.49 Modeling of flow details within zones using either isolated or embedded sub-zone or computational fluid dynamics (CFD) models.

et al., 2002; Mora *et al.*, 2003; Ren and Stewart, 2003; Stewart and Ren, 2003) in an attempt to predict the general structure of these airflow details while properly accounting for their interaction with the *bulk* airflows of the remaining building system. Whereas these *sub-zone* and *embedded sub-zone* models have proven to be computationally attractive, others have questioned their accuracy and have turned instead to *embedding* detailed CFD microscopic models of selected zones within multi-zone models of whole building systems (Li and Holmberg, 1993; Schaelin *et al.*, 1993; Albrecht *et al.*, 2002; Gao, 2002; Lorenzetti *et al.*, 2003; Chen and Wang, 2004).

At the other end of the spectrum, some researchers have attempted to account for the coupled interactions of thermal and/or contaminant mass transport phenomena – *coupled thermal-dispersal-airflow analysis* – that were explicitly ignored in the development of the multi-zone methods introduced above (Figure 4.50). Macroscopic methods of building thermal and contaminant dispersal (including moisture transport) analysis are invariably based on idealizations of building systems as collections of control volumes linked by airflow and other mass and energy transport paths. Consequently, the coupling of multi-zone airflow analysis programs with both multi-zone thermal and contaminant dispersal analysis programs has been realized by a number of researchers, albeit not without practical and theoretical challenges (Walton, 1982; Axley and Grot, 1989 and 1990; Clarke and Hensen, 1990; Hensen and Clarke, 1990 and 1991; Kendrick, 1993; Haas, 2000; Vuolle and Sahlin, 2000; Woloszyn *et al.*, 2000; Bring *et al.*, 2003; Transsolar, 2004).

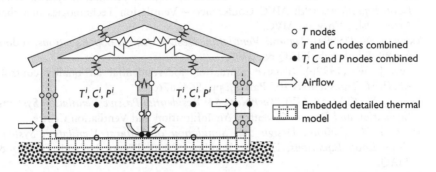

Figure 4.50 Hypothetical multi-zone idealization for combined thermal, airflow and contaminant dispersal analysis to account for the interaction of zone temperatures T^i, zone contaminant concentrations C^i and zone pressures P^i.

These recent and emerging developments as well as new efforts underway to combine *embedded sub-zone* methods with *coupled thermal-dispersal-airflow methods* to address the challenge of natural and hybrid ventilation system design and the concern for chemical/biological attacks on building ventilation systems are currently shaping the future of the discipline. Regrettably, however, they fall outside the scope of this section. The interested reader is, therefore, directed to a recent publication of the author of this section for additional detail (Axley, 2006).

References

Albrecht, T., Gritzki, R., Grundmann, R., Perschk, A., Richter, W., Rösler M. and Seifert J. (2002). *Evaluation of a Coupled Calculation for a Hybrid Ventilated Building From a Practical and Scientific Point of View*. P. Heiselberg (ed). Aalborg, Denmark, Hybrid Ventilation Centre, Aalborg University.

Allard, F., ed. (1998). *Natural Ventilation in Buildings: A Design Handbook*. London, James & James (Science Publishers) Ltd.

Anderson, J. D. (1995). *Computational Fluid Dynamics. The Basis with Applications*. McGraw-Hill, Inc, New York.

ASHRAE (2005). *2005 ASHRAE Handbook – Fundamentals*. Atlanta, GA, ASHRAE.

Awbi, H. B. (1991). Chapter 3: Air infiltration and natural ventilation. *Ventilation of Buildings*. London, Spon Press, pp. 60–98.

Awbi, H. B. (1998). 'Calculation of convective heat transfer coefficients of room surfaces for natural convection'. *Energy and Buildings* 28: pp. 219–227.

Awbi, H. B. (2003). *Ventilation of Buildings – Second Edition*. London, Spon Press.

Axley, J. (1998). *Introduction to the Design of Natural Ventilation Systems Using Loop Equations.* 19th AIVC Conference – Ventilation Technologies in Urban Areas, Oslo, Norway, AIVC.

Axley, J. W. (1999a). *Natural Ventilation Design Using Loop Equations.* Indoor Air 99, Edinburgh, ISIAQ & AIVC.

Axley, J. W. (1999b). 'Passive ventilation for residential air quality control'. *ASHRAE Transactions* 105(Part 2): pp. 864–876.

Axley, J. W. (2000a). *AIVC TechNote 54: Residential Passive Ventilation Systems: Evaluation and Design.* Coventry, Air Infiltration and Ventilation Centre.

Axley, J. W. (2000b). *Design and Simulation of Natural Ventilation Systems Using Loop Equations.* Healthy Buildings 2000, Espoo, Finland, FiSIAQ & ISIAQ.

Axley, J. (2001). *Application of Natural Ventilation for U.S. Commercial Buildings – Climate Suitability, Design Strategies & Methods, Modeling Studies.* Gaithersburg, MD, NIST, 150 pages.

Axley, J. W. (2006). Chapter 2: Analytical methods and computing tools for ventilation. *Building Ventilation – The State of the Art.* M. Santamouris and P. Wouters (eds). London, Earthscan (James & James) Publications, 352 pages.

Axley, J. W. and Chung D. H. (2005a). 'POWBAM0: Mechanical power balances for multi-zone building airflow analysis'. *International Journal of Ventilation* 4(No. 2): pp. 95–112.

Axley, J. W. and Chung D. H. (2005b). *Mechanical Energy Power Balances for Building Airflow Analysis.* 2nd Workshop on Natural Ventilation, Tokyo.

Axley, J. W. and Chung D. H. (2006). 'Well-posed models of porous buildings for macroscopic ventilation analysis.' *International Journal of Ventilation* 5(No. 1): pp. 89–104.

Axley, J. and Grot R. (1989). *The Coupled Airflow and Thermal Analysis Problem in Building Airflow System Simulation.* ASHRAE Symposium on Calculation of Interzonal Heat and Mass Transport in Buildings, Vancouver, B. C., ASHRAE.

Axley, J. and Grot R. (1990). *Coupled Airflow and Thermal Analysis for Building System Simulation by Element Assembly Techniques.* RoomVent '90: Engineering Aero- and Thermodynamics of Ventilated Rooms Second International Conference, Oslo, Norway, NORSK VVS: Norwegian Association of Heating, Ventilating and Sanitary Engineers.

Axley, J., Emmerich, S. J., Dols S. and Walton G. (2002a). *An Approach to the Design of Natural and Hybrid Ventilation Systems for Cooling Buildings.* Indoor Air 2002, Monterey, CA.

Axley, J., Wurtz E. and Mora L. (2002b). *Macroscopic Airflow Analysis and the Conservation of Kinetic Energy.* Room Vent 2002: Air Distribution in Rooms – Eight International Conference, Copenhagen, Denmark.

Aynsley, R. (1999). *Unresolved Issues in Natural Ventilation for Thermal Comfort.* HybVent Forum '99, Sydney, Australia.

Aynsley, R. M., Melbourne W. and Vickery B. J. (1977a). *Architectural Aerodynamics.* London, Applied Science Publishers Ltd.

Aynsley, R. M., Melbourne W. and Vickery B. J. (1977b). Chapter 1: Historical developments in architectural aerodynamics. *Architectural Aerodynamics.* London, Applied Science Publishers Ltd.

Bassett, M. (1990). *Technical Note AIVC 27: Infiltration and Leakage Paths in Single Family Houses – A Multizone Infiltration Case Study*. Coventry, England, Air Infiltration and Ventilation Centre.

Baturin, V. V. (1972). *Fundamentals of Industrial Ventilation*. Oxford, UK, Pergamon Press.

Bird, R. B., Stewart W. E. and Lightfoot E. N. (1960). *Transport Phenomena*. New York, John Wiley & Sons, Inc.

Bird, R. B., Stewart W. E. and Lightfoot E. N. (2002). *Transport Phenomena – Second Edition*. New York, John Wiley & Sons, Inc.

Bjerg, B., Svidt, K., Morsing, S., Zhang G. and Johnsen J. O. (2002). 'Modeling of a wall inlet in numerical simulation of airflow in livestock buildings'. *The CIGR Journal of Agricultural Engineering Scientific Research and Development* 4, http://eigr-journal.tamu.edu/volume 4.html

Blevins, R. D. (2003). *Applied Fluid Dynamics Handbook*. Melbourne, Florida, Krieger Publishing Company.

Bouia, H. and Dalicieux P. (1991). *Simplified Modeling of Air Movements Inside Dwelling Room*. Building Simulation '91, Sophia-Antipolis, Nice, France, IBPSA (International Building Performance Simulation Association).

Bring, A., Sahlin P. and Vuolle M. (2003). *IDA Indoor Climate and Energy 3.0*. Stockholm, EQUA.

Casey, M. and Wintergerste T. (ed.) (2000). *Best Practice Guidelines*. ERCOFTAC Special Interest Group on Quality and Trust in Industrial CFD.

CEN, E. C. f. S. (1999). *European Standard: Ventilation for Buildings – Calculation Methods for the Determination of Air Flow Rates in Dwellings* (English Version). Brussels, European Committee for Standardization, 44 pages.

Chapman, D. R. (1979). 'Computational aerodynamics development and outlook'. *American Institute of Aeronautics and Astronautics Journal*, 17: pp. 1293–1313.

Chen, Q. (1988). Indoor airflow, air quality and energy consumption of buildings. Doctoral Dissertation, Technical University of Delft.

Chen, Q. (1996). Prediction of Room Air Motion by Reynolds-Stress Models. *Building and Environment*, 31(No. 3): pp. 233–244.

Chen, Q. and Jiang Z. (1992). 'Significant questions in predicting room air motion'. *ASHRAE Transactions*, 98 (Pt. 1): pp. 929–939.

Chen, Q. and Srebric J. (2002). 'A procedure for verification, validation and reporting of indoor environment CFD analysis'. *HVAC & R Research* 8(No. 2): pp. 201–216.

Chen, Q. Y. and Wang L. (2004). *Coupling of Multizone Program CONTAM with Simplified CFD Program CFD0-C*. West Lafayette, IN, School of Mechanical Engineering, Purdue University, 121 pages.

Chiu, Y. H. and Etheridge D. W. (2002). 'Calculations and Notes on the Quadratic and Power Law Equations for Modelling Infiltration'. *International Journal of Ventilation* 1(No. 1): pp. 65–77.

Chorin, A. J. and Marsden J. E. (1993). *A Mathematical Introduction to Fluid Mechanics*. New York, Springer-Verlag.

Clarke, J. A. and Hensen J. L. M. (1990). *An Approach to the Simulation of Coupled Heat and Mass Flows in Buildings*. Ventilation System Performance, 11th AIVC Conference, Belgirate, Italy, AIVC.

Courant, R., Isaacson E. and Ress M. (1952). 'On the solution of non-linear hyperbolic differential equations by finite differences'. *Communications on Pure and Applied Mathematics* 5: p. 243.

Dascalaki, E. and Santamouris M. (1996). Chapter 9: Natural ventilation. *Passive Cooling of Buildings*. M. Santamouris and D. Asimakopoulos (eds). London, James & James Ltd., pp. 220–306.

de Gids, W. F. (1978). 'Calculation method for the natural ventilation of buildings'. *Verwarming Vent* (no. 7): pp. 552–564.

Dick, J. B. (1949). 'Experimental studies in natural ventilation of houses'. *Journal of the Institution of Heating and Ventilating Engineers* (December, 1949): pp. 420–466.

Dick, J. B. (1950). 'The fundamentals of natural ventilation of houses'. *Journal of the Institution of Heating and Ventilating Engineers* 18: pp. 123–124.

Dick, J. B. (1951). 'The principles of natural ventilation of buildings'. *Building Research Station Digest* (No. 34): 6 pages.

Dols, W. S. (2001a). *NIST Multizone Modeling Website*. Gaithersburg, MD, NIST, 22 pages.

Dols, W. S. (2001b). 'A tool for modeling airflow & contaminant transport'. *ASHRAE Journal* 43(No. 3): pp. 35–42.

Dols, W. S. and Emmerich S. J. (2003). *LoopDA – Natural Ventilation Design and Analysis Software*. Gaithersburg, MD, US DOC NIST, 39 pages.

Dols, W. S. and Walton G. N. (2000). *CONTAMW 2.0 User Manual: Multizone Airflow and Contaminant Transport Analysis Software*. Gaithersburg, MD, U.S. NIST, 165 pages.

Dols, W. S., Walton G. N. and Denton K. R. (2000). *CONTAMW 1.0 User Manual: Multizone Airflow and Contaminant Transport Analysis Software*. Gaithersburg, MD, U.S. NIST, 133 pages.

Etheridge, D. W. (1998). 'A note on crack flow equations for ventilation flow modeling'. *Building and Environment* 33(No. 5): pp. 325–328.

Etheridge, D. W. (2002). 'Nondimensional methods for natural ventilation design'. *Building and Environment* 37(No. 11): pp. 1057–1072.

Etheridge, D. W. (2004). 'Natural ventilation through large openings – measurements at model scale and envelope flow theory'. *The International Journal of Ventilation* 2(No. 4): pp. 325–343.

Etheridge, D. and Sandberg M. (1996). *Buildng Ventilation: Theory & Measurement*. Chichester, John Wiley & Sons.

Feustel, H. E. (1990). *The COMIS Air Flow Model A Tool for Multizone Applications*. Indoor Air '90 The 5th International Conference on Indoor Air Quality and Climate, Toronto, Canada Mortgage and Housing Corporation.

Feustel, H. E. and Raynor-Hoosen, A. eds. (1990). *COMIS – Fundamentals*. Berkeley, CA, Lawrence Berkeley Lab.

Feustel, H. E. and Smith, B. V. eds. (1997). *COMIS 3.0 – User's Guide*. Berkeley, CA, Lawrence Berkeley National Laboratory.

Fried, E. and Idelchik I. E. (1989). *Flow Resistance: A Design Guide for Engineers*. New York, Hemisphere Publishing Corporation, New York.

Fürbringer, J.-M., Dorer, V., Huck F. and Weber A. (1993). *Air Flow Simulation of the LESO Building Including a Comparison with Measurements and Sensitivity Analysis*. Indoor Air '93: The 5th International Conference on Indoor Air Quality and Climate, Helsinki, Finland, Helsinki University of Technology.

Fürbringer, J. M., Roulet C. A. and Borchiellini, R. eds. (1996a). *Annex 23: Multizone Air Flow Modeling: Evaluation of Comis*. Lausanne, Swiss Federal Institute of Technology, Institute of Building Technology.

Fürbringer, J. M., Roulet C. A. and Borchiellini, R. eds. (1996b). *Annex 23: Multizone Air Flow Modeling: Evaluation of Comis: Appendices*. Lausanne, Swiss Federal Institute of Technology, Institute of Building Technology.

Gao, Y. (2002). *Coupling of Computational Fluid Dynamics and a Multizone Airflow Analysis Program for Indoor Environmental Design*. Cambridge, MA, Massachusetts Institute of Technology.

Gerhart, P. M. and Gross R. J. (1985). *Fundamentals of Fluid Mechanics*. London, Addison Wesley.

Ghiaus, C., Allard F. and Axley J. (2003). Chapter 6: Natural ventilation in an urban context. *Solar Thermal Technologies for Buildings – The State of the Art*. M. Santamouris (ed.). London, James & James.

Girault, P. and Spennato B. (1999). *The Impact of Wind Turbulence on the Precision of a Numerical Modeling Study*. Indoor Air 99, Edinburgh, ISIAQ & AIVC.

Gosman, A. D., Nielsen, P. V., Restivo A. and Whitelaw J. H. (1980). 'The flow properties of rooms with small ventilation openings'. *Transactions of the ASME*, J. Fluid Eng., 102, 316–323.

Guffey, S. E. and Fraser D. A. (1989). 'A power balance model for converging and diverging flow junctions'. *ASHRAE Transactions* 95(Part 2): pp. 1661–1669.

Haas, A. (2000). *COMIS 3.1, EMPA*, – Swiss Federal Laboratories for Materials Testing and Research.

Haghighat, F., Brohus H. and Rao J. (2000). 'Modelling air infiltration due to wind fluctuations– a review'. *Building and Environment* 35(No. 5): pp. 377–385.

Hensen, J. L. M. and Clarke J. A. (1990). *A Fluid Flow Network Solver for Integrated Building and Plant Energy Simulation*. The 3rd International Conference on System Simulation in Buildings, University of Liège, Belgium, pp. 151–167.

Hensen, J. L. M. and Clarke J. A. (1991). *A Simulation Approach to the Evaluation of Coupled Heat and Mass Transfer in Buildings*. Building Simulation '91, Sophia-Antipolis, Nice, France, IBPSA (International Building Performance Simulation Association).

Holmes, M. J. and McGowan S. (1997). *Simulation of a Complex Wind and Buoyancy Driven Building*. Building Simulation '97 – Fifth International IBPSA Conference, Prague, IBPSA.

Huang, P. G., Launder B. E. and Leschziner M. A. (1985). 'Discretization of nonlinear convection processes: a broad-Range comparison of four schemes'. *Computer Methods, in Applied Mechanics and Engineering* 48(No. 1): pp. 1–24.

Hunt, G. R. and Syrios K. (2004). 'Roof-mounted ventilation towers – design criteria for enhanced buoyancy-driven ventilation'. *International Journal of Ventilation* 3(No. 3): pp. 193–208.

Idelchik, I. E. (1994). *Handbook of Hydraulic Resistance – 3rd Edition*. Boca Raton, CRC Press.

Inard, C. and Buty D. (1991a). *Simulation of Thermal Coupling Between a Radiator and Room with Zonal Models*. Building Simulation '91, Sophia-Antipolis, Nice, France, IBPSA (International Building Performance Simulation Association).

Inard, C. and Buty D. (1991b). *Simulation of Thermal Coupling Between a Radiator and Room with Zonal Models.* 12th AIVC Conference: Air Movement and Ventilation Control Within Buildings, Ottawa, Canada, Air Infiltration and Ventilation Centre, Coventry, Great Britain.

Irving, S. and Uys E. (1997). *CIBSE Applications Manual: Natural Ventilation in Non-domestic Buildings.* London, CIBSE.

Jackman, P. J. (1970). *Air Movement in Rooms with Sidewall Mounted Grilles – A Design Procedure.* Laboratory Report No. 65, HVRA (now BSRIA), Bracknell, UK.

Jensen, R. L. (2005). *Modellering af naturlig ventilation og natkøling-ved hjlp af ringmetoden* (Modeling of Natural Ventilation and Night Cooling – by the Loop Equation Method). Department of Building Technology and Structural Engineering. Aalborg, Denmark, Aalborg University, 262 pages.

Jensen, J. P., Heiselberg P. and Nielsen P. V. (2002a). *Numerical Simulation of Airflow Through Large Openings, International Energy Agency – Energy Conservation in Buildings and Community Systems*, Annex 35: Hybrid Ventilation in New and Retrofitted Office Buildings, 19 pages.

Jensen, J. P., Heiselberg P. and Nielsen P. V. (2002b). *Numerical Simulation of Airflow Through Large Openings.* RoomVent 2002 – 8th International Conference on Air Distribution in Rooms, Copenhagen, The Technical University of Denmark & Danvak.

Jeppson, R. W. (1976). *Analysis of Flow in Pipe Networks.* Ann Arbor, MI, Ann Arbor Science.

Jones, P. J. and Whittle G. E. (1992). 'Computational fluid dynamics for building air flow prediction – current status and capabilities'. *Building and Environment* 27(No. 3) pp. 321–338.

Karava, P., Stathopoulos T. and Athienitis A. K. (2004). 'Wind-driven flow through openings – a review of discharge coefficients'. *International Journal of Ventilation* 3(No. 3): pp. 255–266.

Kato, S. (2004). 'Flow network model based on power balance as applied to cross-ventilation'. *The International Journal of Ventilation* 2(No. 4): pp. 395–408.

Kato, S., Murakami S. and Kondo Y. (1994). 'Numerical simulation of two-dimensional room airflow with and without buoyancy by means of ASM'. *ASHRAE Transactions*, 100(No. 1): pp. 238–255.

Kato, S., Murakami, S., Shoya, S., Hanyu F. and Zeng J. (1995). 'CFD analysis of flow and temperature fields in atrium with ceiling height of 130 m'. *ASHRAE Transactions*, 101(Part 2) pp. 1144–1157.

Kelley, C. T. (1995). *Iterative Methods for Linear and Nonlinear Equations.* Philadelphia, Soc for Industrial & Applied Math.

Kelley, C. T. (2003). *Solving Nonlinear Equations With Newton's Method (Fundamentals of Algorithms).* Philadelphia, Soc for Industrial & Applied Math.

Kendrick, J. (1993). *AIVC Technical Note 40: An Overview of Combined Modeling of Heat Transport and Air Movement.* Coventry, UK, AIVC.

Klote, J. H. and Milke J. A. (1992). *Design of Smoke Management Systems.* Atlanta, GA, ASHRAE.

Knoll, B. and Phaff J. C. (1996). *Two Unique New Tools for the Prediction of Wind Effects on Ventilation, Building Construction, and Indoor Climate.* Delft, TNO Building Construction & Research.

Knoll, B., Phaff J. C. and de Gids W. F. (1997). *Pressure Simulation Program*. Delft, TNO Building Construction & Research.

Kobayashi, T., Sagara, K., Yamanaka, T., Kotani H. and Sandberg M. (2006). 'Wind driven flow through openings – an analysis of the stream tube'. *International Journal of Ventilation* 4(No. 4): pp. 323–336.

Kofoed, P. and Nielsen P. V. (1990). *Thermal Plumes in Ventilated Rooms*. Proceedings of the International Conference on Engineering Aero- and Thermodynamics of Ventilated Rooms, ROOMVENT'90, Oslo.

Kunugiyama, N. (2002). *Private Communication*. Earth Simulator Center, Japan.

Kurabuchi, T., Ohba, M., Arashiguchi A. and Iwabuchi T. (2000). *Numerical Study of Airflow Structure of a Cross-ventilated Model Building*. RoomVent 2000: Ventilation for Health and Sustainable Environment, University of Reading, Elsevier.

Kurabuchi, T., Ohba, M., Endo, T., Akamine Y. and Nakayama F. (2004). 'Local dynamic similarity model of cross-ventilation, Part 1 – Theoretical framework'. *The International Journal of Ventilation* 2(No. 4): pp. 371–381.

Launder, B. E. (1989). 'Second-moment closure: present. . . and future?' *International Journal of Heat and Flow* 10: pp. 282–300.

Launder, B. E. and Spalding D. B. (1974). 'The numerical computation of turbulent flow'. *Computer Methods in Applied Mechanics and Engineering* 3: pp. 269–289.

Leonard, B. P. (1979). 'A stable and accurate convective modeling procedure based on quadratic upstream interpolation'. *Computer Methods in Applied Mechanics and Engineering* 19: pp. 59–98.

Li, Y. (1993). *Predictions of Indoor Air Quality in Multi-Room Buildings*. Indoor Air '93: The 5th International Conference on Indoor Air Quality and Climate, Helsinki, Finland, Helsinki University of Technology.

Li, Y. and Holmberg S. (1993). *General Flow- and Thermal-Boundary Conditions in Indoor Air Flow Simulation*. Indoor Air '93: The 5th International Conference on Indoor Air Quality and Climate, Helsinki, Finland, Helsinki University of Technology.

Liddament, M. (1986a). 'Air infiltration and ventilation calculation techniques'. *Air Infiltration Review* 8(No. 1): pp. 6–7.

Liddament, M. (1986b). *Air Infiltration Calculation Techniques – An Applications Guide*. Coventry, England, Air Infiltration and Ventilation Centre.

Liddament, M. W. (1996). *A Guide to Energy Efficient Ventilation*. Coventry, AIVC.

Liddament, M. and Allen C. (1983). *Technical Note AIVC 11: The Validation and Comparison of Mathematical Models of Air Infiltration*. Air Infiltration and Ventilation Centre.

Lorenzetti, D. M. and Sohn M. D. (2000). *Improving the Speed and Robustness of the Comis Solver*. RoomVent 20000, University of Reading, Elsevier.

Lorenzetti, D., Jayaraman, B., Gadgil, A., Hong S. and Chyczewski T. (2003). *Results from a Coupled CFD + Multizone Simulation*. NIST Workshop on Non-uniform Zones in Multizone Network Models, Indoor Environment Dept., Lawrence Berkeley National Laboratory.

Mansouri, Y. (2003). *Conception des enveloppes de bâtiments pour le renouvellement d'air par ventilation naturelle en climats tempérés Proposition d'une méthodologie de conception*. Mécanique Thermique et Génie Civil. Nantes, France, Ecole polytechnique de l'Université de Nantes, 320 pages.

Mansouri, Y. and Allard F. (2003). *Methods and Methodological Tools for the Elaboration of Natural Ventilation Strategy*. Healthy Building 2003 7th International Conference of Energy-Efficient Healthy Buildings, Singapore.

Marques da Silva, F. and Saraiva J. G. (2005). 'A simplified model to estimate natural ventilation flows for simple dwelling layouts'. *International Journal of Ventilation* 3(No. 4): pp. 353–362.

MathSoft (2004). MathCAD11.

MathWorks (2000). *MATLAB® – The Language of Technical Computing m- Using MATLAB Version 6*. Natick, MA, The MathWorks.

McRee, D. I. and Moses H. L. (1967). *The Effect of Aspect Ratio and Offset on Nozzle Flow and Jet Reattachment. Advances in Fluidics*. The 1967 Fluidics Symposium, ASME.

Mora, L., Gadgil, A. J., Wurtz E. and Inard C. (2002). *Comparing Zonal and CFD Model Predictions of Indoor Airflows Under Mixed Convection Conditions to Experimental Data*. EPIC Conference – Third European Conference on Energy Performance and Indoor Climate in Buildings, Lyon, France.

Mora, L., Gadgil A. J. and Wurtz E. (2003). 'Comparing zonal and CFD model predictions of isothermal indoor airflows to experimental data'. *Indoor Air* 13: pp. 77–85.

Moser, A., Off, F., Schälin A. and Yuan X. (1995). 'Numerical modeling of heat transfer by radiation and convection in an atrium with thermal inertia'. *ASHRAE Transactions*, 101(Part 2) pp. 1136–1143.

Murakami, S. (1993). 'Comparison of various turbulence models applied to a bluff body'. *Journal of Wind Engineering and Industrial Aerodynamics* 46 & 47: pp. 21–36.

Murakami, S., Kato, S., Akabayashi S. and Mizutani K. (1991). 'Wind tunnel test on velocity-pressure field of cross-ventilation with open windows'. *ASHRAE Transactions Symposia* NY-91-5-4: pp. 525–538.

Murakami, S., Kato S., and Ooka R. (1994). 'Comparison of numerical predictions of horizontal nonisothermal jet in a room with three turbulence models – k-ε EVM, ASM, and DSM'. *ASHRAE Transactions*, 100(No. 2), pp. 697–706.

Nielsen, P. V. (1973). 'Berechnung der Luftbewegung in einem zwangsbelüfteten Raum'. *Gesundheits-Ingenieu r* 94(H10): pp. 299–302.

Nielsen, P. V. (1991). *Models for the Prediction of Room Air Distribution*. 12th AIVC Conference, Ottawa, Canada, ISBN 0 946075 53 0.

Nielsen, P. V. (1992a). *Velocity Distribution in the Flow from a Wall-Mounted Diffuser in Rooms with Displacement Ventilation*. Presented at ROOMVENT'92. Third International Conference on Air Distribution in Rooms, Aalborg, Denmark.

Nielsen, P. V. (1992b). 'The description of supply openings in numerical models for room air distribution'. *ASHRAE Transactions* 98(Part 1).

Nielsen, P. V. (1995). 'Air flow in an exposition pavilion studied by scale-model experiments and computational fluid dynamics'. *ASHRAE Transactions* 101(Part 2) pp. 1118–1126.

Nielsen, P. V. (1998a). Ventilation in commercial and residential buildings. *Ventilation System and Air Quality*. Lecture Series; 1998-07, von Karman Institute for Fluid Dynamics, Bruxelles p. 26.

Nielsen, P. V. (1998b). 'The selection of turbulence models for prediction of room airflow'. *ASHRAE Transactions* 104(Part 1B) pp. 1119–1127, ISSN: 0001-2505.

Nielsen, P. V. (2000). 'Velocity distribution in a room ventilated by displacement ventilation and wall-mounted air terminal devices'. *Energy and Buildings* 31(No. 3), pp. 179–187.

Nielsen, P. V. and Möller Å. T. A. (1985). *Measurement of the Three-Dimensional Wall Jet from Different Types of Air Diffusers.* 'Clima 2000' World Congress on Heating, Ventilating and Air Conditioning, Copenhagen.

Nielsen, P. V. and Tryggvason T. (1998). *Computational Fluid Dynamics and Building Energy Performance Simulation.* Proceedings of ROOMVENT '98: Sixth International Conference on Air Distribution in Rooms, Stockholm, Sweden, June 14–17. Vol. 1, pp. 101–107.

Nielsen, P. V., Dam, H., Srensen, L.C., Svidt K. and Heiselberg P. (2000). *Characteristics of Buoyant Flow from Open Windows in Naturally Ventilated Rooms.* Proceedings of ROOMVENT 2000, Reading.

Nitta, K. (1994). Calculation method of multi-room ventilation. *Memoirs of the Faculty of Engineering and Design.* Kyoto Institute of Technology Vol. 42, pp. 59–94.

Ohba, M., Kurabuchi, T., Endo, T., Akamine, Y., Kamata M. and Kurahashi A. (2004). 'Local dynamic similarity model of cross-ventilation. Part 2 – Application of local dynamic similarity model'. *The International Journal of Ventilation* 2(No. 4): pp. 383–393.

Olesen, B. W. (1979). *Draught and Air Velocity Measurements.* DISA Information, No. 24, Disa Elektronik A/S, Denmark.

Orme, M. (1999). *AIVC Technical Note 51: Applicable Models for Air Infiltration and Ventilation Calculations.* Coventry, AIVC, 66 pages.

Orme, M., Liddament M. W. and Wilson A. (1998). *Numerical Data for Air Infiltration & Natural Ventilation Calculations.* Coventry, AIVC.

Patankar, S. V. (1980). *Numerical Heat Transfer and Fluid Flow.* Hemisphere Publishing Corporation, New York.

Pelletret, R. Y. and Keilholz W. P. (1997). *COMIS 3.0 – A New Simulation Environment for Multizone Air Flow and Pollutant Transport Modeling.* Building Simulation '97 – Fifth International IBPSA Conference, Prague, IBPSA.

Persily, A. K. (1994). *Manual for Ventilation Assessment in Mechanically Ventilated Commercial Buildings,* Gaithersburg, MD, US NIST.

Persily, A. K. and Ivy E. M. (2001). *Input Data for Multizone Airflow and IAQ Analysis.* Gaithersburg, NIST, 33 pages.

Press, W. H., Teukolsky, S. A., Vetterling W. T. and Flannery B. P. (1992). *Numerical Recipes in C: The Art of Scientific Computing – Second Edition.* Cambridge, Cambridge University Press.

Rajaratnam, N. (1976). *Turbulent Jets.* Elsevier, Amsterdam.

Ren, Z. and Stewart J. (2003). 'Simulating air flow and temperature distribution inside buildings using a modified version of COMIS with sub-zonal divisions'. *Energy & Buildings* 35: pp. 257–271.

Rodriguez, E. A., Alvarez S. and Cáceres I. (1993). *Prediction of Indoor Temperature and Air Flow Patterns by means of Simplified Zonal Models.* ISES Solar World Conference, Budapest, Hungarian Energy Society.

Roulet, C.-A., Fürbringer J.-M. and Borchiellini R. (1996). *Evaluation of the Multizone Air Flow Simulation Code Comis.* 5th International Conference on Ar Distribution in Rooms – ROOMVENT '96.

Russell, M. B. and Surendran P. N. (2000). 'Use of computational fluid dynamics to aid studies of room air distribution: a review of some recent work'. *Proceedings of CIBSE* A: *Building Services Engineering Research and Technology* 21(No. 4): pp. 241–247.

Saleh, J., ed. (2002). *Fluid Flow Handbook*. New York, McGraw-Hill.

Sandberg, M. (2002). *Airflow Through Large Openings – A Catchment Problem?* RoomVent 2002 – 8th International Conference on Air Distribution in Rooms, Copenhagen, The Technical University of Denmark & Danvak.

Sandberg, M. (2004). 'An alternative view on the theory of cross-ventilation'. *The International Journal of Ventilation* 2(No. 4): pp. 395–408.

Santamouris, M., ed. (2003). *Solar Thermal Technologies for Buildings – The State of the Art*. London, James & James.

Santamouris, M. and Dascalaki E. (1998). Chapter 3: Prediction methods. *Natural Ventilation in Buildings: A Design Handbook*. F. Allard (ed.). London, James & James (Science Publishers) Ltd., pp. 63–157.

Saraiva, J. G. and Marques da Silva F. (1999). *Atmospheric Turbulence Influence on Natural Ventilation Air Change Rates*. Indoor Air 99, Edinburgh, ISIAQ & AIVC.

Sawachi, T. (2002). *Detailed Observation of Cross Ventilation and Airflow Through Large Openings by Full Scale Building in Wind Tunnel*. RoomVent 2002 – 8th International Conference on Air Distribution in Rooms, Copenhagen, The Technical University of Denmark & Danvak.

Schaelin, A., Dorer, V., van der Maas J. and Moser A. (1993). 'Improvement of multizone model predictions by detailed flow path values from CFD calculations'. *ASHRAE Transactions* 99(Pt. 2): pp. 709–720.

Schälin, A. and Nielsen P. V. (2004). 'Impact of turbulence anisotropy near walls in room air flow. Indoor air'. *International Journal of Indoor Environment and Health* 14(No. 3), pp. 159–168.

Schild, P. (1997). *Accurate Prediction of Indoor Climate in Glazed Enclosures*. PhD Thesis, Department of Refrigeration and Air Conditioning, Norwegian University of Science and Technology (NTNU), Norway.

Seifert, J., Gritzki, R., Rösler M. and Grundmann R. (2000). *Contribution to Flows Through Large Openings*. 4th IEA-ECBCS-Annex 35 Expert Meeting, Athens, Greece, IEA-ECBCS-Annex 35.

Seifert, J., Li, Y., Axley J. and Rösler M. (2004). *The Effect of Wall Porosity on the Flow Rate in a Building Ventilated by Cross Wind*. RoomVent 2004 – 9th International Conference on Air Distribution in Rooms, Coimbra, Portugal, ADAI, Universidade de Coimbra.

Shaw, W. N. (1907). *Air Currents and the Laws of Ventilation*. Cambridge, University Press.

Sirén, K. (1997). *A Modification of the Power-Law Equation to Account for Large-Scale Wind Turbulence*. 18th AIVC Conference – Ventilation and Cooling, Athens, Greece, AIVC.

Skåret, E. (1986). *Ventilation by Displacement – Characterization and Design Implications*. Ventilation '85, H. D. Goodfellow (ed.), Elsevier Science Publishers B. V., Amsterdam.

Spalding, D. B. (1972). 'A novel finite-difference formulation for differential expressions involving both first and second derivatives'. *International Journal for Numerical Methods in Engineering* 4: p. 551.

Solomons, D. E. W. (1990). *BREEZE 5.0 User Manual*. London, Building Research Establishment.

Stewart, J. and Ren Z. (2003). 'Prediction of indoor gaseous pollutant dispersion by nesting sub-zones within a multizone model'. *Building and Environment* 38: pp. 635–643.

Sørensen, D. N. and Nielsen P. V. (2003). 'Quality control of computational fluid dynamics in indoor environments'. *International Journal of Indoor Environment and Health* 13(No. 1): pp. 2–17.

Thomas, D. A. and Dick J. B. (1953). 'Air infiltration through gaps around windows'. *Journal of the Institute of Heating and Ventilation Engineers* 21: pp 85–97.

Transsolar (2004). *TRNFLOW – A module of an air flow network for coupled simulation with TYPE 56 (multi-zone building of TRNSYS) – Version 1.1 – 19.02.2004*. Stuttgart, Transsolar Energietecnik.

True, J. J., Sandberg, M., Heiselberg P. and Nielsen P. V. (2003). 'Wind driven cross-flow analysed as a catchment problem and as a pressure driven flow'. *International Journal of Ventilation* 1(Hybrid Ventilatioin Special Edition): pp. 88–102.

van Leer, B. (1974). 'Towards the ultimate conservative difference scheme. Monotonicity and conservation combined in a second order scheme'. *Journal of Computational Physics* 14: pp. 361–370.

van der Maas, J., ed. (1992). *Air Flow Through Large Openings in Buildings*. Lausanne, Switzerland, Laboratoire d'Energie Solaire et de Physique du Batiment, Ecole Polytechnique Fédérale de Lausanne.

van der Maas, J., Hensen J. L. M. and Roos A. (1994). *Ventilation and Energy Flow Through Large Vertical Openings in Buildings*. 15th AIVC Conference – The Role of Ventilation, Buxton, Great Britain, AIVC.

Versteeg, H. K. and Malalasekera W. (1995). *An Introduction to Computational Fluid Dynamics*. Harlow, Essex, England, Longman Ltd.

Vuolle, M. and Sahlin P. (2000). *IDA Indoor Climate and Energy – A New-Generation Simulation Tool*. Healthy Building 2000, Espoo, Finland, SIT Indoor Air Information Oy, Helsinki, Finland.

Walker, I. S. and Wilson D. J. (1994). *Practical Methods for Improving Estimates of Natural Ventilation Rates*. 15th AIVC Conference – The Role of Ventilation, Buxton, Great Britain, AIVC.

Walton, G. N. (1982). 'Airflow and multiroom thermal analysis'. *ASHRAE Transactions* 88(Pt. 2): Paper No. 2704, 11 p.

Walton, G. N. (1984). *A Computer Algorithm for Predicting Infiltration and Inter-room Airflows*. ASHRAE Transactions, AT-84-11 No. 3, Atlanta, ASHRAE.

Walton, G. (1988). *AIRNET: A Computer Program for Building Airflow Network Modeling*, Gaithersburg, MD, National Institute of Standards and Technology.

Walton, G. (1989). *Airflow Network Models for Element-Based Building Airflow Modeling*. ASHRAE Symposium on Calculation of Interzonal Heat and Mass Transport in Buildings, Vancouver, B.C., ASHRAE.

Walton, G. N. (1994). *CONTAM93 User Manual*. National Institute of Standards and Technology Gaithersburg, MD, NIST.

Walton, G. (1997). *CONTAM96 User Manual*. Gaithersburg, MD, NIST.

Walton, G. (1998). *DRAFT: Notes on Simultaneous Heat and Mass Transfer* [CONTAM97]. Gaithersburg, NIST.

Wolfram_Research (2004). MATHEMATICA-5.

Woloszyn, M., Duta, A., Rusaouen G. and Hubert J. L. (2000). *Combined Moisture, Air and Heat Transport Model ing Methods for Integration in Building Simulation Codes*. Room Vent 2000 – 7th International Conference on Air Distribution in Rooms, University of Reading, UK, Elsevier.

Wood, D. J. and Funk J. E. (1993). Hydraulic analysis of water distribution systems. *Water Supply Systems, State of the Art and Future Trends*. E. Cabrera and F. Martinez (eds), Southampton, Boston, Computational Mechnaics Publications, pp. 43–68.

Wray, C. P. (1990). *Development and Application of a Multizone Airflow and Contaminant Dispersal Analysis Microcomputer Program for Simulating Radon and Radon Progeny Transport and Accumulation in Soils and Houses with Basements*. Department of Mechanical Engineering, University of Manitoba.

Wray, C. P. (1993). *Development of Usable Techniques for Designing, Commissioning, and Testing Smoke Control Systems*. Winnipeg, Manitoba, G.K. Yuill and Associates (B.C.) Ltd.

Wray, C. P. and Yuill G. K. (1993). 'An evaluation of algorithms for analyzing smoke control systems'. *ASHRAE Transactions* 99(Part 1): pp. 160–174.

Yaglou, C. P., Riley E. C. and Coggins D. I. (1936). 'Ventilation requirements'. *ASHVE Transactions* 42: pp. 133–162.

Yaglou, C. P. and Witheridge W. N. (1937). 'Ventilation requirements (Part 2)'. *ASHVE Transactions* 43: pp. 423–436.

Nomenclature for macroscopic methods

Latin symbols

A	area, surface or cross-sectional	m^2
A_L	equivalent leakage area	m^2
a_0, a_1, a_2	fan curve coefficients	–
C	thermal capacity, constant	JK^{-1}, varies
C_d	discharge coefficient, taking account for friction losses	–
C_p	wind pressure coefficient	–
c	specific heat capacity	$Jkg^{-1}K^{-1}$
D	diameter (or average size) of a duct	m
D_h	hydraulic diameter of a two-port control volume 2(area/wetted perimeter)	m
$\dot{E}_d$	rate of viscous energy dissipation	W
$F(\tau)$	probability function	–
$F_b()$	boundary condition function	Pa or ms^{-1}
$F_d(\Delta P, \vartheta)$	*forward* dissipation function for element flow	kgs^{-1}
$F_f()$	zone field assumption function	Pa
$F_m()$	mass balance function	kgs^{-1}
f	Fanning friction factor	–

$f(\tau)$	probability density function	s^{-1}
$G_d(\dot{m}, \vartheta)$	*inverse* dissipation fnction for element flow	W or Pa
g	acceleration of gravity (9.806)	ms^{-2}
H	heat loss coefficient	WK^{-1}
h	enthalpy	Jkg^{-1}
Δh	difference between liquid column heights	m
I	electric current	A
i, j, k, l, m, n	integer indices	–
K	constant or element flow resistance	– or $kg^{-1}m^{-1}$
K_{tot}^l	total kinetic energy within control volume l	J
K^P	flow resistance of a parallel flow circuit	$kg^{-1}m^{-1}$
K^S	flow resistance of a serial flow circuit	$kg^{-1}m^{-1}$
l	distance, length	m
M	average molecular weight of the air (0.02894)	$kgmol^{-1}$
m	mass, mass of air within a control volume	kg
$\dot{m}$	mass flow rate	kgs^{-1}
N	number, rotational speed	–, s^{-1}
n	air change rate, integer index	s^{-1} or h^{-1}, –
p	pressure	Pa
Δp	pressure differential	Pa
P	modified pressure $(p + \rho g z)$	Pa
Q	energy	J
R	universal gas constant (8.31696)	$Jmol^{-1}K^{-1}$
R^l	measured porous media resistance	$Pakg^{-1}s$
Re	Reynolds number	–
$\underline{r}$	vector locating a point in the space	m
T	absolute temperature	K
t	time	s
Δt	time interval	s
U	voltage	V
V	volume	m^{-3}
$\dot{V}$	volumetric flow rate	m^3s^{-1}
v	air velocity, wind speed	ms^{-1}
$\bar{v}_{ref}$	time-averaged reference wind speed	ms^{-1}
x	dummy variable	
z	vertical position with reference to the ground or a datum	m

Greek symbols

α	velocity distribution correction factor	–
β	pressure-work distribution correction factor	–
Φ	power	W
γ	gain/loss ratio	–
δy	error in the variable y	
ε_a	coefficient of air change performance $= 2\eta_a$	–
ε_c	local ventilation effectiveness or pollutant removal effectiveness	–
ε_d	air distribution effectiveness	–
ϑ^l	model parameters that establish the size of a flow component l	varies
η	efficiency	–
η_a	air change efficiency	–
Π_{tot}^l	total geopotential energy within control volume l	J
θ	temperature	°C
μ	viscosity	$kgs^{-1}m^{-1}$
ρ	density	kgm^{-3}
σ_x	standard deviation of the variable x	
τ	time constant	s
τ_a	air change time	s
τ_n	nominal time constant of the room	s
τ_r	local age of air at point $\underline{r}$	s
τ_r	mean age of air at location $\underline{r}$	s
Ω	angular velocity	s^{-1}
υ	air velocity resultant (speed)	ms^{-1}
ω	specific humidity	–
ζ	friction loss factor	–
$\langle x \rangle_t$	average of the variable x over time period t	

Embellishments, index and matrix notation

$\hat{x}$	surface average of x (e.g. averaged over port plane area)
$\bar{x}$	time-smoothed average of x
$\dot{x}$	time rate of change of x
x_i	value of x associated with port plane i (Chapter 4)
$x_{[k]}$	k^{th} iterative estimate of x
x^l	value of x associated with control volume l (Chapter 4)

x^* specified numerical value for variable x

$\{x\}$ vector $\mathbf{x}$ with n elements defined as:

   $\{\mathbf{x}\} = \{x_1, x_2, \ldots x_n\}^T$ for element indexed to n port plane numbers

   $\{\mathbf{x}\} = \{x^1, x^1, \ldots x^m\}^T$ for element indexed to m control volume numbers

$[X]$ matrix $\mathbf{X}$ with $n \times m$ elements defined as: $[\mathbf{X}] =$

$$\begin{bmatrix} X_{1,1} & X_{1,2} & \cdots & X_{1,n} \\ X_{2,1} & X_{2,2} & \cdots & X_{2,n} \\ \vdots & \vdots & \ddots & \vdots \\ X_{m,1} & X_{m,2} & \cdots & X_{m,n} \end{bmatrix}$$

Chapter 5

Air distribution
system design

Peter V. Nielsen and Hazim B. Awbi

5.1 Introduction

A building requires supply of fresh air and removal of heat, gases and particulates that are emitted in the building. Heat is generated by persons in the building, different types of equipment (PCs, display screens, lighting, etc.) and by solar radiation into the room. Vapour, gases and particulates are also generated inside the building and emitted to the room air. Moisture is generated by people and drying processes, and various gases and volatile organic compounds (VOC) are emitted from building materials, furniture and office equipment.

It is not sufficient to supply clean air to an individual room in a building with the correct temperature and flow rate, but it is also necessary to design an air distribution system in the room in such a way that occupants experience high air quality and thermal comfort.

Chapters 1 and 4 introduced different flow elements as free jet, wall jet, thermal plumes, etc. They are all elements of the air distribution in the room, and they form the basis for different design procedures. This chapter describes the design procedures for air distribution systems that involve the calculation of velocity and temperature gradients in the occupied zone of a room. The design procedure is given for different types of systems: mixing ventilation with an end-wall-mounted supply opening; mixing ventilation with a ceiling-mounted radial diffuser and a ceiling-mounted diffuser with swirl. Design methods for displacement ventilation with wall-mounted low velocity diffusers or floor-mounted diffusers are also given. In addition, newly developed high momentum displacement ventilations systems are introduced giving comparisons with standard displacement systems. The chapter is closed by a discussion of the design procedure for vertical ventilation from ceiling mounted diffusers with low momentum flow as for example textile terminals.

5.2 The room and the occupied zone

The occupied zone is that part of the room where people are situated. Therefore, it is important that the occupied zone has the optimum climate with respect to values of air temperature, air velocity, temperature and velocity gradients, mean radiant temperature and asymmetric radiant temperature. It is also important that the supply air reaches all parts of the occupied zone without the presence of stagnant zones. Areas of the room outside the occupied zone may have higher velocity levels and higher or lower temperatures than the occupied zone without compromising comfort. This area is often used for the location of air supply units as these generate draught in the vicinity of their openings. Table 5.1 and Figure 5.1 show some typical boundaries of the occupied zone in a ventilated room.

The distance from a diffuser located in the lower part of the room to the occupied zone could be larger than those indicated in Table 5.1 for other elements, and this distance is considered in the design procedure. Diffusers should be located in walking areas (e.g. corridors) outside the occupied zone, i.e. outside areas of sedentary activities.

It is well known that the highest velocity in a room with displacement ventilation is very close to the floor. This is the reason for including the floor within the occupied zone. New measurements of the velocity levels in a room with other air distribution systems also show a high velocity close to the floor, so in all situations, it is recommended to define the occupied zone down to the floor.

An important element of the design procedure for an air distribution system is to ensure an acceptable low air velocity (typically $0.2\,\mathrm{ms^{-1}}$) where the flow passes the imaginary surfaces defining the occupied zone. The design value at the upper horizontal surface of the occupied zone can have higher

Table 5.1 Elements in a room and distance from the elements to the boundary of the occupied zone

Element	Distance from the inner surface of the elements [m]	
	Typical range	Default value (CR 12792)
External windows, doors and radiators	0.5–1.5	1.0
External and internal walls	0.25–0.75	0.5
Floor (lower boundary)	0–0.2	0[c]
Floor (upper boundary)	1.30[a]–2.0[b]	1.8
Wall mounted diffuser	Dependent on the dimensions of diffuser	–

[a]Mainly seated occupants.
[b]Mainly standing occupants.
[c]Deviation from CR 12792.

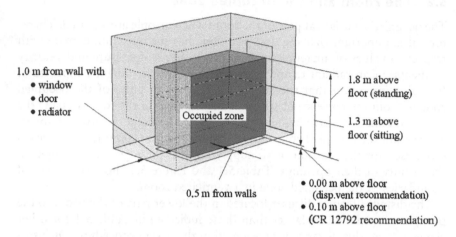

1.0 m from wall with
- window
- door
- radiator

Occupied zone

1.8 m above
floor (standing)

1.3 m above
floor (sitting)

0.5 m from walls

- 0.00 m above floor
  (disp.vent recommendation)
- 0.10 m above floor
  (CR 12792 recommendation)

Figure 5.1 The occupied zone in a room.

values, depending on the type of air distribution system used. The thermal plume above a person has an upward velocity of approximately $0.25\,\mathrm{ms}^{-1}$, and this flow often prevents draught at head height. Figure 5.2 shows the thermal boundary layer around a person illustrated by a smoke release from the surface of the manikin for different downward flow velocities. The boundary layer is preserved up to a downward velocity of $0.25\,\mathrm{ms}^{-1}$.

A downward jet of limited cross-section does not normally create draught at head height, but it penetrates the occupied zone between the plumes and creates draught at floor level. A restriction of the velocity at ankle level can be associated with a certain maximum velocity at the upper imaginary surface of the occupied zone for a given air distribution system as discussed in Section 5.8.

0.0   0.15   0.20   0.25   0.30 m/s

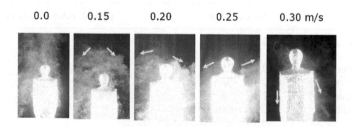

Figure 5.2 A manikin located in a room with downward air flow. The boundary layer around the manikin is preserved at head height up to a downward velocity of $0.25\,\mathrm{ms}^{-1}$.

5.3 The q_o–ΔT_o relationship for the design of air distribution systems in rooms

Figure 5.3 describes the idea behind a design chart for air distribution in rooms. The chart is based on the minimum and maximum allowable flow rate (q_o) to the room and also on the maximum temperature difference between return and supply (ΔT_o). It is expressed graphically as q_o–ΔT_o. The figure indicates that it is necessary to have a minimum flow rate of fresh air into the room to achieve a given air-quality level. It is also expected that there must be a limit for the maximum flow rate to avoid draught in the room as well as a maximum flow rate for the design of the system. It is common for air distribution systems based on the supply of higher momentum flow that draught is present when the flow rate is above a certain value. These systems also generate a mixing in the occupied zone, which is important for creating uniform conditions in the room when the heat load is high. The temperature difference ΔT_o between return and supply is also restricted. A too high temperature difference may either cause draught in the occupied zone or create a large temperature gradient in the room. Furthermore, there is a restriction on the temperature difference ΔT_o with regard to the design of energy efficient cooling systems.

Figure 5.3 indicates an area for q_o and ΔT_o, which supplies sufficiently fresh air and ensures a draught free air movement in the occupied zone plus a limited vertical temperature gradient (white area in the figure). The limits around the area in the design chart can be defined for a maximum air velocity in the occupied zone of $0.15\,\text{ms}^{-1}$, a maximum vertical temperature gradient of $2.5\,\text{Km}^{-1}$ and a minimum flow rate of $10\,\text{ls}^{-1}$ per person, but other reference values can also be selected.

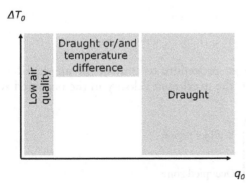

Figure 5.3 Design chart showing the limits on the flow rate q_0 and on the temperature difference ΔT_0 between return and supply owing to draught, temperature gradient and air quality.

The design chart can also be given as a u_o–ΔT_o chart, where u_o is the supply velocity of the diffuser in the room. The design chart was introduced by Nielsen (1980).

5.4   Design of mixing ventilation

Figure 5.4 shows a typical situation in a room ventilated by mixing ventilation system. A wall jet is generated in the upper part of the room outside the occupied zone. The jet penetrates the occupied zone and produces a mixing flow with the effect that heat and contaminant are diluted in the room air and then removed by the exhaust air flow.

Design methods are often based on flow elements, see Section 4.2.1. Two flow elements are considered in the wall jet region. One flow element is the isothermal flow from the diffuser. This element of the wall jet determines the maximum velocity u_{rm} that exists in the occupied zone. Another flow element is the non-isothermal wall jet, which determines the penetration length (x_s) of the flow into the occupied zone. A design method can also be based on the value of velocity in the occupied zone (u_{ocz}) which can be tolerated as the cold jet crosses the upper boundary of the occupied zone, or on the velocity at head height (u_{11}) or ankle height (u_{01}) of persons in the room (see Nielsen *et al.*, 2001).

5.4.1   Mixing ventilation with end-wall-mounted diffuser

A number of design methods will be discussed in this section. The first method is based on the *throw of the isothermal jet.* A length l_{Th}, called the throw, is defined as the distance from the opening to the location where the maximum velocity is equal to a given reference value u_{Th}. The following expression can be obtained from Equation (4.3):

$$l_{Th} = \frac{u_o K_a \sqrt{a_o}}{u_{Th}} - x_o \,(\text{m}) \tag{5.1}$$

It is the purpose of the design procedure to control the air distribution in the room in such a way that the maximum velocity in the occupied zone

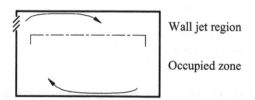

Figure 5.4 A room showing the wall jet region and the occupied zone.

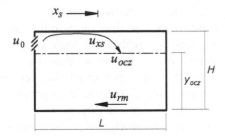

Figure 5.5 Penetration length of the wall jet and definitions used in a design method based on a maximum velocity in the occupied zone and a design method based on the penetration length of a cold jet as well as a design method based on a permitted velocity entering in the occupied zone.

$u_{rm} \leq 0.15\,\mathrm{ms}^{-1}$. Experience shows that this is achieved when the throw l_{Th} is equal to the room length, L, and the reference velocity u_{Th} is equal to 0.2 or $0.25\,\mathrm{ms}^{-1}$.

Another method is based on the *maximum velocity in the occupied zone*. The maximum velocity in the reverse flow u_{rm} is located close to the floor at a distance of $\sim 2/3\,L$ from the supply opening (see Figure 5.5) where L is the length of the room. The velocity u_L is the velocity in an undisturbed wall jet at a distance L from the actual diffuser. u_L contains information on the supply velocity, distance from inlet and geometrical details around the initial flow, for example type of diffuser, adjustable blades and distance from the ceiling. Experiments with isothermal flow show that u_{rm} is a simple function of the velocity u_L, and the ratio u_{rm}/u_L and can therefore be used as an element in a design procedure.

The maximum velocity in the occupied zone, u_{rm}, can be obtained from Equation (4.3) when the variation of u_{rm}/u_L is known for the given diffuser and geometry.

$$u_{rm} = u_o \left(\frac{u_{rm}}{u_L} \right) k_a \frac{\sqrt{a_o}}{L + x_o}\,(\mathrm{ms}^{-1}) \tag{5.2}$$

Two design methods consider a non-isothermal flow. The first method is related to the *penetration length* of a non-isothermal wall jet. Cases with a short penetration length are undesirable because the jet will have a high velocity and a low temperature as it enters the occupied zone. An acceptable length is dependent on the type of supply opening, but it is often recommended that the penetration length x_s/L be equal to or larger than 0.5. A restriction on the minimum penetration length to a given value has the consequence that the specific heat load of the room is restricted and related to the design variable as $Q \sim u_{rm}^3/K_a$ (see Nielsen, 1991).

It is desirable to have an air distribution system that can handle a high-heat load. Equation (5.3) shows that this is possible if a high maximum velocity in the occupied zone is tolerated, and it also shows that the ability to handle heat load is proportional to the third power of this velocity. Therefore, it is important that the design procedure is able to produce a system that gives a velocity u_{rm} close to the design velocity.

The K_a-value is another important design parameter when only one diffuser is used in the room. Equation (5.3) shows that it is efficient to use an air terminal devise with a low K_a-value. A low K_a-value corresponds to a high initial diffusion, and it is partly achieved by a semi-radial or radial flow in the wall jet below the ceiling.

Another non-isothermal design method is based on the principle that the jet should penetrate the upper boundary of the occupied zone with a restricted velocity u_{ocz}. The method is based on the value of *permitted velocity entering the occupied zone*. An expression can be estimated from the wall jet velocity decay equation, Equation (4.3), where the length x is equal to $x_s + x_o + H - y_{ocz}$. The length x_s is the distance from the diffuser to the location where the jet drops into the occupied zone, and $H - y_{ocz}$ is the distance from the ceiling to the upper part of the occupied zone, i.e.:

$$u_{ocz} \sim u_o \frac{K_a \sqrt{a_o}}{x_s + x_o + H - y_{ocz}} (\mathrm{ms^{-1}}) \tag{5.3}$$

The design conditions are represented by the values of u_o and ΔT_o that restrict u_{ocz} to a given value.

A more direct expression can be found from equations of the form:

$$u_{ocz} = u_o \cdot f(Ar) \, (\mathrm{ms^{-1}}) \tag{5.4}$$

$$u_{01} = u_o \cdot f(Ar) \, (\mathrm{ms^{-1}}) \tag{5.5}$$

$$u_{11} = u_o \cdot f(Ar) \, (\mathrm{ms^{-1}}) \tag{5.6}$$

$$u_{18} = u_o \cdot f(Ar) \, (\mathrm{ms^{-1}}) \tag{5.7}$$

where u_{ocz} is the actual velocity found by measurements at the location where the jet penetrates the room at y_{ocz} (see Figure 5.5), and u_{01}, u_{11} and u_{18} are the air velocities at the height of 0.1, 1.1 and 1.8 m in the occupied zone close to persons, respectively. The use of Equations (5.5–5.7) represents a design that is based on the *maximum velocity in the occupied zone for non-isothermal flow*, and the reference velocity could typically be 0.15 ms⁻¹. The maximum supply velocity obtained from Equations (5.5–5.7) could be expressed as a maximum u_{ocz} design velocity (see Section 5.8).

The design of air distribution system in a room with an end-wall-mounted diffuser is now discussed. The room that will be considered is a typical small office. The room was the standard room for the International Energy Agency Annex 20 project (see Nielsen *et al.*, 2001), and it has the dimensions length, width and height equal to 4.2, 3.6 and 2.5 m, respectively.

Figure 5.6 shows the heat load location in the room. The heat load consists of two PCs, two lamps and two manikins giving a total load of 460 W. Figure 5.7 shows the 'Annex 20' diffuser for mixing ventilation mounted on the end wall.

Figure 5.8 shows the design chart that is based on the temperature differences ΔT_o between return and supply and the air flow rate q_o to the room. The curves show the combination of ΔT_o and q_o that encloses an area that fulfils thermal comfort requirements. A maximum velocity u_{rm} of 0.2 ms^{-1} restricts the flow rate to the room to 0.036 m^{3}s^{-1}, and the requirement for a penetration length of $x_x/L = 0.5$ is expressed by the dotted curve in

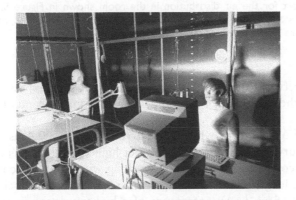

Figure 5.6 Office room with two persons (manikins).

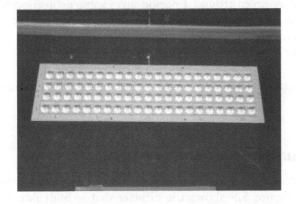

Figure 5.7 Diffuser located on the end wall 0.2 m below the ceiling.

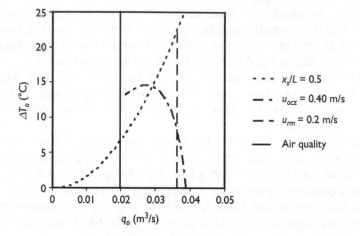

Figure 5.8 Design chart for the air distribution in the room shown in Figure 5.6. The curves show different limitations for the velocity levels and the temperature gradient, which ensure thermal comfort in the room.

Figure 5.8. The area between the curve and the vertical line thus expresses an area that permits the variation of ΔT_o and q_o in a room with mixing ventilation (see Nielsen *et al.*, 2003).

The combination of ΔT_o and q_o, which gives a velocity $u_{ocz} = 0.4\,\mathrm{ms^{-1}}$, is shown by another curve in Figure 5.8. The rather high velocity seems to give the same level of comfort as obtained by working with flow elements for the jet penetration length, and it corresponds to a maximum velocity at ankle level u_{01} of $0.15\,\mathrm{ms^{-1}}$. An acceptance of a higher velocity in the vertical downward direction is discussed in Section 5.2, where it is shown that the thermal boundary layer around a person gives some protection for downward directed flow. The acceptance of higher velocity is also confirmed by Toftum *et al.* (1997).

Finally, the air quality requires a certain flow rate for example $0.02\,\mathrm{m^3\,s^{-1}}$ as shown in Figure 5.8.

All the results are of course dependent on the diffuser design room dimensions and room load.

5.4.2 *Mixing ventilation with radial ceiling diffuser*

A dimensioning procedure with a radial ceiling diffuser is discussed in the following paragraphs. Figure 5.9 shows the diffuser that is built into the ceiling with the radial opening located flush with the ceiling surface. The velocity decay is described by Equation (4.5).

Figure 5.9 Radial ceiling diffuser.

Figure 5.10 shows the airflow pattern in a test room with a radial ceiling diffuser. The airflow rate is $3\,h^{-1}$ in the upper figure and up to $10\,h^{-1}$ in the lowest figure. The load is in all cases equal to 480 W. The middle figure shows the situation when the air change rate is between 5 and $6\,h^{-1}$. The jet does not usually flow directly down into a person's head region. The boundary layer flow around people prevents draught at head height, and draught at the floor will often be a determining design parameter, see Section 5.2 for the discussion of the influence of a person's thermal boundary layer.

Figure 5.11 shows the design chart for a radial ceiling diffuser based on the *maximum velocity in the occupied zone for non-isothermal flow* (Section 5.4.1). Experiments with the radial diffuser in Figure 5.9 show that $u_{01} = 0.15\,ms^{-1}$ corresponds to a u_{ocz} value close to $0.3\,ms^{-1}$ and a maximum ΔT_o value of 10 K. It can thus be concluded that $u_{ocz} = 0.3\,ms^{-1}$ combined with $\Delta T_o = 10\,K$ should be the maximum velocity entering the occupied zone and the maximum temperature difference for a design with a radial diffuser, see Nielsen *et al.* (2006). The results are to some extent dependent on diffuser design, size of the room, and the load.

5.4.3 Mixing ventilation with ceiling swirl diffuser

Figure 5.12 shows a typical ceiling swirl diffuser. It is built into the ceiling with its surface flush with the ceiling surface.

Figure 5.13 shows the air distribution generated by a typical ceiling swirl diffuser. The general air distribution in the room is largely uninfluenced by the airflow rate. The supply flow moves in a thick layer (0.5 m) with swirl towards the side wall and flows into the occupied zone along the walls. The air distribution shown in the figure is typical of air change rates between 3 and $9\,h^{-1}$.

The flow from a diffuser with swirl can be described by Equation (4.2) if the air moves down into the room, and described by Equation (4.5) if the air

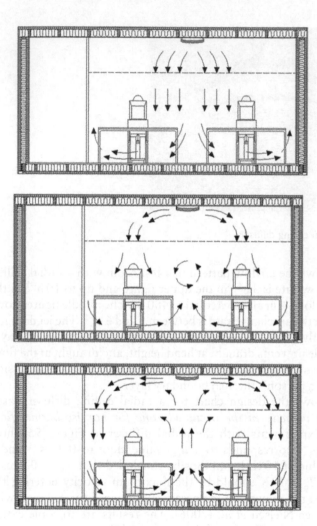

Figure 5.10 Air distribution in a room with a radial ceiling diffuser. All three situations cover different air flow rates and a heat load of 480 W in the "Annex 20" room shown in Figure 5.6.

moves parallel to the ceiling. Flow elements such as a free jet (Equation 4.2) and a wall jet (Equation 4.5) are not extensively described in the literature when they have a high content of swirl, and the dimensioning of the system is in the case of Annex 20 room is based on the *maximum velocity in the occupied zone for non-isothermal flow* (see Section 5.4.1).

Figure 5.14 shows the design chart for the ceiling swirl diffuser. The experiments with this diffuser show that $u_{01} = 0.15\,\mathrm{ms}^{-1}$ corresponding to

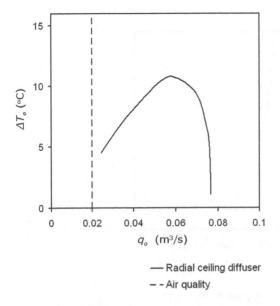

Figure 5.11 Design chart for an air distribution system with a radial ceiling diffuser.

Figure 5.12 Ceiling swirl diffuser and a smoke visualization of the flow from the diffuser.

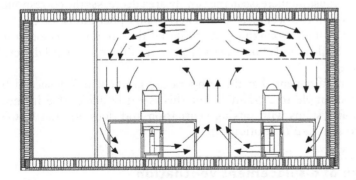

Figure 5.13 Air distribution in a room with a ceiling swirl diffuser. The general air flow pattern is uninfluenced by the air flow rate.

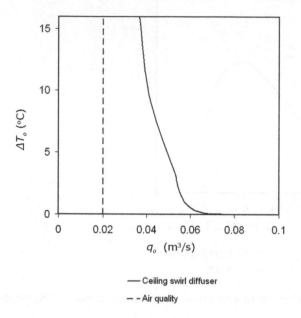

Figure 5.14 Design chart for an air distribution system with ceiling swirl diffuser.

a u_{ocz} value in the range of 0.3–0.4 ms^{-1}. It can thus be concluded that $u_{ocz} = 0.35$ ms^{-1} should be the maximum velocity entering the occupied zone for a design with ceiling mounted diffuser with swirl when the air distribution is optimized for the Annex 20 room shown in Figure 5.6. The results are to some extent dependent on the diffuser design, the size of the room and the load.

The design area for the diffuser with swirling flow corresponds to lower airflow rates than the design area for the radial diffuser. A swirl diffuser with large supply openings could, if necessary, move the design area towards large airflow rates because an increase in the supply area decreases the corresponding velocity level in the occupied zone. There is no upper limits for ΔT_o for this diffuser. The high mixing rate in the initial flow removes temperature differences in the flow and eliminates a large part of the downward directed buoyancy force.

It is also a requirement that the temperature gradient dT/dy should be restricted for example to 2.5 Km^{-1}, but this is easily achievable in most air distribution systems with mixing ventilation and does not add further restrictions to the q_o–ΔT_o relationship.

5.5 Design of displacement ventilation

For many years, buoyancy-driven displacement ventilation has been used in industrial areas with high thermal load as for example in hot process buildings in the steel industry. The displacement ventilation system has also grown popular during the last 20 years as comfort ventilation in rooms with low thermal loads, for example in offices. The supply openings are located at a low level in case of displacement ventilation and the air flows direct into the occupied zone. Free convection from heat sources creates a vertical air movement in the room, and the heated air is removed by return openings located in the ceiling or just below the ceiling. It is thus the free convection or the buoyancy that controls the flow in the room while the momentum flux from the diffusers is low and without any practical importance for the general flow in the room.

The vertical temperature gradient is characteristic of a room with displacement ventilation. Heat from heat sources is supplied to the room as free convection and radiation. Free convection increases the ceiling temperature compared with the surroundings, and radiation from the ceiling increases the temperature of the floor that is also cooled by the cold supply flow from the low-level diffuser. The total effect is a vertical temperature gradient, which is rather similar at different locations because of the stratification in the room. It is therefore possible to remove exhaust air from the room with temperatures that are several degrees above the temperature in the occupied zone. This allows an efficient use of energy because the supply temperature can be higher than the supply temperature used by mixing ventilation. It is, on the other hand, necessary to use a higher airflow rate in displacement ventilation to obtain a large volume of fresh air in the occupied zone and to avoid too low temperatures in this zone. This leads to a duct system that is slightly larger than the duct system used in mixing ventilation.

Displacement ventilation is only used for cooling. It can also be used as a fresh air supply system combined with a radiator heating system in rooms that require heating.

A displacement ventilation system can have high ventilation effectiveness if the sources are both heat and contaminant sources. The vertical temperature gradient implies that both fresh air and contaminated air are separated and that the most contaminated air can be found above the occupied zone.

Figure 5.15 shows the main principle of displacement ventilation. The airflow q_o is supplied into the occupied zone at low velocity from a wall-mounted diffuser. The plumes from heat sources, such as equipment and persons, entrain air from the surroundings in an upward movement, and cold downdraught transports air down into the occupied zone. A stratification zone is located over a height where the flow q_2-q_3 is equal to q_o in the situation shown in Figure 5.15.

The diffuser can either be a wall-mounted low velocity diffuser or floor-mounted diffusers. The flow from a diffuser moves downwards to the floor because of gravity, and it flows across the room in a thin layer, which is typical of stratified flow. The flow along the floor is the only air movement that influences the comfort of the occupants, and it is therefore important

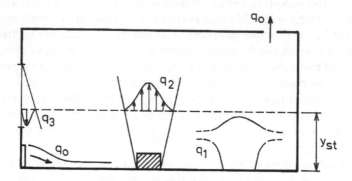

Figure 5.15 Room with displacement flow and natural convection.

to have a design procedure that can support the selection of diffusers. The velocity distribution close to the floor is given by Equation (4.11).

Floor-mounted diffusers generate a high entrainment close to the openings. They should therefore mainly be distributed in secondary areas of the occupied zone.

The design of a displacement ventilation system involves the calculation of the vertical flow in plumes from different heat sources (see Equations 4.8 and 4.10). The stratification height y_{st} is determined from the supply flow rate and heat sources, and the height of fresh air generated should ideally be comparable with the height of the occupied zone. It is possible to estimate the vertical temperature gradient in the room. This gradient is important for thermal comfort because of both the temperature difference between head and ankle height and to asymmetric radiation from the ceiling.

Thermal flows from heat sources and cold surfaces can be treated independently of other flows in the room. Such flow is, for example, independent of the stratified flow along the floor, and it is often supposed to be independent of the vertical temperature gradient. A practical design procedure for displacement ventilation is based on this simplified theory.

The stratification height is unimportant in cases with low contaminant emission in the room, and the flow rate can, in this situation, be given from considerations of the temperature gradient and the draught at the floor. When the principle of a clean low zone is to be applied, a stratification height of minimum 1.1 m can be recommended in a room where the occupants mainly have a sedentary working position. Furthermore, it is possible to work with a stratification height, which is lower than the breathing zone. The free convection around a person generates a flow of clean air in the breathing zone, although the stratification height is smaller than the height of the breathing zone in undisturbed surroundings (see for example Brohus and Nielsen, 1996; Xing et al., 2001; Skistad et al., 2002). Figure 5.16 shows measurements of this effect made by Stymne et al. (1991).

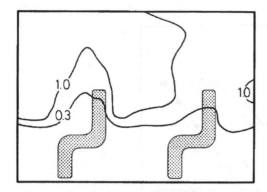

Figure 5.16 Normalized concentration distribution c/c_R around two persons. The ventilation rate is 0.011 m³s⁻¹ per person. The source of tracer gas is located in the upper zone close to the ceiling.

The air movement in the upper zone is often well mixed, which means that the normalized concentration c/c_R, where c_R is the concentration in the return opening, is close to 1.0. A normalized concentration distribution of 1.0 is the concentration distribution that can be obtained by a mixing air distribution system, and it can therefore be concluded that even cases with concentration of 1.0 in the breathing zone should be acceptable in most situations.

The vertical temperature gradient is a characteristic of a room with displacement ventilation. The temperature gradient can best be given in a dimensionless form. Figure 5.17 shows that the variations between non-dimensional gradients are relatively small compared with the variations that are found for gradients in a dimensioned form. T_o and T_R are the supply and return temperature, respectively.

A design of a displacement ventilation system involves *inter alia* the prediction of vertical temperature gradient. If this gradient is known, it is possible to calculate the percentage of dissatisfied because of temperature gradient and asymmetric radiation from the ceiling. Measurements indicate that it is possible to make the simplified assumption that the temperature varies linearly with the height from a minimum temperature at floor level T_f to a maximum temperature at ceiling level. The ceiling level temperature is equivalent to the return temperature T_R.

$$T = \frac{y}{H}\left(T_R - T_f\right) + T_f \tag{5.8}$$

Comparison with Figure 5.17 shows that the minimum temperature at floor level varies little when it is given in a dimensionless form, which is

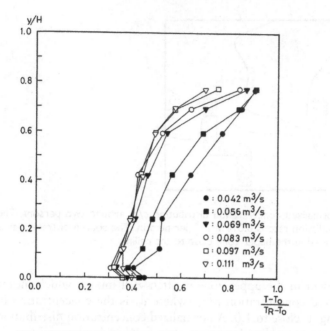

Figure 5.17 Vertical temperature distribution for different air flow rates (Nielsen *et al.*, 1988).

suitable for application in a design method based on a floor temperature $(T_f - T_o)/(T_R - T_o)$ of 0.5, called the '50 % rule', Skistad *et al.* (2002).

The dimensionless air temperature near the floor $(T_f - T_o)/(T_R - T_o)$ is to some extent a function of the type of heat source in the room. A concentrated heat load, for example a small furnace in an industrial environment, can have a value of 0.3. Ceiling lighting gives a vertical temperature gradient with a floor temperature of 0.5, which is due to radiation from the light source. When persons are the primary heat source, the dimensionless floor temperature is about 0.58, and for an evenly distributed heat sources, it is about 0.65. These different temperatures are shown in Figure 5.18. If many different heat sources are present in the room, it is recommended that a dimensionless floor temperature of 0.5 is used.

5.5.1 Displacement ventilation with wall-mounted diffuser

Displacement ventilation has a high air velocity in the stratified flow at the floor. A comfortable velocity in the occupied zone is therefore obtained by restricting the velocity in the stratified flow at the entrance to the occupied zone in front of the diffuser. This is expressed by the length of the adjacent zone l_n, which is the distance from the diffuser to a given value of velocity

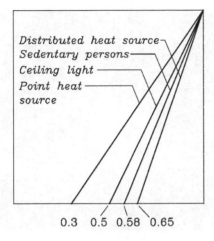

Figure 5.18 Vertical temperature distribution for different types of heat loads (Nielsen, 1996).

in the stratified flow (see Figure 5.19). The vertical temperature gradient is also important in displacement ventilation, and it should be restricted to a certain value. The flow from a wall-mounted diffuser for displacement ventilation is given by Equation (4.11), and it is further described by Nielsen (2000) and Skistad *et al.* (2002), while the vertical temperature gradient can be predicted from Equation (5.8).

Figure 5.20 shows a typical wall-mounted low velocity diffuser used for displacement ventilation. The smoke visualization shows how the air flow from the diffuser accelerates down to floor level because of gravity effect and moves in a radial pattern along the floor.

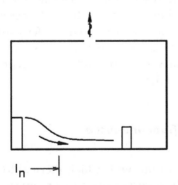

Figure 5.19 Room with displacement ventilation and wall mounted diffuser.

Figure 5.20 Wall mounted low velocity diffuser and smoke visualization of the stratified flow from the diffuser.

The design chart for the displacement ventilation case shown in Figure 5.21 is based on an entering velocity into the occupied zone of $0.2\,\mathrm{ms}^{-1}$ at a distance of 1 m from the diffuser. It is further based on a maximum vertical temperature gradient of $2.5\,\mathrm{K\,m}^{-1}$. The minimum airflow rate is $0.02\,\mathrm{m}^3\mathrm{s}^{-1}$ (air quality considerations) corresponding to the value used in case of full mixing in the room. This means that a slightly better air quality is obtained when the persons in the office are in sedentary positions without large movements because it is possible to utilize the boundary layer flow around the persons to achieve improved air quality of the inhaled air. The results in Figure 5.21 are to some extent dependent on the diffuser design (Figure 5.20), the size of the room and the load.

It became clear from this work that the results for displacement ventilation are very dependent on an efficient low velocity diffuser. Generally, Figures 5.8 and 5.11 indicate that constant air volume systems (CAV) work well for mixing ventilation, whereas variable air volume systems (VAV) are easy to control in the case of displacement ventilation and mixing ventilation with a swirl diffuser (Figures 5.21 and 5.14).

5.5.2 Displacement ventilation with floor-mounted diffusers

Floor-mounted diffusers are often used in rooms with a high thermal load. The supply area is small, and a high supply velocity generates a momentum flow in a vertical upward direction $(2 \sim 4\,\mathrm{ms}^{-1})$.

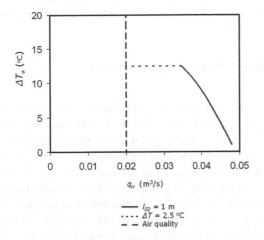

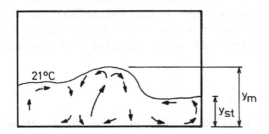

Figure 5.21 Design chart for an air distribution system with wall mounted diffusers for displacement ventilation.

Figure 5.22 Flow in a room with displacement ventilation and a floor mounted diffuser (Fitzner, 1988).

The velocity decay in vertical direction can be obtained from Equation (4.3), both when it is a free jet and when it is a jet with swirl.

The 21°C isotherm in Figure 5.22 indicates the boundary between the lower and upper zones in a room with a floor-mounted supply opening. The recirculation around the cold jet generates a secondary flow in the lower part of the room with a movement towards the jet in a layer corresponding to the stratification height. The stratification height y_{st} is calculated for the room as discussed in Section 5.5 and will not be influenced by the jet from the floor-mounted diffuser. The penetration height of the jet y_m is given by the following equation:

$$\frac{y_m}{\sqrt{a_o}} = 3.33 \left[\frac{K_a u_o^2}{(T_{oc} - T_o) \sqrt{a_o}} \right]^{0.5} \tag{5.9}$$

where the supply temperature T_o is lower than the temperature in the occupied zone T_{oc} (Helander *et al.*, 1953). y_m is the maximum penetration of the jet.

5.6 Design of high momentum displacement ventilation

The displacement ventilation systems described in Section 5.5 operate on the principle of low momentum jets and high buoyancy from heat sources at low level in the room. Although these are very popular systems, particularly in Europe, because of their having higher ventilation effectiveness than mixing systems and also more energy efficient because of the inherent large vertical temperature gradient in the room, they are not free from shortcomings. The most obvious of these are the inability of displacement systems to be used for heating, and also, they are not effective in delivering fresh air long distances from the supply diffuser as the fresh supply air is consumed by the heat sources present before it is able to reach far places in the room.

New methods of air distribution that are based on high momentum jets and upward displacement in the room have recently been developed. The most promising of these are the impinging jet and the confluent jet ventilation systems. The principle of these two air distribution methods, their performance and comparison with both mixing and standard displacement systems will be described here.

5.6.1 Impinging jet ventilation

This method is based on the principle of supplying a jet of air with high momentum downwards onto the floor (see Figure 5.23). As the jet impinges onto the floor, it spreads over a large area causing the jet momentum to recede but still has a sufficient force to overcome the heat sources present in its path and reach long distances. Unlike the displacement system that 'floods' the floor with supply air, the flow from an impinging jet ventilation system (IJV) is in the form of a very thin layer of air over the floor.

The impingement of a turbulent jet on a flat plate has many applications and has been widely studied. The flow characteristics of the impinging jet and its application in ventilation is described by Karimipanah (1996) and Karimipanah and Awbi (2002). The maximum velocity u_r in the radial direction on the floor at a distance x from the centreline of a vertical impinging jet is given by (Karimipanah and Awbi, 2002):

$$u_r = K_v u_o \left(\frac{x}{\sqrt{a_o}} \right)^{-1.10} \tag{5.10}$$

where u_o is the supply velocity, $K_v = 2.45$ is the throw constant and a_o is the supply outlet area. In an ideal situation, the exponent in Equation (5.10)

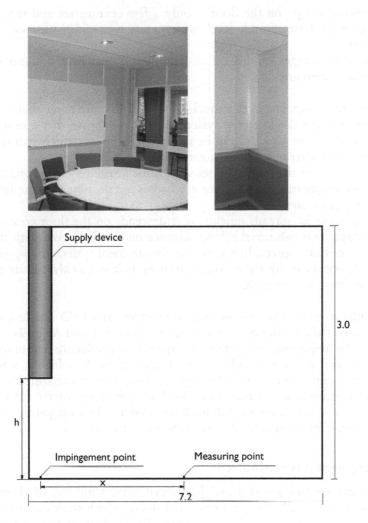

Figure 5.23 Impinging jet ventilation system.

should be −1.0 and the lower value given in the equation is due to loss of momentum caused by frictional losses, i.e. the momentum cannot be conserved.

The height of the supply diffuser above the floor, h in Figure 5.23, has little influence on the maximum velocity on the floor for $h = 300–950\,\text{mm}$ (Karimipanah and Awbi, 2002). Hence, Equation (5.10) is applicable to this range of h, which is most commonly used in practice. The equation can be used by the designer to predict the maximum velocity close to the floor to avoid draft at foot/ankle level in the room. Experiments show that the thickness of the jet on the floor is only a few centimetres and that is unlikely to cause draft at ankle level, which must be avoided for comfort considerations.

The main advantages of this air distribution system over a standard displacement system are:

- the air can be supplied over a much wider temperature range than can be tolerated in a displacement system. Tests have shown that a supply temperature down to 15 °C can be tolerated as the jet quickly entrains room air at higher temperature as it leaves the diffuser.
- the ability to use the impinging jet system for heating by supplying air at higher temperature than room air, which is not possible in a standard displacement system.
- the impinging jet spreads quickly as it descends on the floor creating a thin layer that can travel a long distance on the floor even with the presence of heat sources. In a standard displacement system, the supply air does not normally travel long distances as it is quickly consumed by heat sources in its path.

The results from tests in environmental chambers and CFD simulations (Karimipanah et al., 2000; Cho et al., 2002; Karimipanah and Awbi, 2002) show that the impinging jet system is capable of producing acceptable indoor environment over a wide range of cooling loads, which a standard displacement system cannot although a mixing system can with lower ventilation effectiveness. Furthermore, the local mean age of air for the impinging jet and displacement systems have similar values suggesting that the ventilation effectiveness for the two systems are similar.

5.6.2 Confluent jets ventilation

Confluent jets are free jets that are discharged from a number of closely spaced openings (usually from a perforated duct), which coalesce further downstream to form a single jet (see Awbi, 2003). A jet that is formed in this way produces a lower diffusion than a single jet whilst entrains room air. Confluent jets can be supplied at low-level in a room to produce displacement flow over the floor, see Figure 5.24. Before the single jet is supplied to the floor, it is usually diffused over a large surface, such as a wall that then descends to the floor as single thin wall jet. The spreading rates and other characteristics of the wall confluent jets can be found in Cho et al. (2004, 2007).

The maximum velocity in the wall confluent jets region (close to the floor), u_r, can be expressed as (Cho et al., 2007):

$$u_r = \frac{K_v u_o}{l_c} \qquad (5.11)$$

(a) Typical confluent jets installations

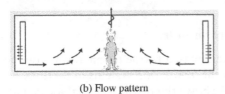

(b) Flow pattern

Figure 5.24 The confluent jets ventilation system (courtesy of Fresh AB and Dr. T. Karimipanah).

where u_o is the nozzle outlet velocity, $l_c = \sqrt{(r/h)}$ is the characteristic length, r is the distance from the diffuser, h is the diameter of the jet nozzle, and K_v is the throw constant of the jet that is for the floor region $= 5.2$. This value of K_v compared with 3.5 for a plane wall jet (Rajaratnam, 1976) suggests that the decay of the confluent jets is slower than a plane wall jet, i.e. the momentum is more conserved.

The performance of the confluent jet system has been compared with that of a standard displacement system in environmental chambers as well as using CFD simulations (Cho *et al.*, 2005; Karimipanah *et al.*, 2007). The two systems give similar performance in terms of indoor air quality and energy efficiency at low cooling loads but the confluent jets ventilation (CJV) system is superior for cooling loads $> 35\,\mathrm{Wm^{-2}}$. The CJS can provide acceptable indoor environment with a supply temperature as low as that which can be used with a mixing system, i.e. down to about 13°C (Karimipanah *et al.*, 2007). It is a known fact that the cooling capacity of a displacement system is very limited as when the room load is larger than about $35\,\mathrm{Wm^{-2}}$, the supply air is quickly consumed and a mixing flow results in the room. This creates zones of stagnant air in the room. As was the case for the impinging jet system, the CJS creates a thin layer of fresh air on the floor with a relatively high momentum; hence, the system can also be used for heating the room.

5.7   Design of vertical ventilation

Vertical ventilation is a distribution system where the supply air is led into the room from large areas in the ceiling or from the whole ceiling. It is normally used in special rooms such as operating rooms in hospitals and clean rooms in industry, and the design in these rooms require a very high air change rate (50–$100\,h^{-1}$). This section, however, discusses vertical ventilation in typical comfort rooms such as offices. Two designs are considered, namely textile terminals, which only cover a small part of the ceiling, and vertical supply ceiling, which covers large parts of the ceiling or the whole ceiling.

5.7.1   Ventilation with textile terminals

The test of an air distribution system with a textile terminal was conducted in the same office as that used for the ventilation systems in Sections 5.4.1, 5.4.2, 5.4.3 and 5.5.1. Figure 5.25 shows the textile terminal located at the ceiling. The textile terminal is designed as half a cylinder ($d = 315\,mm$), and the whole surface serves as supply area. The textile terminal is not normally used in the heating mode because the low supply velocity results in a low momentum flow and a strong stratification in the room. A special design of the diffuser with an additional opening generating horizontal supply jets parallel to the ceiling increases the mixing in the room and makes the heating mode possible. This diffuser can be used for inlet temperatures up to 3–5°C above room temperature.

Figure 5.25 Textile terminal for low impulse air distribution system (Nielsen et al., 2005).

Figure 5.26 shows the location of a textile terminal in the room. It can either be located above the heat loads (Figure 5.26a) with return at the floor, or it can be located asymmetrical to the heat loads (Figure 5.26b) as for example above corridors in an office. The return opening is located at the ceiling the other side of the supply in Figure 5.26(b).

The air distribution in the room is mainly controlled by buoyancy forces from the heat sources with an upward flow above the computer, the lamps and the manikins (the people), and a downward flow between the plumes from the textile terminal. A location of the textile terminal directly above the two work stations therefore creates a large mixing of air in the two opposite plumes as indicated in Figure 5.27a. The layout of the air distribution system with the textile terminal located close to a side-wall (see Figure 5.27b) does, on the other hand, generate a room air movement with displacement flow in large areas of the room. The high location of the return opening and

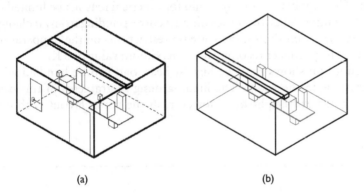

(a)   (b)

Figure 5.26 Office ventilation with a textile terminal. The terminal has a symmetrical location in (a), and an asymmetrical location in (b).

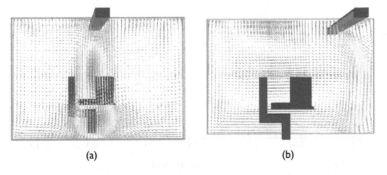

(a)   (b)

Figure 5.27 CFD simulation of the flow in the room when (a) the textile terminal is located in the centre plane and (b) when the textile terminal is located close to the side wall.

the two vertical flows from the supply and the heat sources support the displacement effect.

Figure 5.28 shows the design chart for the low impulse system. Experiments in the Annex 20 room (Figure 5.6) with one and two persons were carried out in a layout where the textile terminal is located above the heat sources as in Figure 5.27a. It is obvious that draught is dependent on the source geometry (number and location of people and other heat loads).

The comfort requirements (draught and temperature gradients) do not have restrictions on the level of the supply flow rate q_o or on the temperature difference ΔT_o as shown in the Figure 5.28. The acceptance of a high flow rate q_o can be explained by the fact that a small temperature difference gives a high flow rate, but still a low velocity from the diffuser because of the large supply area. The acceptance of a high temperature difference ΔT_o can be explained by the fact that a large temperature difference gives such a low inlet velocity that the downward buoyancy acting on the flow is not sufficient to generate draught. The supply flow rate q_o could for practical reasons be limited to $0.024 \, m^3 s^{-1}$ per metre length of the textile duct corresponding to an air change rate of 10. As the graph shows, there are no restrictions on the temperature difference, but the producer recommends a maximum value of 6 K.

The design graph shows that the highest heat load can be handled with two workstations in the room. The limiting design parameter in this case is the velocity at head height u_{11}. However, in the room layout with one

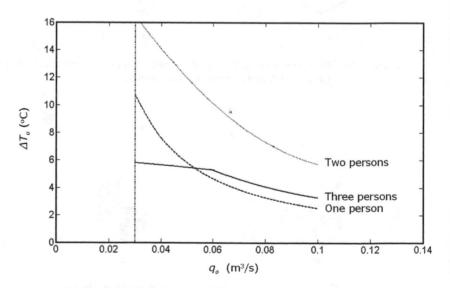

Figure 5.28 Design chart for an air distribution system with a ceiling mounted textile terminal (see Figure 5.25).

workstation, the velocity at ankle height u_{01} is the limiting velocity, and the maximum heat load is decreased.

The room with three persons corresponds to a hospital ward with a healthcare worker and two bedside patients, and it is shown in Figure 5.29 where the textile terminal is located close to the side wall (see also Figure 5.26b). The design chart of Figure 5.28 is related to the healthcare worker standing between the two beds.

The flow rate to the room should first and foremost be larger than $0.03 \, \text{m}^3\text{s}^{-1}$ because of air quality requirements when three persons are in the room. The velocity at ankle level is the limiting velocity at high air-flow rates, and it is seen that unlike for the case of one or two occupants, there is a restriction on the temperature difference ΔT_o here. The velocity at head height is the limiting velocity for a flow of $0.03–0.06 \, \text{m}^3\text{s}^{-1}$, which restricts the temperature difference to 5–6 K, as shown in Figure 5.28. When the healthcare worker moves to the area below the diffuser, there will be more draught at his ankle level, and the area below the diffuser should be considered to be outside the occupied zone. The results in Figure 5.28 are of course dependent on the diffuser design and the size of the room.

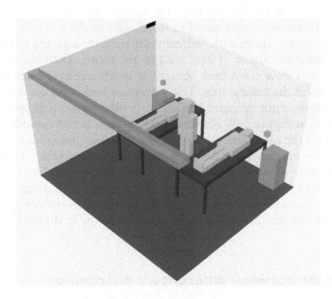

Figure 5.29 A room layout with three persons: a healthcare worker and two patients in a hospital ward. The textile terminal is located asymmetrical close to the side wall to produce displacement flow in the room.

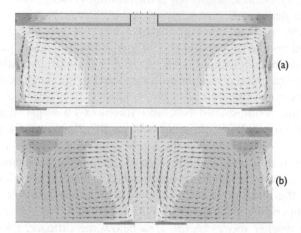

Figure 5.30 Air distribution in a room with vertical ventilation through a ceiling panel. The heat load in the room is close to the side wall in case (a) and close to the centre line in case (b) (Jacobsen et al., 2004).

5.7.2   Vertical ventilation through ceiling panels

Vertical ventilation through ceiling panels is used in offices and in the ventilation of livestock buildings. Figure 5.30 shows a CFD prediction of the air distribution in a livestock building. The supply area covers most of the ceiling with a return opening located in the middle of the room.

The air distribution in a room with vertical ventilation through the ceiling is controlled by the distribution of heat load in the room. The buoyancy effect on the thermal flow from heat sources is much stronger than the momentum flow from the ceiling area. The advantage is that the air distribution system itself does not generate draught as in other systems as shown in the design chart in Figure 5.3. However, this system is not capable of producing mixing of the room and supply air, and therefore, horizontal and vertical temperature gradients at high heat loads are unavoidable. Those gradients may result in a low level of comfort at high heat loads. The influence of the heat load on the air distribution in the room is illustrated in Figures 5.30(a and b), where a relocation of the heat load in the occupied zone changes the whole air movement in the room to a flow in the opposite direction.

5.8   Comparison between different air distribution systems

It is possible to make a direct comparison between several of the different air distribution systems described in this chapter because the experiments are made in the same room with the same office equipment; two computers, two desk lamps and two manikins. The thermal load is in all cases equal to 480 W. The room geometry is shown in Figure 5.6.

The air distribution systems are:

- mixing ventilation with a wall-mounted diffuser
- mixing ventilation with a ceiling-mounted radial diffuser
- mixing ventilation with a ceiling-mounted swirl diffuser
- displacement ventilation with a wall-mounted low velocity diffuser
- vertical ventilation with a ceiling-mounted textile terminal.

Table 5.2 gives a summary of the different systems.

Figure 5.31 shows the design chart for all five air distribution systems described before. Air quality considerations require a flow rate that is larger than $0.02 \, \text{m}^3\text{s}^{-1}$ for all systems.

Table 5.2 Five different air distribution systems

System	Supply position	Return position	Diffuser
Mixing ventilation	End wall mounted	Below supply terminal	
Mixing ventilation	Radial ceiling diffuser	End wall mounted below ceiling	
Mixing ventilation	Ceiling swirl diffuser	End wall mounted below ceiling	
Displacement ventilation	End wall mounted	End wall mounted below ceiling	
Vertical ventilation	Ceiling mounted low impulse textile terminal	End wall mounted at floor level or at ceiling level	

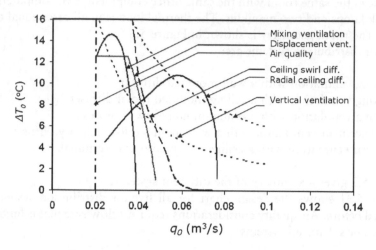

Figure 5.31 Design chart for the five different air distribution systems.

The maximum penetration velocity through the occupied zone (design velocity, u_{ocz}) is equal to $0.4\,\mathrm{ms}^{-1}$ in the case of mixing ventilation from an end-wall-mounted diffuser. The rather high velocity gives the same level of comfort as obtained by the textile terminal for a u_{ocz} design velocity of $0.2\,\mathrm{ms}^{-1}$.

Figure 5.31 indicates that the room to some extent has the same level of comfort (in terms of maximum velocity and temperature gradient) in the case of mixing ventilation with a wall-mounted diffuser or displacement ventilation with a wall-mounted low velocity diffuser. The figure shows also that the vertical ventilation (low impulse) system is close to both mixing ventilation based on a wall-mounted diffuser and to displacement ventilation. In connection with vertical ventilation, it should be noticed that the layout with two workstations in a small room may result in $q_o - \Delta T_o$ curve, which is more favourable than that for a large room with vertical ventilation. (The lower curve in Figure 5.31 for vertical ventilation is obtained for one manikin in the room, corresponding to two persons in a larger room.)

Mixing ventilation by ceiling-mounted diffusers is capable of handling a higher heat load than any of the other systems. The design velocity u_{ocz} is $0.3\,\mathrm{ms}^{-1}$ for the radial diffuser combined with a maximum value of $\Delta T_o = 10\,\mathrm{K}$, whereas the design velocity u_{ocz} for the diffuser generating a flow with swirl is $0.35\,\mathrm{ms}^{-1}$.

It should also be emphasized that the design chart to some extent is dependent on room size, room layout and design of the terminal units (number, location, etc.). It is possible to use the approach behind the design chart in rooms larger than the test room used in this chapter by using flow elements in the design procedure (wall jet, penetration length, stratified flow, etc.) which correspond to the actual room dimension.

The design methods in this chapter can only give the limits for the operation of the air distribution system. It is necessary to consider thermal comfort for all flow rates if the system has to be optimized. The thermal environment often shows temperature gradients, velocity gradients, different turbulence levels and an asymmetrical radiant temperature distribution. The local discomfort, resulting from such an environment, is found from measurements of the local values of air temperature, air velocity, turbulence level and surface temperatures or asymmetric radiant temperatures (see Fanger and Langkilde, 1975; Olesen *et al.*, 1979; Fanger *et al.*, 1989; Toftum *et al.*, 1997).

The number of dissatisfied because of draught, the draught rating (DR), is used as a measure of local discomfort. The DR is defined as:

$$DR = e^{d(t_a - 24)}(34 - t_a)(u - 0.05)^{0.62}(0.37uTu + 3.14) \qquad (5.12)$$

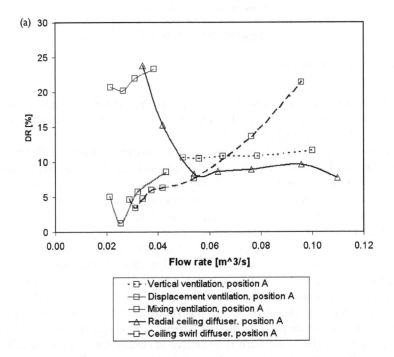

Figure 5.32 Draught rating (DR) versus flow rate q_0 measured at (a) position A for five different air distribution systems and (b) position B for five different air distribution systems.

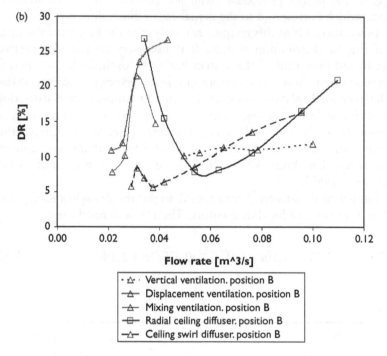

Figure 5.32 (Continued).

where t_a, u and Tu are ambient air temperature, air velocity and turbulence intensity, respectively (see Toftum *et al.*, 1997). d is a factor that is dependent on direction of the air flow.

Equation (5.12) is especially based on a situation where the velocity is exposed to the back of the neck of the subjects. In this chapter, it was observed that the highest velocity is often found at ankle level, and the sensitivity at this position is slightly lower than that for the back of the neck. The draught rating given from Equation (5.12) (shown in Figure 5.32) is therefore representing a conservative estimate.

Figure 5.32(a and b) shows the *DR* for the occupants at different positions in the room. Position A is close to the end wall with the diffuser, and position B is opposite that wall in the IEA Annex 20 room (see Figure 5.6) in which the experiments are made. The draught rating is dependent on the location of the workstation in mixing ventilation with a wall-mounted diffuser as well as for displacement ventilation; however, it is not dependent on the position in the room with the vertical ventilation system. Generally, the best results are obtained for displacement ventilation with a workstation position far away from the diffuser, whereas mixing ventilation with a wall-mounted diffuser has a high draught rating. The vertical ventilation system gains from the high level of buoyant flow in the room, which in this case protected the persons from draught. Mixing ventilation with a ceiling-mounted diffuser generating flow with swirl has a draught rating close to the best result obtained by displacement ventilation at position A. The draught rating for the mixing ventilation with a ceiling-mounted radial diffuser also shows a low value at both positions under the optimal flow conditions of around $0.055 \, \mathrm{m^3 s^{-1}}$. All the systems with ceiling-mounted diffusers (radial diffuser, swirl diffuser, textile terminal) are superior to systems with end-wall-mounted diffusers.

Measurements show that the temperature gradient and asymmetric radiation are only important for displacement ventilation.

References

Awbi, H. B. (2003). *Ventilation of Buildings*, Second Edition, ISBN 0415270553. Taylor and Francis, London.

Brohus, H. and Nielsen, P. V. (1996). Personal exposure in displacement ventilated rooms. *Indoor Air: International Journal of Indoor Air Quality and Climate* 6(3), 157–167.

CEN Report. (1997). CR 12792 *Ventilation for Buildings – Terminology, Symbols and Units*.

Cho, Y., Awbi, H. B. and Karimipanah, T. (2002). A comparison between four different ventilation systems, Proceedings of the 8th International Conference on Air Distribution in Rooms (Roomvent 2002), pp 181–184, Denmark, ISBN 87-989117-0-8.

Cho, Y., Awbi, H. B. and Karimipanh, T. (2004). The characteristics of wall confluent jets for ventilated enclosures, Proceedings of the 9th International Conference on Air Distribution in Rooms (Roomvent 2004), pp 6, 5–8 September, Coimbra, Portugal.

Cho, Y., Awbi, H. B. and Karimipanah, T. (2005). Comparison between wall confluent jets and displacement ventilation in aspects of the spreading ratio on the floor, Paper 4.4-12, Proceedings of 10th International Conference in Indoor Air Quality and Climate (*Indoor Air* 2005), Beijing, China, ISBN 7-89494-830-6.

Cho, Y., Awbi, H. B. and Karimipanah, T. (2007). Theoretical and experimental investigation of wall confluent jets ventilation and comparison with wall displacement ventilation. *Building and Environment*, 10.1016/j.buildenv.2007.02.006.

Fanger, P. O. and Langkilde, G. (1975). Interindividual differences in ambient temperature preferred by seated persons. *ASHRAE Transactions* 81(2), 140–147.

Fanger, P. O., Melikov, A. K., Hanzawa, H. and Ring, J. (1989). Turbulence and draft. *ASHRAE Journal*, April, 18–23.

Fitzner, K. (1988). Impulsarme Luftzufuhr durch Quellüftung. Heizung Lüftung/-Klima Haustechnik, No. 4.

Helander, L., Yen, S. M. and Crank, R. E. (1953). Maximum downward travel of heated jets from standard long radius ASME nozzles. *ASHVE Transactions*, No. 1475.

Jacobsen, L., Nielsen, P. V. and Morsing, S. (2004). Prediction of Indoor Airflow Patterns in Livestock Buildings Ventilated through a Diffuse Ceiling, Roomvent 2004, 9th International Conference on Air Distribution in Rooms, Coimbra.

Karimipanah, T. (1996). Turbulent jets in confined spaces, PhD thesis, Royal Institute of Technology, Sweden.

Karimipanah, T. and Awbi, H. B. (2002). Theoretical and experimental investigation of impinging jet ventilation and comparison with wall displacement ventilation. *Building and Environment* 37, 1329–1342.

Karimipanah, T., Sandberg, M. and Awbi, H. B. (2000). A comparative study of different air distribution systems in a classroom, air distribution in rooms: ventilation for health and sustainable environment, Proceedings of ROOMVENT 2000, H. B. Awbi (ed.), Vol. 2, pp 1013–1018, Elsevier, Oxford.

Karimipanah, T., Awbi, H. B., Sandberg, M. and Blomqvist, C. (2007). Investigation of air quality, comfort parameters and effectiveness for two floor-level air supply systems in classrooms. *Building and Environment* 42, 647–655.

Nielsen, P. V. (1980). The influence of ceiling-mounted obstacles on the air flow pattern in air-conditioned rooms at different heat loads. *Building Services Engineering Research and Technology* 1, 55–71.

Nielsen, P. V. (1991). Models for the Prediction of Room Air Distribution, 12th AIVC Conference, Ottawa, Canada.

Nielsen, P. V. (1996). Temperature distribution in a displacement ventilated room, in ROOMVENT'96, Proceedings of the 5th International Conference on Air Distribution in Rooms, Yokohama, Japan, July 17–19, Murakami, S. (ed.), Vol. 3, pp. 323–330. ISBN: 4-924557-01-3.

Nielsen, P. V. (2000). Velocity distribution in a room ventilated by displacement ventilation and wall-mounted air terminal devices. *Energy and Buildings* 31(3), 179–187.

Nielsen, P. V., Heby, T. and Moeller-Jensen, B. (2006). Air distribution in a room with ceiling-mounted diffusers – comparison with wall-mounted diffuser, vertical ventilation and displacement ventilation. *ASHRAE Transactions* 112(Part 2): pp. 498–504.

Nielsen, P. V., Hoff, L. and Pedersen, L. G. (1988). Displacement Ventilation by Different Types of Diffusers, 9th AIVC Conference on Effective Ventilation, Ghent, Belgium.

Nielsen, P. V., Jensen, R. L., Pedersen, D. N. and Topp, C. (2001). Air distribution in a room and design considerations of mixing ventilation by flow elements, Proceedings of the 4th International Conference on Indoor Air Quality, Ventilation & Energy Conservation, Vol. II, pp. 1055–1062, Hunan University.

Nielsen, P. V., Larsen, T. S. and Topp, C. (2003). Design methods for air distribution systems and comparison between mixing ventilation and displacement ventilation, Proceedings of Healthy Buildings 2003, Singapore.

Nielsen, P. V., Topp, C., Snnichsen, M. and Andersen, H. (2005). Air distribution in rooms generated by a textile terminal – comparison with mixing ventilation and displacement ventilation. *ASHRAE Transactions* 111 (Part 1): 733–739.

Olesen, B. W., Scholer, M. and Fanger, P. O. (1979). Vertical air temperature differences and comfort. In: *Indoor Climate*, P.O. Fanger and O. Valbjoern (eds). Danish Building Research Institute, Copenhagen, pp. 561–579.

Rajaratnam, N. (1976). *Turbulent Jets*. Elsevier, Amsterdam.

Skistad, H., Mundt, E., Nielsen, P. V., Hagström, K. and Railio, J. (2002). Displacement Ventilation in Non-Industrial Premises. REHVA Guidebook No 1.

Stymne, H., Sandberg, M. and Mattsson, M. (1991). Dispersion pattern of contaminants in a displacement ventilated room, Proceedings of the 12th AIVC Conference, ISBN 0-946075-53-0, Warwick.

Toftum, J., Zhou, G., Melikov, A. (1997). *Effect of Airflow Direction on Human Perception of Draught.* Clima 2000, Brussels.

Xing, H., Hatton, A. and Awbi, H. B. (2001). A study of the air quality in the breathing zone in a room with displacement ventilation. *Building and Environment* 36(7), 809–820.

Chapter 6

Characteristics of mechanical ventilation systems

Claude-Alain Roulet

6.1 Introduction

There are many types of mechanical ventilation systems, each type being adapted to particular needs (ventilation, contaminant extraction, heating, cooling, full air-conditioning, etc.) and to constraints such as available space, cost, etc.

The aim of this chapter is to characterize the various types of mechanical systems, to present the advantages and disadvantages of each type and to provide indication on how to characterize them by measurements.

6.2 Types of ventilation systems

Ventilation systems can be classified according to:

- their function – ventilation only or air-conditioning including heating, cooling, drying or humidifying the air;
- the way the task is fulfilled – ventilation by supply, extract or both, and temperature control using dual duct, local coils, variable air volume (VAV), etc.

Detailed description of various systems and their variants can be found in the literature (ASHRAE, 2004). Therefore, only short descriptions of the type of systems will be given below.

6.2.1 Natural ventilation

Natural ventilation is a system where the air is moved through ventilation openings by natural forces such as wind pressure and stack effect.

In many buildings, these openings are doors, windows, and leakage in the building envelope and partition walls and decks. Because this does not allow a good control of the ventilation rate, it is however recommended that one should install ventilation openings designed for this purpose. The

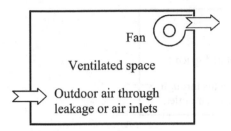

Figure 6.1 Principle of extract-only ventilation system.

location and size of these openings should be chosen to allow the required airflow rate and to favour an appropriate airflow pattern in the room.

This type of ventilation is described in more detail in Chapter 7.

6.2.2 Extract-only mechanical ventilation

This very simple mechanical ventilation system extracts the air from the ventilated space with ducts and fans (Figure 6.1). The purpose of such systems is to extract contaminants as close as possible to the source. To protect the fan and keep duct clean, one may install a filter upstream of the fan. Examples are kitchen or laboratory hoods, painting canopies, small extracting fans in toilets or bathrooms, etc.

The extracted air is replaced by outdoor air, the internal pressure being lower than the external one but remaining close to it. In some cases, the air enters by infiltration through leakage in the building envelope. However, it is recommended that air inlets are installed at appropriate places on the building envelope, so that the incoming air is directed to where it is needed to reduce the risk of creating draughts.

6.2.3 Supply-only mechanical ventilation

The supply-only system blows the air into the ventilated space (Figure 6.2). The purpose of such systems is to introduce conditioned air into the ventilated space. This air could be only filtered, but also heated, cooled, humidified or dried.

This system introduces a slight overpressure in the ventilated space. Therefore, indoor air leaves the space by exfiltration through leakage in the building envelope. In some cases, air outlets are installed to favour an appropriate airflow pattern.

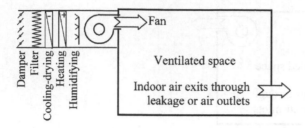

Figure 6.2 Principle of supply-only ventilation system.

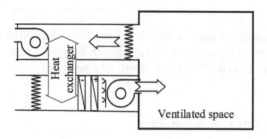

Figure 6.3 Principle of balanced system.

6.2.4   *Supply and exhaust or balanced mechanical*
ventilation systems

This system uses two fans, one to supply air and the other to extract vitiated air (Figure 6.3). This allows a proper balancing of pressure and a better control of the airflow pattern. In addition, a heat exchanger could be installed to recover heat from the extract air and give it to the supply air. The contrary is also possible during the cooling season. More information on heat recovery is found in Section 6.5.

6.2.5   *Recirculation*

The air is a poor fluid for transferring heat, because its density and thermal capacity are low (see Chapter 3, Section 3.5.1). Therefore, a larger airflow rate is needed to transfer heat into the ventilated space, and this flow rate may be a multiple of the airflow rate required for providing acceptable indoor air quality. Therefore, recirculation is used in some systems to allow large quantities of conditioned air to mix with a limited outdoor airflow rate. Recirculation may be installed in systems with (Figure 6.4) or without (Figure 6.5) exhaust air duct.

Note that this system has the inconvenience to recirculate pollutants produced at one place in the building back to the whole ventilated space.

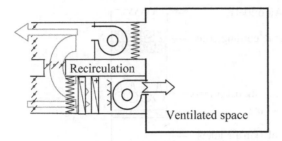

Figure 6.4 Principle of recirculation in a ventilation system.

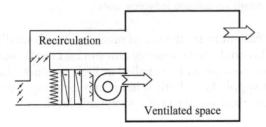

Figure 6.5 Principle of recirculation in a ventilation system without exhaust duct.

6.3 Temperature and humidity control with mechanical ventilation (air-conditioning)

Although air is not an efficient heat transportation medium, it is nevertheless used to heat and cool spaces, as well as to control the room humidity in some cases. Various techniques and systems are used for this purpose, each having its advantages and inconveniences. These are known as air-conditioning systems, which are briefly described below.

6.3.1 Fan coil and induction systems

In these systems, energy is transported to each room as hot or chilled water in a hydronic network. Only fresh air is supplied to the rooms, and extracted if appropriate. In each room, there are heat exchangers to cool and dry or heat the air. The air is forced through these heat exchangers either by a fan (fan coil units) or by induction caused by jets of supply air (induction units) (Figure 6.6). This has the advantage of:

- reducing the airflow rate down to the hygienic level required;
- transporting heat with water (less energy use and less space used to transport heat);
- recirculating air individually in each room, thus avoiding the spread of pollutants from one room to another.

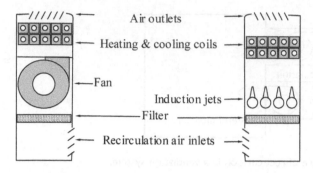

Figure 6.6 Schematic cut through fan coil and induction units.

The disadvantages of fan coil units are the use of many but much smaller fans than in central units. This could be less energy efficient and needs more maintenance. Induction units need air to be supplied at high pressure, thus increasing the energy use of supply fans. Both systems may be noisy, but it is technically possible to limit their noise to fully acceptable levels.

6.3.2 Constant and variable air volume systems

If heat is transported with air, the room temperature can be controlled either by varying the temperature of the supply air delivered at a nearly constant rate (constant air volume or CAV systems) or by varying the airflow rate of cold air and hot air that are both at constant temperature (VAV systems).

CAV systems are convenient where a constant airflow rate is required. An advantage is that the fan always operates at its maximum efficiency. When this system is applied to cool down several zones with different loads, the air is centrally cooled at the lowest required temperature, and then reheated locally with heating coils for each zone. This is of course not energy efficient.

VAV systems are used where thermal loads and airflow rate requirements vary together, which is the case in assembly halls, schools, etc., where the occupants change with time and are the main heat load and main pollutant source. In this system however, the simultaneous control of temperature, humidity and airflow rate may be difficult to achieve. The variable airflow rate should be controlled by changing the fan speed, using a variable frequency controller. Dampers should not be used for this purpose; their pressure drops when nearly closed, strongly reducing the energy efficiency of the system.

6.3.3 Single-duct system

The air is first conditioned in an air-handling unit (AHU) to the required temperature and humidity. During the cooling season, the air should be colder and dryer than the room air, and vice versa during the heating season.

The air is then distributed through a single network. In CAV systems, the temperature and humidity are controlled in the AHU to maintain the required room climate when the load changes. In VAV systems, the fan speed or dampers are adjusted so that there is just enough supply airflow rate to cope with the room loads.

A simpler variant of this is found in tropical countries, where the main requirement is to dry the air. Each building zone is equipped with a simple supply-only unit with recirculation (Figure 6.7). Wet and hot outdoor air is mixed with drier and colder indoor air coming from recirculation ducts. This mix is cooled down and dried through a cooling coil. The dew point is controlled by adjusting the chilled water flow rate so that the off-coil air temperature is close to a set point. This air is supplied to the ventilated space at such a rate that it compensates internal and solar gains to reach a comfortable room temperature. For this, a thermostat located in the ventilated space controls the fan speed.

In such a system, the recirculation rate is generally very large (around 95 per cent) and outdoor airflow rate depends more on thermal loads than on indoor air quality. In addition, mixed air reduces the drying efficiency of the cooling coil.

Therefore, a dual-fan system was developed (Sekhar *et al.*, 2004) to control separately the outdoor and recirculation airflows (Figure 6.8).

This system allows separate control of the airflow rates and dew points of both outdoor and recirculation air, so that a good indoor environment

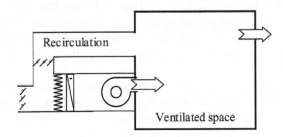

Figure 6.7 Principle of simple 'tropical' units.

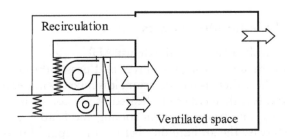

Figure 6.8 Principle of dual-fan unit.

can be obtained with good energy efficiency. This system can easily be used with a dual-duct air distribution, thus allowing individual climate control of several zones.

6.3.4 Dual-duct systems

In such systems, part of the supply air is cooled and another part is heated separately. Cold and hot air are distributed to conditioned spaces in two separate duct networks and mixed close to each ventilated space in proper proportions to compensate for the space load. These systems can also control the room climate either by changing the supply air temperature and humidity or by providing more or less supply air at a given temperature and humidity.

6.3.5 Advantage and inconveniences

In most buildings, the heating or cooling load is not uniform. It may even happen that some rooms need cooling, whereas others need heating. Therefore, it is important to be able to control the temperature of each room individually. This is straightforward in fan coil or induction systems, but not in centralized systems.

To ensure good control, one needs to install one single-duct CAV system for each room or zone with a given load, which is expensive. In single-duct VAV systems, it is in principle possible to control and vary the airflow rates in each room by using dampers, but in this case, the pressure in the main duct should be large enough to avoid a strong interaction between the airflow rates (when a damper closes, this increases the airflow rates into other rooms and vice versa).

The dual-duct system allows controlling the temperature in each room just with a simple mixing box. However, it needs transporting both hot and cold air on long distances, introducing important heat losses. This system cannot simultaneously control humidity, is expensive and uses room space for the two duct networks.

6.4 Components of air-handling units

Figure 6.9 shows schematically a supply and exhaust AHU.

The outdoor air enters the unit through a louver and passes through a first filter. In some cases, where there is recirculation, it is mixed with return air and passes a finer filter. It is then either pre-heated or pre-cooled by the heat recovery system (if any) and passes through cooling and heating coils. A humidifier may then increase the air humidity before it is supplied to the air distribution ducts.

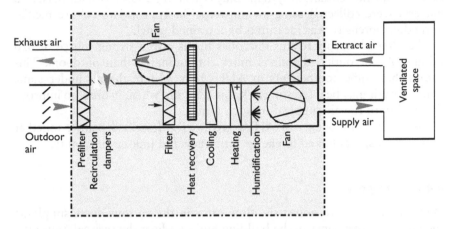

Figure 6.9 Schematics of a supply and exhaust air-handling unit. The location of fans may differ from the one illustrated here, and some elements (in particular heater, cooler, humidifier and heat recovery) are not installed in every unit.

The air extracted from the ventilated spaces is exhausted after passing through the heat recovery unit, if any. Supply- or exhaust-only systems simply use only one half of this scheme and do not have heat recovery.

Several field studies have shown that the ventilation system is often a source of contaminants (Fanger, 1988; Bluyssen *et al.*, 2000b). Figure 6.10 shows the presence of contaminants at several locations in the supply duct

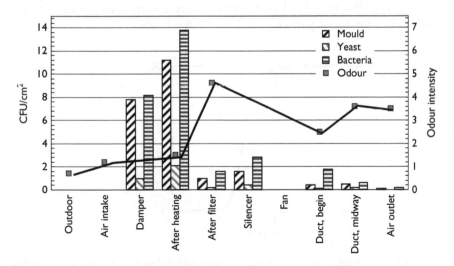

Figure 6.10 Colony-forming units (CFU) of mould, yeast and bacteria, as well as odour intensity at several locations in an air-handling unit.

of a mechanical ventilation system (Bluyssen *et al.*, 2003). Mould, yeast and bacteria were collected using an impactor on appropriate culture media, and odour intensity was measured by a trained panel.

The filter obviously retains microbes but is also a strong source of bad odours. At air outlet, the air is more contaminated than outdoors. This shows that some components in AHUs may pollute the air under some conditions. It may be interesting to know why and when, in order to improve the quality of the air supplied by such units.

The main components of AHUs are briefly described below, together with some aspects linked to energy efficiency and indoor air quality.

6.4.1 Dampers

Dampers are used to adjust the flow rates in the various ducts supplying the air to various zones in the building and to adjust the recirculation rate, if any.

Dampers should be avoided wherever possible because they generate additional pressure drops and therefore more energy use for fans, and they also accumulate dirt and can generate noise.

6.4.2 Filters

The atmosphere contains various liquid and solid contaminants in the form of particles (aerosols), some of which are so small that they remain suspended in the air for long periods, which are then entrained by the airflow. As shown in Figure 6.11, larger particles are filtered in the nose or the trachea, but the finest particles easily enter the lungs.

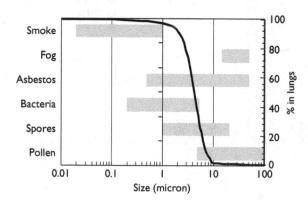

Figure 6.11 Sizes of various atmospheric aerosols and percentage of these particles entering the lungs [Extracted from ACGIH (American Conference of Governmental Industrial Hygienists)].

Figure 6.12 Bag filter viewed from downstream.

Filters reduce the aerosol content of the air, mainly to avoid dust accumulation in the AHU and in the ducts. In some units, fine filters are installed to retain the smallest particles such as microbes or smoke. An important characteristic of filters is their ability to remove particles from the airflow. The percentage of removed particles is the efficiency of the filter. This efficiency depends on the structure of the filter and on the type and size of the captured particles.

There are mechanical, electrostatic and biological filters. Mechanical filters collect dust on fibre mats. Some mats are coated with a viscous fluid on which impinging particles stick. Dust accumulates on media unit filters such as bag filters illustrated in Figure 6.12 until the filter is so dirty that the pressure drop through it reaches a limit at which the filter should be either cleaned or changed. In renewable media filters, a fresh mat constantly replaces loaded mat, generally by unrolling it into the airstream, so that pressure drop remains constant.

Filters are sorted into classes according to their capturing efficiency of various particle sizes. Figure 6.13 shows the efficiency of three types of filters (according to European classes) versus particle size. Comparing this figure with Figure 6.11, it is shown that filters of class EU 13 are needed to eliminate breathable particles from the airstream.

Dirty filters pollute the air because dirt generally emits volatile compounds. New filters may emit pollutant too, as illustrated in Figure 6.14,

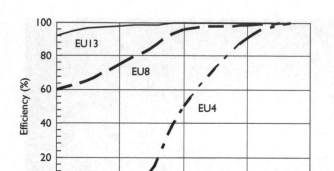

Figure 6.13 Efficiency of three types of filters as a function of particle size.

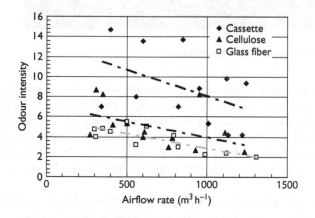

Figure 6.14 Olfactive pollution of various new filters as a function of airflow rate, from (Bluyssen *et al.*, 2000a, 2003).

because some materials used for filters and their frames emit volatile organic compound (VOCs).

Filters are one of the main sources of sensory pollution in ventilation systems (Bluyssen *et al.*, 2000a and 2003). The filter material had a significant influence on the initial pollution emission from new filters. The pollution of new filters decreases after some time of use. When filters get older, i.e. are in use for some time, the pollution increases again. The reason for the increase of pollution after use is however still unclear. It seems that microorganisms may not be the only pollution source on a filter.

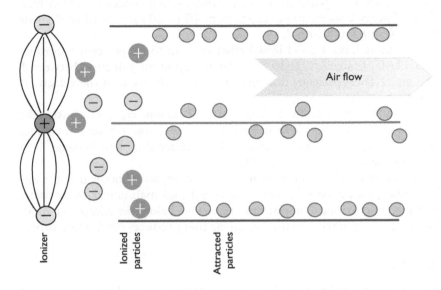

Ionizer

Ionized particles

Attracted particles

Air flow

Figure 6.15 Cross-section of a part of an electrostatic filter (principle). The complete filter is made of several such layers in parallel.

Environmental conditions such as airflow (amount or frequency, i.e. intermittent/continuous) and temperature do not seem to have an influence on the pollution effect.

It is important to keep filters dry because wet media filters are perfect supports for microbial growth, and microbes may also emit dangerous pollutants and bad odours. Filters may be moistened either by snow, driving rain or fog entering the outdoor air inlet, or by water droplets spread by some humidifiers or found in airflows downstream of the cooling coils.

Electrostatic filters electrically charge the dust particles and attract them onto plates maintained at a high voltage (Figure 6.15). These filters have the great advantage of not passing the air through the accumulated dirt and have a lower pressure drop than media filters.

In the ionization section of these filters, a high direct voltage (6–25 kV) is maintained between suspended thin wires. The electric field ionizes a few atoms that charge the particles in the airflow. These charged particles are then attracted on a series of plates parallel to the airflow, between which a high direct voltage of 4–10 kV is maintained. The particles are then neutralized but remain on the plates by intermolecular adhesion forces. The retention effectiveness can be improved by coating the plates with adhesives. The retention plates should be cleaned periodically because the efficiency of the filter decreases with the dust load.

A solid state unit powered by the electrical grid provides the direct high voltage. Energy consumption ranges from 40 to $80 \, \mathrm{Jm^{-3}}$ of filtered air or 10 to $20 \, \mathrm{W}$ for $1{,}000 \, \mathrm{m^3 h^{-1}}$.

These filters have a good initial efficiency (up to 98 per cent according to ASHRAE Standard 52.2-1999, Filter rating) at air velocities between 1 and $2 \, \mathrm{ms^{-1}}$. This efficiency decreases at higher velocity or if the velocity is not uniform.

Particles that pass an inefficient filter and remain charged transfer the electrical charge further downstream. This may charge surfaces in contact with the airflow, and these surfaces then attract dust and become dirty faster than without electrostatic filter.

Any ionizer produces some ozone from the oxygen contained in the air. The ozone production of properly designed and maintained electrostatic filters results in ozone concentration that is far below the acceptable limit. However, a dirty filter may present arcing that produces much more ozone.

6.4.3 Fans

Fans should move the air at the required airflow rate, taking into account the pressure drop in ducts and through the various elements of the system. To avoid draught resulting from infiltration through the building envelope, one should maintain the building at a pressure slightly above the outdoor atmospheric pressure. For this, a supply flow rate slightly higher than the extract is used.

Poor fan efficiency not only wastes costly electrical energy but also hinders efficient cooling. The cooling power of the air delivered by the fan is

$$\Phi_c = \rho c \dot{V} \Delta \theta \qquad (6.1)$$

where

ρ is the density of air $(\mathrm{kgm^{-3}})$
c is the heat capacity of air $(\mathrm{kJkg^{-1}K^{-1}})$
$\Delta \theta$ is the temperature difference between exhaust air and supply air (K)
$\dot{V}$ is the volume flow rate $(\mathrm{m^3 s^{-1}})$.

The kinetic energy given to the air by the fan is, sooner or later, degraded into heat by viscosity and friction on the surfaces of ducts, room walls and furniture. Because the fan motor is usually in the airflow, its heat loss is also delivered to the air. Therefore, nearly all the energy given to the fan ends as heat in the indoor air. This corresponds to a heating power equal to the electrical power consumed by the fan motor, Φ_e. Hence

$$\Phi_e = \frac{\dot{V} \Delta p}{\eta_f} \qquad (6.2)$$

where

Δp is the pressure difference (Pa)
η_f is the fan efficiency.

During the cooling season, the heating power should be small when compared with the cooling power. Therefore, the ratio

$$\frac{\Phi_c}{\Phi_e} = \eta_f \frac{\rho c \Delta \theta}{\Delta p} \tag{6.3}$$

should be as large as possible. This means that the fan efficiency should be as close as possible to one (or 100 per cent). In addition, the pressure differential should be as small as possible.

Another way to look at this issue is to calculate the air temperature increase resulting from heat gain:

$$\Delta \theta_h = \frac{\Phi_e}{\rho c \dot{V}} = \frac{\Delta p}{\eta_f \rho c} \tag{6.4}$$

This should be as small as possible, so again, the fan efficiency should be large and the pressure differential should be as small as possible.

Figure 6.16 shows the temperature increase and pressure differential across fans measured in several AHUs.

As far as indoor environment quality is concerned, fans do not emit contaminants, as long as the motor does not overheat and fan belts are in good state. Efficient fan blades are silent because noise is the result of

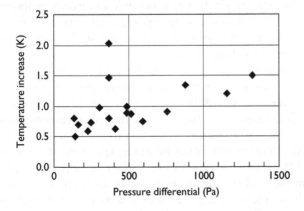

Figure 6.16 Air temperature increase $\Delta \theta_h$ as a function of pressure differential across various fans.

Figure 6.17 Cooling coil for an air-handling unit.

energy spill to sound waves. This also means that noisy fans are very likely not to be efficient and most likely disturb the indoor environment.

6.4.4   Heating and cooling coils

These devices are water–air or refrigerant–air heat exchangers (Figure 6.17). In heating coils, a controlled flow rate of warm water or steam circulates in the heat exchanger pipes to heat or reheat the air at the required temperature.

In cooling coils, a controlled flow rate of chilled water circulates into the heat exchanger to cool down the air. It is also possible to directly evaporate a refrigerant into the heat exchanger. When its external surface is colder than the dew point of the air, water condenses on the surface and drops into a pan, then to sewer. This dries the air.

Experiments have shown that heating and cooling coils without condensate or stagnating water collecting in the pans are components that have small contributions to the overall odour intensity of the air. On the contrary, cooling coils with condensate in the pans are microbial reservoirs and breading sites that may be major sources of odours to the inlet air.

Some droplets of condensate may also be entrained from the cooling coil surface in the airflow and moist another surface downstream. Such moistened surfaces, in particular if these are filters or fibres in an acoustic attenuator bay, also host microbes and could be a source of pollution. To avoid this, one should install a droplet catcher downstream of the cooling coils.

6.4.5 Humidifier

Humidifiers increase the air moisture by injecting water vapour to it. Some humidifiers directly introduce water vapour, whereas others introduce droplets that evaporate in the air, thus reducing its temperature. The main types of humidifiers are as follows:

- spray humidifier where finely dispersed water droplets are injected into the flowing air.
- steam humidifier in which steam is produced and supplied directly into the airflow.
- ultrasonic humidifier in which water is dispersed into fine water droplets by the effect of an ultrasonic oscillator.

Humidifiers were found to increase the odour intensity in the airflow, which is a sign of pollution (Figure 6.18).

The reasons for pollution found in humidifiers are (Müller *et al.*, 2000) as follows:

- disinfecting additions;
- old water in tanks or dirty tanks;
- microbiological growth;
- water standing in the tank when the humidifier is off;
- desalinization and demineralization devices and agents if used.

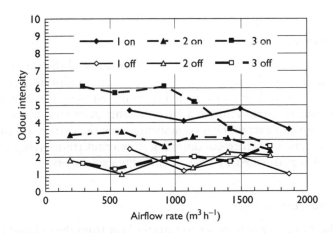

Figure 6.18 Odour intensity in humidifiers when on and off, for three tests.

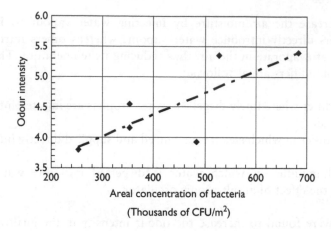

Figure 6.19 Odour intensity and bacteria colony-forming units (CFU) on surfaces in humidifiers (Müller *et al.*, 2000).

A humidifier only pollutes the air significantly if it is not used as recommended or if it is not properly maintained. Frequent cleaning of humidifiers is an absolute must, as is the use of fresh water for top up.

Under normal conditions, for a number of humidifiers, it was found that the airflow has no influence on the odour intensity caused by humidifiers.

A relation was found between the perceived air quality and the concentration of bacteria in humidifier (Figure 6.19). The odour intensity increases with increasing number of bacteria. This was not the case for other locations in a heating, ventilating and air conditioning (HVAC) system. However, similar correlation could not be found for fungi.

6.4.6 Heat recovery

Air supplied into the ventilated space comes from outside and should often be heated, cooled, dried or humidified to be supplied at comfortable conditions. This needs energy. The air extracted from the building is nearly at room conditions, but it is contaminated by pollutants generated indoors. Therefore, it is possible to extract heat or moisture from the extract air and provide it to the supply air before conditioning it, without transferring the contaminants. This is performed by heat or enthalpy exchangers.

Types of heat exchangers

The purpose of heat exchanger is either to transfer heat from the exhaust to the supply part of the AHU (heat recovery heat exchangers), or to transfer

Figure 6.20 Flat plate heat exchanger.

heat from water to air (heating coils) or vice versa in cooling coils. This heat should be transferred in the most possible efficient way, without transferring contaminants.

Water-to-air heat exchangers are in most cases made out of finned tubes in which the water circulates. The fins increase the exchange area between the exchanger surface and the air.

The types of heat exchangers most commonly used for heat recovery are as follows.

PLATE HEAT EXCHANGERS

In plate heat exchangers, the exhaust air is blown in several channels provided by metallic or glass plates (Figure 6.20). The other side of these plates is in contact with inlet air, so that heat can be transferred from the warm side to the other. The heat recovery efficiency of these exchangers ranges from 60 to 80 per cent, depending on the type and size. A variant of this exchanger is the heat pipe exchangers, in which heat pipes are used to transport heat from warm to cold air. Such heat exchangers should not leak air from one side to the other.

New enthalpy plate heat exchangers are now on the market, in which the plates are membranes loaded with salts that transfer not only heat but also air humidity (Kriesi, 2005). Such heat exchangers have efficiencies

Figure 6.21 Rotating heat exchanger.

close to that of rotating heat exchangers (RHEs) without presenting the inconvenience of contaminant transfer from exhaust to supply air.

ROTATING HEAT EXCHANGERS

RHEs are used in larger systems (Figure 6.21). A disk made with a honey-comb structure is placed so as to have half of its area in the exhaust duct and the other half in the supply duct. This disk rotates slowly and is heated by the warmer air, where air moisture also condenses on the surface of the honeycomb. It is then cooled by the colder airstream also evaporating the condensed water, and so on. In this way, sensible and latent heat contained in the warm air is given to the cold air, and the heat recovery efficiency may reach up to 90 per cent. A gasket and a purging sector limit contamination transfer from exhaust air to fresh air side.

A small leakage can be accepted in an RHE, resulting in a recirculation rate of less than four per cent. Reduced leakage is achieved by carefully installing the heat exchanger and by balancing the air pressure between both sides of the exchanger. To achieve this, one should not have supply and exhaust fans on the same side of the heat exchanger (Figure 6.22).

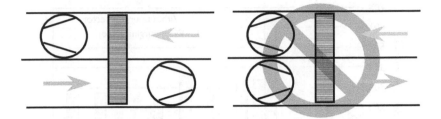

Figure 6.22 Relative position of fans and rotating heat exchangers. Placing both fans on the same side results in a large pressure differential through the RHE, thus increasing leaks.

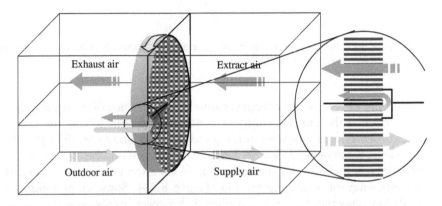

Figure 6.23 Schematics of the purging sector. A part of the outdoor air cleans the honeycomb structure and then is sent back to the exhaust air.

Most RHEs are equipped with a purging chamber, located between the inlet and exhaust air ducts, on the warm side of the wheel (Figure 6.23). This chamber covers a sector of about ±5°, in which the outdoor air passes through the wheel, makes a 180° turn in the purging chamber, passing back into the wheel, and finally leaves the AHU by the exhaust air duct. This cleans the wheel from contaminants accumulated when passing in the extract air, before entering the outdoor air supply.

However, RHEs may transfer contaminants from exhaust to supply air in three ways: with entrained air, through possible leakage around the wheel at the separation wall, and by adsorption–desorption on the inner surfaces of the exchanger's wheel.

Experiments performed in AHUs with RHEs (Roulet *et al.*, 2002) clearly showed that part of the VOCs present in the exhaust air may be recycled to

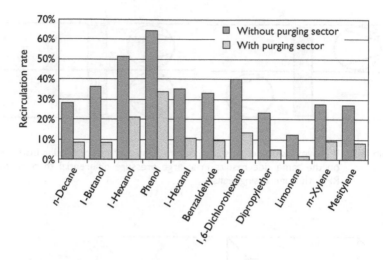

Figure 6.24 Average VOC recirculation rates measured in both units, with and without purging sector.

the supply air by the heat recovery unit in the absence of a purging chamber, or when the purging chamber is not well designed. The low transfer rates observed in a unit with purging chamber for limonene, dipropylether, *m*-xylene, mesitylene, *n*-decane, hexanal, 1-butanol and benzaldehyde confirm the results found for formaldehyde (Andersson *et al.*, 1993) and attest the efficiency of a purging section (Figure 6.24). Some compounds such as dichlorohexane, hexanol and phenol, however, easily pass the purging chamber and are recycled in significant quantities to the supply air. Therefore, RHEs should not be installed where contaminant recirculation are to be avoided.

GLYCOL HEAT EXCHANGER

When the exhaust and inlet ducts are not side by side, heat can be transported by an hydronic circuit with two heat exchangers: the fluid (generally a glycol–water mix) is heated by the air–liquid heat exchanger located in one of the ducts, then pumped to the other exchanger to give heat to the cold air. Such systems have no air leakage from one side to the other.

HEAT PUMP

In exhaust-only systems, the recovered heat cannot be given to the outdoor air supply but to the hydronic heating system or to a hot water boiler. For this, the temperature of the hot side of the heat recovery system is increased using a heat pump, whose cold source is the exhaust air.

Heat exchange efficiency

The efficiency of a heat exchanger has two aspects: the energy (or enthalpy) efficiency and the temperature efficiency.

The first is the ratio of the enthalpy flow delivered to the supply air by the enthalpy flow in exhaust air:

$$\eta_E = \frac{H_{\text{downstream, supply}} - H_{\text{upstream, supply}}}{H_{\text{upstream, exhaust}} - H_{\text{outdoor air}}} \tag{6.5}$$

If the supply air upstream of the heat exchanger (inlet air) has the same properties as the outdoor air, $H_{\text{outdoorair}}$ may be replaced by $H_{\text{upstream,supply}}$. The enthalpy of air is determined by its temperature and moisture content. Therefore, measurement of temperature and moisture content of air both upstream and downstream of the heat exchanger allows the determination of the enthalpy efficiency.

The air enthalpy H is the product of mass airflow rate and specific enthalpy, h:

$$H = \rho \dot{V} h \tag{6.6}$$

where ρ is the density of air.

A good approximation of the specific enthalpy of air, h in (Jkg^{-1}), is given by

$$h = 1004.5\ \theta + \omega(2,500,000 + 1858.4\ \theta) \tag{6.7}$$

where

   θ is the air temperature (°C)
   ω is the specific humidity ratio, i.e. the mass of water vapour to mass
      of dry air (kgkg^{-1}).

The specific humidity can be calculated from the water vapour partial pressure, p, and the atmospheric pressure, p_a:

$$\omega = \frac{0.62198p}{p_a - p} \tag{6.8}$$

The water vapour partial pressure is calculated from relative humidity φ using:

$$p = \varphi p_s \tag{6.9}$$

where p_s is the water vapour pressure at saturation, which depends on the temperature.

The specific humidity can also be derived from the mass concentration of water, C_w, or volume concentration, c_w, using:

$$\omega = \frac{C_w}{1 - C_w} = \frac{\rho c_w}{1 - \rho c_w} \tag{6.10}$$

Also interesting and much simpler to assess is the temperature (energy) efficiency which gives the effectiveness of temperature recovery. This efficiency is simply calculated from the temperature measurements upstream and downstream of the heat exchanger in both the supply and exhaust ducts:

$$\varepsilon_{HR} = \frac{\theta_{\text{downstream, supply}} - \theta_{\text{upstream, supply}}}{\theta_{\text{upstream, exhaust}} - \theta_{\text{upstream, supply}}} \tag{6.11}$$

It is the ratio of the temperature increase (temperature decrease for cooling) of supply air to the temperature difference between indoor and outdoor air.

The efficiency measured in test units and provided by the manufacturer should however not be confused with the global efficiency of the heat recovery system, which is the part of the ventilation heat loss that is recovered in the heat recovery system. This latter figure is always smaller than the nominal efficiency for various reasons explained in detail in Chapter 3 (Section 3.7.2).

Leaks

Some heat exchangers allow some air leakage between both air ducts. In addition, some air is entrained by the rotation of the wheel in RHEs. The amount of air transferred this way can be measured, and the leakage flow rate is one of the results of the measurement of airflow rates in the AHU.

As mentioned in Section 6.5.6, the total recirculation rate can easily be checked by measuring the concentration of a tracer injected in the ventilated space, such as the carbon gas exhaled by occupants. Assuming that there is no reverse recirculation and no leaks in the AHU, the total recirculation rate is

$$R_{xs} = \frac{C_{\text{supply}} - C_{\text{outdoor}}}{C_{\text{exhaust}} - C_{\text{outdoor}}} \tag{6.12}$$

If no recirculation is expected, a significant recirculation rate may result from leakage through the heat exchanger. If more information is required, in particular to check whether it is the exchanger or another part of the AHU that leaks, additional measurements could be performed, as described in Section 6.5.

Pressure differential measurements are useful to explain leakage. In addition, these are easier to perform than leakage measurements and can readily provide information for a diagnosis.

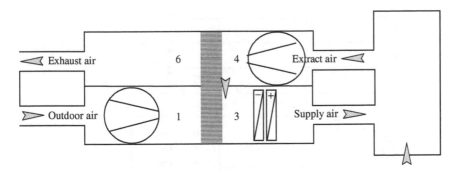

Figure 6.25 Schematics of an AHU, showing location of pressure taps for pressure differentials measurements.

Pressure differentials should be measured between the following locations (Figure 6.25):

- between 1 and 6 on one hand, and 3 and 4 on the other hand. These pressure differentials drive the leakage direction. They should be positive, so that a possible leakage flow goes from supply to exhaust, and not the contrary.
- between 1 and 3 on one hand, and 4 and 6 on the other hand. These pressure differentials increase with clogging. These should compare with the nominal pressure differential given by the manufacturer for the actual airflow rates. If these pressure differentials are significantly larger than the nominal values, the wheel should be cleaned.

6.4.7 Ductwork

Ducts supply the air from the AHU to the ventilated spaces and from the ventilated space back to the unit as necessary. These ducts should be airtight, otherwise considerable amount of air is lost on the way, and does not reach the appropriate ventilated space, thus reducing the indoor air quality and comfort as well as waste energy. Experiments performed in France and Belgium have shown that up to 40 per cent of the air was lost this way.

The ductwork should also be clean, otherwise it may be a source of contaminants. Figure 6.26 shows the increase in odour intensity through ducts having oil residue from the duct during manufacture. It is interesting to note that this increase does not depend on the airflow rate or air velocity in the duct. In other words, increasing the airflow rate does not help in improving the air quality downstream the ducts.

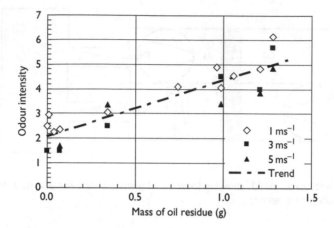

Figure 6.26 Odour intensity downstream ducts with oil residue.

Cleanliness of the ductwork requires not only that ducts be clean when leaving the factory, but also that they should be kept clean on the construction site until they are put in place, and then maintained clean by installing and maintaining appropriate filters at each air inlet.

6.5 Airflow rate measurements in ventilation systems

Measurements of airflow rates in ventilation systems may be useful to check if the air follows the expected paths, to detect potential problems early so they can be corrected and to optimize the performance of the airflow system. This includes, among others, checking that:

- airflow rates are the desired values
- negligible leakage and short circuits are present.

This check should be performed:

- when commissioning a ventilation system, to ensure that the system is built according to the design specifications;
- if there are indoor air quality problems, to help in finding the causes;
- before refurbishing a ventilation system, to accurately identify the potential problems to be cured by the refurbishing.

Diagnosis tools are useful to detect malfunction, preferably before or during commissioning the AHU. Velocity measurements using Pitot tubes

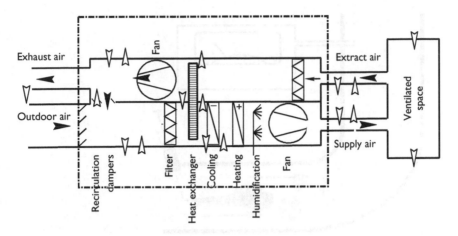

Figure 6.27 Schematics of an air-handling unit showing main (solid arrows) and secondary (open arrows) airflow paths.

or other types of anemometers are often used for that purpose. Such techniques, however, can be applied only to long straight ducts, seldom found in mechanical ventilation systems. In addition, only main airflow rates are measured that way, whereas many other airflow paths may be found in an AHU, as shown in Figure 6.27.

6.5.1 Principle of airflow measurement in ducts

Tracer gas techniques can be used for measuring airflow in ducts. The simplest application of tracer gas dilution technique to the measurement of airflow in ducts is illustrated in Figure 6.28. The tracer is injected at a known, constant flow rate, I (kgs^{-1}). The air is analysed downstream, far enough form the injection port, to ensure a good mixing of the tracer with the air.

If no tracer is lost between the injection and measuring points, the mass balance of tracer gas k is

$$\dot{m}_{tk} = (C_k - C_{k,0})\,\dot{m} \tag{6.13}$$

where

C_k is the concentration of tracer k as obtained by measurement,
$Ck,_0$ the concentration upstream of the injection port (if any) and
$\dot{m}$ the airflow rate in the duct, which is then

$$\dot{m} = \frac{\dot{m}_{tk}}{(C_k - C_{k,0})} \tag{6.14}$$

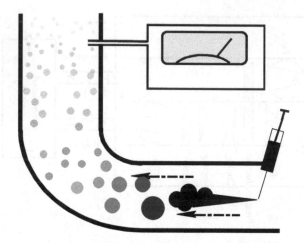

Figure 6.28 Principle of the tracer gas dilution method.

This simple method assumes steady state: both airflow rate and injection flow rate are constant, and the concentration is analysed after a long enough time period from the starting the injection to reach a constant concentration. Note that in many cases, the density of the air does not change much between outdoor and indoor. In these cases, the mass balance can be replaced with a good approximation by a volume balance, mass flow rates by volume flow rates and mass concentrations by volume concentrations.

6.5.2   Measurement of airflow rates in air-handling units

Tracer gas injection

Several tracer gases should be injected at various locations to perform simultaneous measurement of several airflow rates. Tracer gases are injected, most often at a constant flow rate, at carefully chosen locations in the AHU. In principle, the airflow rate in each duct or in each branch of a duct network can be measured by injecting a tracer gas in this duct and analysing the air upstream and downstream the injection port. This requires however many tracer gas injections and many more air samples. Experience has shown that the experiment can be made simpler by identifying the most practical and efficient injection locations, as indicated in Figure 6.29. Two tracer gases suffice in most cases to determine all primary and most secondary airflow rates:

- tracer 1 injected in the outside air duct
- tracer 2 injected in the main return air duct.

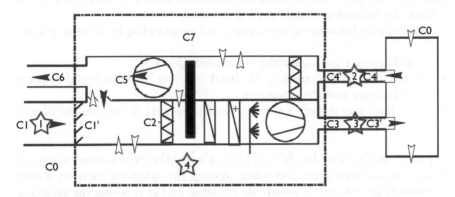

Figure 6.29 Locations of tracer gas injection (stars), and sampling points for concentration measurements (C_i).

Additional injection ports are as follows:

- tracer 3 injected in the main supply air duct, allowing the direct and more accurate determination of supply airflow rate
- tracer 4 injected in the plant room at constant concentration, to determine leakage from the plant room into the AHU.

If several tracer gases are needed but not available, it is possible to use the same tracer gas in several experiments by injecting the tracer successively at different locations. In this case, it is recommended that we start with injection location 2, then 3 and finally 1. This strategy shortens the time required between two experiments to reach zero tracer gas concentration in the system.

The optimal tracer gas injection rate depends on the design airflow rate $\dot{m}$ in the duct and on the desired concentration, C. A good method is to adjust the tracer gas injection flow rate on the basis of the outdoor airflow rate $\dot{m}_{01}$. If C_k is the expected tracer gas concentration of tracer k, then

$$\dot{m}_{tk} = C_k \dot{m}_{01} \tag{6.15}$$

Air–tracer mixture sampling points

Tracer gas concentrations are measured at various locations, in order to obtain sufficient number of equations from the conservation of airflow and tracer gas flows to determine all the wanted airflow rates. It is important that good mixing of tracer gas in the measured airflow be ensured. For this reason, the following criteria shall be fulfilled.

The distance between injection ports and air-sampling location is at least:

- ten diameters (or duct width) in straight ducts
- five diameters if there is a 90° bend or a fan between injection ports and mixture sampling location
- five diameters if there are several injection ports in the same section of the duct.

Proper mixing shall be checked by looking at the variations of measured concentration with time, and when moving the sampling location within the duct (Figure 6.30). If variations are large and at random, the sampling and/or injection points have to change or multiple injection ports are used until variations are within acceptable measurement errors.

If the distances mentioned above cannot be achieved, multiple injection locations must be used (Figure 6.31) and/or obstacles are placed in the airflow path to increase the turbulence and enhance mixing.

Measurement of the tracer concentration upstream of the sampling locations should be avoided. The distance between the sampling location upstream of the injection points and the injection nozzle should be at least one diameter when there is no possibility of backward flow, and larger (three to five diameters) when backward airflow can be suspected (e.g. close to T-junctions).

Note: Never use tubes that were once used for injecting pure tracer for sampling, because some tracer gas is adsorbed in the plastic of the tubes. It is hence recommended to use different colours for injection and sampling tubes.

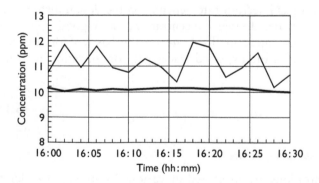

Figure 6.30 Evolution of tracer gas concentration versus time: thick line, good mixing of tracer gas; thin line, poor mixing.

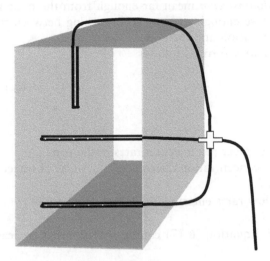

Figure 6.31 Example of a multiple injection device.

6.5.3 Interpretation of measured data

General method

Tracer gas and air mass conservation equations can be written in various ways, leading to different interpretation methods. We have found that some methods are better than others for application to AHUs, and therefore we present these methods below.

The node-by-node method, developed by Roulet and Compagnon (1989) for multi-zone tracer measurements in buildings, is also presented in its general form in Roulet and Vandaele (1991). It can be applied to ductwork and AHUs.

Airflow and tracer gas conservation equations can be rearranged so as to obtain one system of equations per node, giving all airflow rates entering at this node:

$$-\dot{m}_{tik} = \sum_{j=0}^{N} \left(C_{jk} - C_{ik} \right) \dot{m}_{ji} \tag{6.16}$$

where

$\dot{m}_{tik}$ is the injection rate of tracer gas k at (or just upstream) node i
C_{jk} is the concentration of racer gas k at (or just downstream) node j
$\dot{m}_{ji}$ is the airflow rate from node j to node i

Just upstream and *just downstream* mean far enough from the node to ensure good mixing, but close enough to have no branching between the injection port or sampling location and the node.

It can be rewritten in a matrix form as follows:

$$\vec{m}_{t,i} = \underline{C}_i \vec{m}_i \tag{6.17}$$

where

> $\vec{m}_{t,i}$ is the vector for the tracer gas injection rates in the zone i
> $\underline{C}_i$ is the matrix for the concentration differences, $C_{jk} - C_{ik}$, of tracer k between zones j and i
> $\vec{m}_i$ is the vector of airflow rates entering into zone i from zones j.

In zones where $\vec{m}_{t,i} \neq 0$, Equation (6.17) can be solved either by least square fit:

$$\vec{m}_i = \left[\underline{C}^T \underline{C}_i^T \right]^{-1} \underline{C}_i^T \vec{m}_{t,i_i} \tag{6.18}$$

when there are more equations than airflow rates, or reduced first, by elimination or addition of some equations, and the solved in the standard way:

$$\vec{m}_i = \underline{C}_i^{-1} \vec{m}_{t,i} \tag{6.19}$$

In zones where $\vec{m}_{t,i} = 0$, the system of equations can only provide linear combinations of airflow rates, when the determinant $|\underline{C}_i| = 0$.

The airflow rates leaving the zones are determined by mass conservation equations:

$$\dot{m}_{i0} = \sum_{j=0}^{N} (1 - \delta_{ij}) \dot{m}_{ji} - \sum_{j=1}^{N} (1 - \delta_{ij}) \dot{m}_{ij} \tag{6.20}$$

where δ_{ij} is the delta function: $\delta_{ij} = 1$ if $i = j$ and 0 otherwise.

Application to common AHUs

The main airflows as well as leakage and shortcuts in the air-handling system are represented in Figure 6.27. This represents many possible airflows. This representation could however be simplified.

Recirculation may be intentional or could pass through a leak such as those sometimes found in heat exchangers. From the indoor air quality point of view however, there is not much difference between the recirculated air due to a leak between the extract and supply parts of the AHU or from a

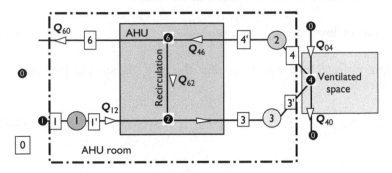

Figure 6.32 A simplified network representing the air-handling unit and ducts. Numbers in black circles represent the nodes of the network, large circles are tracer gas injection locations, and numbered rectangles are air sampling locations. Arrows represent possible airflow rates being measured.

purpose-installed duct. On the other hand, AHUs with RHEs seldom have recirculation duct.

Therefore, the simplified network, as shown in Figure 6.32, is adapted for most investigations.

The possible airflows measured are shown in Table 6.1.

Calculation of airflow rates

As a minimum, it is recommended that the tracer gases are injected in the outdoor and extract air ducts (locations 1 and 2 in Figure 6.32). The airflow rates, calculated from air and tracer gas mass conservation equations are calculated as follows. $\dot{m}_{tk}$ is the injection rate of tracer k, and C_{jk} is the

Table 6.1 Possible airflows in the network represented in Figure 6.32

| | | Going into node | | | | |
		0	1	2	4	6	7
Coming from node	0		$\dot{m}_{01}$		$\dot{m}_{04}$		$\dot{m}_{07}$
	1			$\dot{m}_{12}$			
	2				$\dot{m}_{24}$	$\dot{m}_{26}$	$\dot{m}_{27}$
	4	$\dot{m}_{40}$				$\dot{m}_{46}$	
	6	$\dot{m}_{60}$	$\dot{m}_{61}$	$\dot{m}_{62}$			$\dot{m}_{67}$
	7	$\dot{m}_{70}$		$\dot{m}_{72}$		$\dot{m}_{76}$	

Main airflows are in grey boxes.

steady-state concentration of tracer k at location j.

Intake airflow rate :
$$\dot{m}_{12} = \frac{\dot{m}_{t1}}{C_{1'1} - C_{11}} \tag{6.21}$$

Supply airflow rate, assuming that the AHU is airtight (tracer $k = 2$ recommended):

$$\dot{m}_{24} = \dot{m}_{12} \frac{C_{6k} - C_{1k}}{C_{6k} - C_{3k}} \tag{6.22}$$

Extract airflow rate :
$$\dot{m}_{46} = \frac{\dot{m}_{t2}}{C_{62} - C_{42}} \tag{6.23}$$

Recirculation flow rate :
$$\dot{m}_{62} = \dot{m}_{12} \frac{C_{3k} - C_{1k}}{C_{6k} - C_{3k}} \tag{6.24}$$

Or alternatively
$$\dot{m}_{62} = \dot{m}_{24} \frac{C_{3k} - C_{1k}}{C_{6k} - C_{1k}} \tag{6.25}$$

Another alternative is to calculate $\dot{m}_{62}$ from the recirculation ratio R, if R is assessed using Equation (6.47):

$$\dot{m}_{62} = R\dot{m}_{24} \tag{6.26}$$

Infiltration flow rate, with $k \neq 3$ ($k = 1$ is recommended because this tracer injection location provides the best accuracy for this flow rate):

$$\dot{m}_{04} \cong \dot{m}_{24} \frac{(C_{3'k} - C_{4k})}{(C_{4k} - C_{0k})} = \dot{m}_{12} \frac{(C_{6k} - C_{1k})}{(C_{6k} - C_{3k})} \frac{(C_{3'k} - C_{4k})}{(C_{4k} - C_{0k})} \tag{6.27}$$

Exfiltration flow rate :
$$\dot{m}_{40} = \dot{m}_{04} + \dot{m}_{24} - \dot{m}_{46} \tag{6.28}$$

Exhaust airflow rate :
$$\dot{m}_{60} = \dot{m}_{04} - \dot{m}_{40} + \dot{m}_{01} \tag{6.29}$$

Error analysis of airflow rates

The error analysis is based on the assumption that random and independent errors spoil the measurements of tracer gas concentration and injection rates. In this case, the confidence interval of any result, for example an airflow rate, is:

$$[\dot{m} - \delta\dot{m}; \dot{m} + \delta\dot{m}] \quad \text{with} \quad \delta\dot{m}(x_i) = T(P, \infty) \sqrt{\sum_i \left(\frac{\partial\dot{m}}{\partial x_i}\right)^2 \delta x_i^2} \tag{6.30}$$

where

$T(P, \infty)$ is the Student coefficient for having the actual value within the
confidence interval with probability $1 - P$

x_i is any variable on which the airflow rate $\dot{m}$ depends

δx_i is the standard deviation of the variable x_i, assumed to be a random
variable of mean x_i.and normal distribution.

The confidence intervals of the airflow rates are then as follows.

Intake airflow rate : $\delta \dot{m}_{12} = T(P, \infty) \sqrt{\dfrac{(C_{1'1} - C_{11})^2 \, \delta \dot{m}_{r1}^2 + \dot{m}_{r1}^2 \left(\delta C_{1'1}^2 + \delta C_{11}^2 \right)}{(C_{1'1} - C_{11})^4}}$

$$(6.31)$$

Supply airflow rate: $\delta \dot{m}_{24} = \sqrt{\delta \dot{m}_{12}^2 + \delta \dot{m}_{62}^2} = \dfrac{T(P, \infty)}{(C_{6k} - C_{3k})^2} \sqrt{f_{24}} \qquad (6.32)$

where

$$f_{24} = (C_{6k} - C_{1k})^2 (C_{6k} - C_{3k})^2 \, \delta \dot{m}_{12}^2 +$$

$$+ \dot{m}_{12}^2 \left[(C_{6k} - C_{3k})^2 \, \delta C_{1k}^2 + (C_{6k} - C_{1k})^2 \, \delta C_{3k}^2 + (C_{1k} - C_{3k})^2 \, \delta C_{6k}^2 \right]$$

$$(6.33)$$

Extract airflow rate: $\delta \dot{m}_{46} = T(P, \infty) \sqrt{\dfrac{(C_{62} - C_{42})^2 \, \delta \dot{m}_{r2}^2 + \dot{m}_{r2}^2 \left(\delta C_{62}^2 + \delta C_{42}^2 \right)}{(C_{62} - C_{42})^4}}$

$$(6.34)$$

Recirculation flow rate:

$$\delta \dot{m}_{62} = \dfrac{T(P, \infty)}{(C_{6k} - C_{3k})^2} \sqrt{(C_{6k} - C_{3k})^2 (C_{3k} - C_{1k})^2 \, \delta \dot{m}_{12}^2 + \dot{m}_{12}^2 f_{62}} \qquad (6.35)$$

where $f_{62} = (C_{6k} - C_{3k})^2 \, \delta C_{1k}^2 + (C_{6k} - C_{1k})^2 \, \delta C_{3k}^2 + (C_{3k} - C_{1k})^2 \, \delta C_{6k}^2$

Alternatively $\delta \dot{m}_{62} = \dfrac{T(P, \infty)}{(C_{6k} - C_{1k})^2} \sqrt{(C_{6k} - C_{1k})^2 (C_{3k} - C_{1k})^2 \, \delta \dot{m}_{24}^2 + \dot{m}_{24}^2 f_{62}'}$

$$(6.36)$$

where $f_{62}' = (C_{6k} - C_{3k})^2 \, \delta C_{1k}^2 + (C_{6k} - C_{1k})^2 \, \delta C_{3k}^2 + (C_{3k} - C_{1k})^2 \, \delta C_{6k}^2$

or if calculated from Equation (6.26) : $\delta \dot{m}_{62} = T(P, \infty) \sqrt{R^2 \delta \dot{m}_{24}^2 + \dot{m}_{24}^2 \delta R^2}$

$$(6.37)$$

Infiltration : $\delta \dot{m}_{04} = \dfrac{T(P, \infty)}{(C_{4k} - C_{0k})^2} \sqrt{(C_{3'k} - C_{4k})^2 (C_{4k} - C_{0k})^2 \, \delta \dot{m}_{24}^2 + \dot{m}_{24}^2 f_{04}'}$

$$(6.38)$$

with
$$f_{04} = (C_{3'k} - C_{4k})^2\, \delta C_{0k}^2 + (C_{4k} - C_{0k})^2\, \delta C_{3'k}^2 + (C_{3'k} + C_{0k})^2\, \delta C_{4k}^2$$
(6.39)

Exfiltration :
$$\delta \dot{m}_{40} = \sqrt{\delta \dot{m}_{04}^2 + \delta \dot{m}_{24}^2 + \delta \dot{m}_{46}^2}$$
(6.40)

Exhaust :
$$\delta \dot{m}_{60} = \sqrt{\delta \dot{m}_{04}^2 + \delta \dot{m}_{40}^2 + \delta \dot{m}_{01}^2}$$
(6.41)

6.5.4 Application to large recirculation ratios

In Equations (6.22), (6.24), and (6.27), the difference in concentrations $C_{6k} - C_{3k}$ is in the denominator, and these two concentrations are close to each other in steady state when the recirculation ratio is high. This leads to a large confidence interval of the calculated airflow rates. In this case, it is better to inject the tracer gas at location three instead of location 2. The supply airflow rate can then be determined with a better accuracy, using

$$\dot{m}_{24} = \frac{\dot{m}_{t3}}{C_{3'3} - C_{33}}$$
(6.42)

The confidence interval being calculated, *mutatis mutandis*, using Equation (6.31) or, assuming that the confidence interval is the same for both concentrations:

$$\frac{\delta \dot{m}}{\dot{m}} \cong T(P, \infty)\sqrt{\left(\frac{\delta \dot{m}_t}{\dot{m}_t}\right)^2 + 2\left(\frac{\delta C}{C' - C}\right)^2}$$
(6.43)

The recirculation airflow rate can then be calculated using

$$\dot{m}_{62} = \dot{m}_{24} - \dot{m}_{12}$$
(6.44)

with
$$\delta \dot{m}_{62} = T(P, \infty)\sqrt{\delta \dot{m}_{24}^2 + \delta \dot{m}_{12}^2} \cong T(P, \infty)\sqrt{1 + (1 - R_{xs})^2}\,\delta \dot{m}$$
(6.45)

assuming that the relative error $\delta \dot{m}/\dot{m}$ is the same for both airflow rates, and taking into account that $\dot{m}_{12} = (1 - Rxs)\dot{m}_{24}$. Note that, in this case, $\delta \dot{m}_{62}$ decreases when R_{xs} increases.

The extract airflow rate $\dot{m}_{46}$ cannot be assessed without injecting a tracer gas in the extract duct. However, in AHUs having no exhaust duct (where the air leaves the building by leakages or vents) $\dot{m}_{60} = 0$, hence $\dot{m}_{46} = \dot{m}_{62}$, and $\dot{m}_{40} = \dot{m}_{01} + \dot{m}_{04}$.

Determining the optimum method for a system with large recirculation ratio

The recirculation ratio is defined by

$$R_{xs} = \frac{\dot{m}_{62}}{\dot{m}_{24}} = \frac{\dot{m}_{62}}{\dot{m}_{62} + \dot{m}_{12}} \tag{6.46}$$

Assuming that there is no leak in the AHU, it can be assessed using three different methods:

$$\text{Method A} \qquad R_{xs} = \frac{C_{3k} - C_{1'k}}{C_{6k} - C_{1'k}} \tag{6.47}$$

the subscript k being for any tracer gas except the one injected in inlet duct. The confidence interval is

$$\delta R_{xs} = \frac{T(P, \infty)}{(C_{6k} - C_{1'k})^2} \sqrt{f_R} \tag{6.48}$$

where $f_R = (C_{3k} - C_{6k})^2 \, \delta C_{1'k}^2 + (C_{6k} - C_{1'k})^2 \, \delta C_{3k}^2 + (C_{3k} - C_{1'k})^2 \, \delta C_{6k}^2$ (6.49)

If we assume that the relative error is the same for all concentrations, and taking into account that, for tracers injected at locations 2 and 3, $C_{1'k} \cong 0$ and therefore $C_{3k} \cong RC_{6k}$, we can get a simpler expression for the confidence interval of the recirculation ratio:

$$\delta R_{xs} \cong \frac{T(P, \infty)\delta C}{C} \sqrt{2(R_{xs}^2 - R_{xs} + 1)} \tag{6.50}$$

The recirculation ratio can also be calculated using

$$\text{Method B } R_{xs} = \frac{\dot{m}_{62}}{\dot{m}_{24}} \qquad \text{with} \qquad \delta R_{xs} = \frac{\delta \dot{m}}{\dot{m}} \sqrt{(1 + R_{xs})} \tag{6.51}$$

$$\text{Method C } R_{xs} = 1 - \frac{\dot{m}_{12}}{\dot{m}_{24}} \qquad \text{with} \qquad \delta R_{xs} = \sqrt{2}\frac{\delta \dot{m}}{\dot{m}}(1 - R_{xs}) \tag{6.52}$$

assuming that the relative error $\delta \dot{m}/\dot{m}$ is the same for both airflow rates, and taking into account that $\dot{m}_{12} = (1 - R_{xs})\dot{m}_{24}$.

The three methods for determining R_{xs} and δR_{xs} are compared in Figure 6.33. Method B should be preferred at low recirculation ratio, whereas method C is best at large recirculation ratio. Method A could be applied at low recirculation if method B cannot be applied.

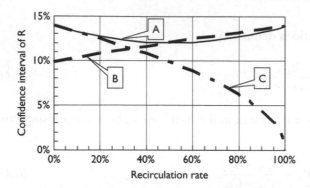

Figure 6.33 Relative confidence interval of the recirculation ratio as function of the recirculation ratio itself, for three assessment methods. For this figure, the relative confidence interval (at 90 per cent) of injection rate and concentrations is 5 per cent.

This error analysis shows that some method of conducting the measurement can provide more accurate results than others and that the best way depends on the type of ventilation unit measured. Therefore, care should be taken to select the most appropriate method for large recirculation ratios. Measurement time could be shortened by fitting the dynamic expression of concentration to the experimental points to assess the steady-state concentration without reaching it.

6.5.5 Unsteady state and assessment of concentrations when the time constant is large

With the constant injection technique, steady state is quickly reached within ducts, where the air velocity is $1\,\mathrm{ms}^{-1}$ or more. The difference in concentration between air sampled downstream (far enough to get complete mixing) and upstream the tracer gas injection port reaches a constant value quickly. However, in the case of high recirculation, it may take a long time to reach steady-state concentrations in the supply duct (location 3) and in the room (location 4), where the air from the room is progressively mixed with outdoor air.

Writing the conservation equation of tracer gas 3 at node 4, in the ventilated space, gives

$$\rho_a V \frac{\partial C_{43}}{\partial t} = \dot{m}_{t3} + \dot{m}_{24} C_{23} + \dot{m}_{04} C_{03} - (\dot{m}_{46} + \dot{m}_{40}) C_{43} \qquad (6.53)$$

Because of the large recirculation ratio, it can be assumed that the concentration is homogeneous in the ventilated space. Dividing this equation by the supply airflow rate $\dot{m}_{24}$ gives

$$\frac{\rho_a V}{\dot{m}_{24}}\frac{\partial C_{43}}{\partial t} = \frac{\dot{m}_{t3}}{\dot{m}_{24}} + C_{23} + \gamma_i C_{03} - (1+\gamma_i)C_{43} \tag{6.54}$$

where γ_i is the infiltration ratio $\dot{m}_{04}/\dot{m}_{24}$. Using the definition of the nominal time constant τ_n, of the recirculation ratio R_{xs}, and using the tracer gas conservation at node 2:

$$\frac{V}{\dot{m}_{24}} = \frac{V}{\dot{m}_{01}}\frac{\dot{m}_{01}}{\dot{m}_{24}} = \tau_n(1-R_{xs}) \quad \text{and} \quad C_{23} = R_{xs}C_{43} + (1-R_{xs})C_{03} \tag{6.55}$$

we get
$$\tau_n(1-R_{xs})\frac{\partial C_{43}}{\partial t} = \frac{\dot{m}_{t3}}{\dot{m}_{24}} - (1-R_{xs}+\gamma_i)(C_{43}-C_{03}) \tag{6.56}$$

The steady-state concentration is

$$C_{43}^\infty = \frac{\dot{m}_{t3}}{\dot{m}_{24}(1-R+\gamma_i)} + C_{03} \tag{6.57}$$

and
$$C_{43}(t) = C_{43}^\infty\left(1-e^{-\frac{t}{\tau}}\right) \quad \text{with} \quad \tau = \frac{\tau_n(1-R_{xs})}{1-R_{xs}+\gamma_i} \tag{6.58}$$

The theoretical exponential can be fitted to the experimental points, allowing the determination of the steady-state concentration and time constant without waiting for equilibrium. Figure 6.34 shows such a fit performed in an actual experiment.

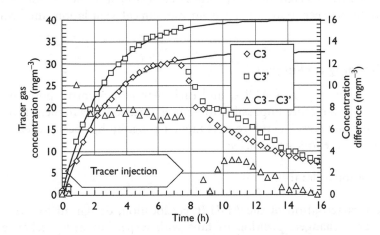

Figure 6.34 Tracer gas concentrations in the supply duct, upstream (3) and downstream (3′) the tracer gas injection port. Dots are measured concentrations, while lines are exponential fits.

6.5.6 Simple measurement

A special case is when only one tracer is injected in the ventilated space. This could be carbon dioxide gas emitted in the ventilated space by occupants. That tracer gas is of great practical interest, because it does not need any injection system. In this case, Equation (6.24) can easily be solved. Assuming that there is no inverse recirculation, and no leaks in the AHU, the global recirculation rate is

$$R_{xs} = \frac{\dot{m}_{62}}{\dot{m}_{12} + \dot{m}_{62}} = \frac{C_{3k} - C_{1k}}{C_{4k} - C_{1k}} \tag{6.59}$$

with $\qquad \delta R_{xs} = \dfrac{T(P, \infty)}{(C_{4k} - C_{3k})^2} \sqrt{f_R}$

where $\qquad f_R = (C_{3k} - C_{4k})^2 \, \delta C_{1k}^2 + (C_{4k} - C_{1k})^2 \, \delta C_{3k}^2 + (C_{3k} - C_{1k})^2 \, \delta C_{4k}^2$

The outdoor airflow rate per occupant is

$$\frac{\dot{m}_{01} + \dot{m}_{04}}{N_{\text{persons}}} = \frac{0.018[\text{m}^3\text{h}^{-1}]}{C_{4k} - C_{0k}} \tag{6.60}$$

assuming that a person exhales $18\,\text{l}\,\text{h}^{-1}$ carbon gas. Airflow rates are given in (m^3h^{-1} per person) if concentrations are as volumetric ratios. With only one tracer injected into the ventilated space, it is not possible to differentiate between outdoor air from mechanical ventilation and from infiltration.

6.5.7 Planning techniques

There are many types of AHUs, and each new measurement poses new problems. It is therefore impossible to write a detailed measurement protocol valid for all types. Therefore, a computer program has been developed that performs the following tasks:

1   Request input data:

    a   Characterisation of the AHU (type, location, design airflow rates, heat exchanger, position of fans with respect to heat exchanger, etc.);

    b   Tracer gas(es) used, injection location(s) and design concentration(s);

c   Characterisation of building (approximate volume, number of occupants, over-pressurized or not, etc.;

d   Airflows that certainly cannot occur in the investigated system.

2   Evaluate the risk of poor tracer gas mixing between the injection and sampling locations and from devices (e.g. fans, bends, filters and dampers) placed in-between.

3   Prepare a printed measurement protocol containing injection and sampling locations, injection rates of tracer gases and system of equations.

4   Request measured tracer gas concentrations and actual injection flow rates or read them in a file.

5   Solve the system of equations and prepare a measurement report.

Such a computer program is available from the website: www.e4tech.com.

6.5.8 Examples

Measurements in one single unit

Sulphur hexafluoride (SF_6) was injected as tracer 1, and nitrous oxide as tracer 2 in an AHU without planned recirculation, but equipped with a RHE. The resulting concentrations are shown in Figure 6.35 and measurement results in Figure 6.36.

Leaks in the heat exchanger, as well as in the return air channel, were detected with this measurement. Measurement in three other identical units

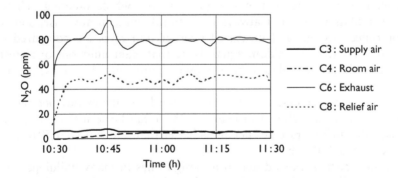

Figure 6.35 Concentrations at locations shown in Figure 6.29 resulting from injection of SF_6 as tracer 1 and N_2O as tracer 2, in a leaky air-handling unit. Short-circuiting through the heat exchanger dilutes exhaust air, thus decreasing the relief air concentration. Presence of this tracer gas in the supply air results from parasitic recirculation.

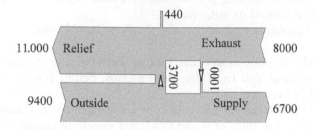

Figure 6.36 Measured airflow rates in a leaky air-handling unit. Design airflow rates were $13,000\,m^3h^{-1}$ for both supply and return, and zero for recirculation.

in the same office did not show any short-circuiting. However, measured outdoor airflow rates were between 55 and 60 per cent of the design value.

Results from several units

In order to show the usefulness of measurements, some results of investigations performed on several AHUs and various buildings are shown below. It should be emphasized that these ventilation units were not selected for having problems. These are units routinely investigated within the frame of two European research programs in Switzerland (Roulet *et al.*, 1999).

Air recirculation is sometimes designed for a purpose, e.g. in order to distribute heat or cold without conditioning too much outdoor air. This may however decrease the global indoor air quality, because this disseminates the contaminants generated at one place in the whole building. Therefore, recirculation may not always be desirable and, in any case, should be controlled. Design and measured recirculation rates are compared in Figure 6.37. These are seldom equal. Note that eight units out of sixteen designed without recirculation have shown significant – but unexpected – recirculation.

The comparison of design and measured outdoor airflow rate per person in twelve buildings is shown in Figure 6.38. It can be seen that in several buildings the airflow rate per person is larger than the expected $50\,m^3h^{-1}$, and overpasses $200\,m^3h^{-1}$.

Measured airflow rates differ from design ones in many buildings. The percentage differences between measured and design outdoor airflow rates in thirty-four AHUs are shown in Figure 6.39. These range from −76 per cent to +54 per cent. Only four units are within the ±10 per cent range.

In principle, supply and exhaust airflow rates are balanced, or the building is exposed to a slight positive pressure. When the envelope is not airtight, and when the difference between supply and exhaust air is too large, air

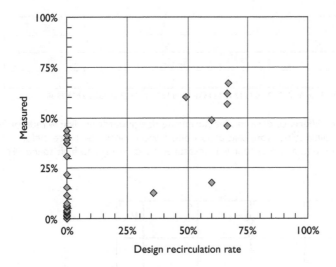

Figure 6.37 Comparison between design and measured recirculation rate in nineteen air-handling units.

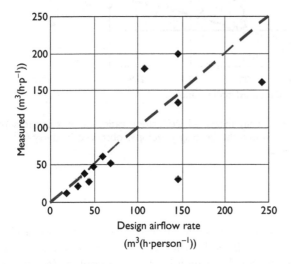

Figure 6.38 Design and measured outdoor airflow rate per person in twelve buildings.

leaks through the envelope. This has not much influence on indoor air quality but may strongly decrease the efficiency of the heat recovery when used. In some buildings, as much as 70 per cent of the supply air is lost that way (Figure 6.40).

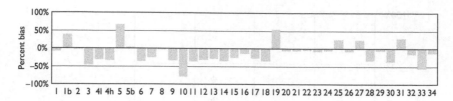

Figure 6.39 Relative difference between measured and design outdoor airflow rate for 34 air-handling units. Note that units I and 5 were measured twice, before and after retrofit, and that four was measured at low (4 I) and high speed (4 h).

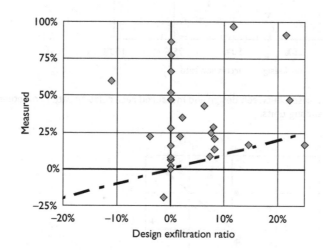

Figure 6.40 Design and actually measured exfiltration ratio, i.e. part of the supply air-leaking through the building envelope, observed in thirty-four air-handling units.

6.6 Summary

In addition to natural ventilation systems, there are several types of mechanical ventilation systems that differ by the way the air is ducted to and from the ventilated space and the principles used to control (or not) the temperature and the humidity in this space.

AHUs include several components to control the airflow rate, to clean, heat, cool, dry or humidify the air. When poorly designed, installed or maintained, these components may be sources of pollution. Therefore, all components should be clean when installed, and the whole installation should be kept clean by periodic maintenance. This is especially valid for filters and humidifiers.

Air leakages and short circuits in mechanical ventilation systems as well as airflow rates larger or smaller than planned cannot be easily detected. There are however ways to measure the main and parasitic airflow rates and to check whether the systems function as expected or not. This is particularly useful when commissioning a new system.

References

Andersson, B., Andersson, K., Sundell, J. and Zingmark, P.-A. (1993), Mass transfer of contaminants in rotary enthalpy exchangers. *Indoor Air*, v. 3, p. 143–148.

ASHRAE, (2004), *Handbook – HVAC Systems and Equipment*. Atlanta, ASHRAE.

Bluyssen, P. M., Cox, C., Souto, J., Müller, B., Clausen, G. and Bjrkroth, M. (2000a), Pollution from filters: what is the reason, how to measure and to prevent it? *Healthy Buildings*, p. 251.

Bluyssen, P. M., Björkroth, M., Müller, B., Fernandes, E. d. O. and Roulet, C.-A. eds., (2000b), Why, when and how do HVAC-systems pollute? Characterisation of HVAC systems related pollution. *Healthy Buildings 2000*, vol.2, Espoo (SF), SIY Indoor Air Information Oy, Finland, p. 233–238.

Bluyssen, P. M., Cox, C., Seppänen, O., Fernandes, E. d. O., Clausen, G., Müller, B. and Roulet, C.-A. (2003), Why, when and how do HVAC-systems pollute the indoor environment and what to do about it? The European AIRLESS project. *Building and Environment*, v. 38, p. 209–225.

Fanger, P. O. (1988), Introduction of the olf and decipol units to quantify air pollution perceived by human indoors and outdoors. *Energy and Buildings*, v. 12, p. 1–6.

Kriesi, R. (2005), Feuchte zurückgewinnen – Gründe für und Massnahmen gegen zu trockene Luft Faktor SWKI, p. 32–34.

Müller, B., Fitzner, K. and Bluyssen, P. M. (2000), Pollution from humidifiers: what is the reason; how to measure and to prevent it. *Healthy Buildings 2000*, p. 275.

Roulet, C.-A. and Compagnon, R. (1989), Multizone gas tracer infiltration measurement – interpretation algorithms for non-isothermal cases. *Energy and Environment*, v. 24, p. 221–227.

Roulet, C.-A. and Vandaele, L. (1991), Airflow patterns within buildings – measurement techniques: AIVC technical note 34, v. 34: Bracknell. Order at inive@bbri.be, AIVC, p. 265

Roulet, C.-A., Foradini, F. and Deschamps, L. (1999), Measurement of air flow rates and ventilation efficiency in air handling units. *Indoor Air'99*, p. 1–6.

Roulet, C.-A., Pibiri, M.-C., Knutti, R., Pfeiffer, A. and Weber, A. (2002), Effect of chemical composition on VOC transfer through rotating heat exchangers. *Energy and Buildings*, v. 34, p. 799–807.

Sekhar, S. C., Maheswaran, U., Tham, K. W. and Cheong, K. W. (2004), Development of energy efficient single coil twin fan air-conditioning system with zonal ventilation control. *ASHRAE Transactions*, v. 110, Part 2, p. 204–217.

Nomenclature

Symbols and units

C_{ik}	mass concentration of tracer k at node i	–
C	matrix of mass concentration differences $C_{jk} - C_{ik}$, of tracer k between zones j and i	–
c	volume concentration	–
c	specific heat capacity	$J(kg \cdot K)^{-1}$
f	factor	–
h	specific enthalpy	Jkg^{-1}
H	enthalpy flow	W
m	mass	kg
$\dot{m}$	mass flow rate	kgs^{-1}
$\vec{m}_i$	vector of mass airflow rates entering into zone i from zones j	
$\vec{m}_{t,i}$	vector for the tracer gas injection rates in the zone i	
N	number	–
p	pressure	Pa
R_{xs}	recirculation ratio in an AHU	–
$T(P, n)$	Student coefficient for probability $1 - P$ and n degree of freedom	–
V	volume	m^3
$\dot{V}$	volume flow rate	m^3s^{-1}
x	humidity ratio	–
δx	half confidence or uncertainty interval of variable x	
δ_{ij}	delta function: $\delta_{ij} = 1$ if $i = j$ and 0 otherwise	
$\Delta\theta$	temperature difference	°C
Δp	pressure differential	Pa
Φ	power	W
γ	ratio	–
η	efficiency, utilization factor	–
φ	relative humidity	–
ρ	density	kgm^{-3}
τ	time constant	s

Subscripts

0	zero, reference		k	numbering the tracer gas
a	air		n	nominal
e	electricity		o	outdoor
f	fan		s	saturation, supply
h	heating, heat		t	tracer
i	internal		T	(exponent) transposed
i	infiltration		w	water vapour
i, j	numbering zones		x	extract

Chapter 7

Characteristics of natural and hybrid ventilation systems

Per Heiselberg

7.1 Introduction

Natural ventilation in buildings is caused by the pressure difference created between inlets and outlets of the building envelope, as a result of natural driving forces such as wind and/or stack effect. Natural ventilation can provide fresh air for the occupants, necessary to maintain acceptable indoor air quality levels, and to cool the building, if the climatic conditions allow it.

The effectiveness of natural ventilation is determined by the prevailing outdoor conditions – microclimate (wind speed, temperature, humidity and surrounding topography) and the building itself (orientation, number of windows or openings, their size and location).

In hybrid ventilation, mechanical and natural driving forces are combined in a two-mode system. The operating mode can vary according to the season and/or within individual days. The active mode takes maximum advantage of ambient conditions, and a control system switches automatically between natural and mechanical modes in order to minimise energy consumption and optimise indoor air quality and thermal comfort.

The expectations of natural and hybrid ventilation performance will vary because of climate variations, differences in energy prices and other factors. In cold climates, natural and hybrid ventilation can avoid the trend to use air conditioning in new buildings, which has occurred in response to higher occupant expectations, requirements of codes and standards, and in some cases higher internal heat gains and changes in building design. In warm climates, it can reduce the reliance on air conditioning and reduce the cost, energy penalty and consequential environmental effects of full year-round air conditioning.

7.1.1 Benefits of natural and hybrid ventilation

There are multiple motivations for the interest in natural and hybrid ventilation. The most obvious are higher indoor environmental quality and user satisfaction and lower energy use and environmental impact.

Indoor environmental quality and user satisfaction

Buildings with natural ventilation are associated with less SBS symptoms, than buildings with traditional ventilation systems (Seppänen and Fisk, 2002) and are well accepted by occupants. A comparative study by Hummelgaard *et al.* (2005) in five mechanically ventilated buildings and in four naturally ventilated buildings indicated that occupants in naturally ventilated offices had a lower prevalence of symptoms than those in mechanically ventilated offices, and although the room air temperature and the concentration of CO_2 were higher in the naturally ventilated offices, their occupants were generally more satisfied with the indoor environment. More occupants in the naturally ventilated offices (60 per cent compared with 41 per cent for mechanically ventilated) did not prefer a change of the thermal environment.

Natural and hybrid ventilation systems are well accepted by occupants and often result in high user satisfaction because of the high degree of individual control of the indoor climate (adaptive comfort) as well as the direct and visible response to user interventions (Rowe, 2002).

Energy use and environmental impact

The environmental impacts of energy production and consumption have provided an increased awareness of the energy used by ventilation and air conditioning systems, and expectation of a reduction in annual energy cost is also an important driving force for the development of natural and hybrid ventilation strategies.

Available data from case studies provided in the international research project IEA ECBCS-Annex 35 (Heiselberg, 2002) show that substantial energy savings have been achieved in a number of buildings, mainly because of a very substantial reduction in energy use for fans and cooling.

In Table 7.1, the energy use, energy cost and environmental impact expressed as CO_2 emission for heating and transport of ventilation air in a typical Danish office building is shown for different ventilation strategies. The results are given per $m^3 s^{-1}$ transported air in the ventilation system during the hours of occupation. The calculations are based on the following assumptions:

Heating is provided by district heating. The balanced mechanical ventilation system has heat recovery with an efficiency of 0.7. No heat recovery for natural ventilation or mechanical exhaust.

Specific fan power is $0.4 \, kJ m^{-3}$ for the mechanical exhaust and $2.5 \, kJ m^{-3}$ for the balanced mechanical system.

The hybrid ventilation system consists of a balanced mechanical system in the heating season and a natural ventilation system outside the heating season. The change over happens at an outdoor temperature of 15° C.

Typical data for energy sources in Denmark (year 2003) is shown in Table 7.2.

Table 7.1 Energy use, energy cost and environmental impact (CO_2 emission) for heating and transport of ventilation air using different ventilation principles in a typical office building in Denmark heated by district heating. (Aggerholm et al., 2006)

	Natural ventilation	Mechanical exhaust	Balanced mechanical ventilation	Hybrid ventilation
Net energy use in MWh per m^3s^{-1} per annum				
Heat	23.5	23.5	3.2	4.0
Electricity	0.0	0.4	5.5	3.8
Energy cost in US$ per m^3s^{-1} per annum				
Heat	1,600	1,600	220	270
Electricity	0	90	1,100	770
Total	1,600	1,690	1,320	1,040
CO_2-emission in tonne per m^3s^{-1} per annum				
Heat	3.1	3.1	0.4	0.5
Electricity	0.0	0.3	3.3	2.3
Total	3.1	3.4	3.7	2.8

Table 7.2 Typical data for energy sources in Denmark (Year 2003) (Aggerholm et al., 2006)

	Cost US$ per kWh	CO_2-emission kg CO_2 per kWh	Boiler efficiency
Gasoil	0.084	0.265	0.85
Natural gas	0.084	0.205	0.95
District heating	0.067	0.130	–
Electricity	0.202	0.600	–

The calculations in Figure 7.1 show that it is both economically and environmentally beneficial to use a combination of balanced mechanical ventilation with heat recovery in the heating season and natural ventilation outside the heating season. The optimal point of change between mechanical and natural ventilation varies between outdoor temperatures of 7 and 13°C (see Figure 7.2). For comfort reasons (reduce the risk of draught), it will usually be best to use a setpoint of change of 12–15°C.

7.1.2 Drawbacks and design constraints

Natural and hybrid ventilation also have some drawbacks and design constraints. Air humidity is the most important limiting factor for the application of natural and hybrid ventilation techniques. High levels of humidity

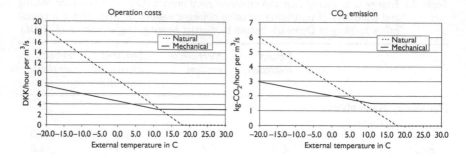

Figure 7.1 Energy cost and environmental impact expressed as CO_2 emission for heating and transport of ventilation air in a typical Danish office building for natural and balanced mechanical ventilation, respectively (Aggerholm *et al.* 2006).

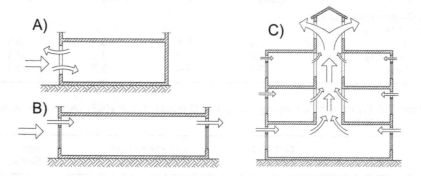

Figure 7.2 Natural ventilation principles: (A) single-sided ventilation, (B) cross ventilation and (C) stack ventilation.

have a negative influence on thermal comfort. As a result, in regions with high relative humidity levels during the summer, the use of conventional air-conditioning systems is necessary in order to remove water vapour from indoor air. Under such circumstances, natural ventilation during day- or night-time hours should be avoided.

Another major disadvantage of natural ventilation systems is the uncertainty in performance, which results in an increased risk of draught problems and/or low indoor air quality in cold climates and a risk of unacceptable thermal comfort conditions during summer periods. Hybrid ventilation systems have access to both ventilation modes and therefore allow the best ventilation mode to be chosen depending on the circumstances.

Estimating the initial cost of hybrid ventilation systems in buildings can be quite difficult as the installation often consists of both mechanical installations and building elements. Part of the investment in mechanical equipment is often shifted towards a larger investment in the building itself: increased room air volume per person, a shape favourable to air movement, a more intelligent façade/window system, etc. On the other hand, the building might provide more useable (rentable) space, as space for plant rooms, stacks for ventilation channels, etc., is not needed. Recently, a method for calculation of life cycle costs (LCC) of natural ventilation systems has been developed (Vik, 2003), which takes all these issues into consideration. This method can also be applied to buildings with hybrid ventilation systems.

In ECBCS-Annex 35, the reference cost range provided by the participants was used to compare the initial costs of hybrid ventilation systems and buildings with the initial cost of traditional systems and buildings (van der Aa, 2002a). The LCC for hybrid ventilated buildings were often lower than that for reference buildings, but the relationship between initial, operating and maintenance costs was different.

The effectiveness of natural ventilation also depends greatly on the design process. Mechanical ventilation systems can be designed separately from the design of the building in which they are installed. They can also be installed in existing buildings after a few modifications. In contrast, ventilation systems using only natural forces such as wind and thermal buoyancy need to be designed together with the building, because the building itself and its components are the elements that can reduce or increase air movement as well as influence the air content (dust, pollution etc.). Architects and engineers need to acquire qualitative and quantitative information about the interactions between building characteristics and natural ventilation in order to design buildings and systems consistent with a passive low-energy approach.

7.2   Ventilation concepts

The design principles of natural and hybrid ventilation in buildings are relatively few and straightforward, relying on wind, thermal buoyancy, fans or a combination as driving forces. However, to characterise the natural and hybrid ventilation concept in a particular building is more difficult as different ventilation principles are often applied and because a whole range of subtle and sophisticated ways are used to take advantage of the natural driving forces to promote the ventilation principles applied.

7.2.1   Characterisation of concepts

To define and describe different natural and hybrid ventilation concepts, five essential aspects can be used as suggested by Kleiven (2003). These are ventilation driving force, ventilation principle, air distribution principle, ventilation elements, building geometry and the supply and exhaust air paths. The complete set of characteristic aspects and parameters is shown in Table 7.3.

Table 7.3 Characteristic aspects and parameters for natural ventilation concepts (Kleiven, 2003)

Characteristic aspect	Characteristic parameter
Ventilation driving force	Buoyancy
	Wind
	Fan
	Combination of forces
Natural ventilation principle	Single-sided
	Cross
	Stack
Air distribution principle	Mixing
	Displacement
Ventilation element	Ventilation openings in the façade
	Wind tower
	Wind scoop
	Chimney
	Low pressure ducts
	Atrium
	Low pressure fan
	Ventilation chamber
	Embedded duct
Building geometry	Low-, medium-, or high-rise
	Compact
	Distributed
Supply and exhaust air paths	Local
	Central

7.2.2 Ventilation driving force

The ventilation driving force can be wind, buoyancy, fans or a combination of these. A natural ventilation system will often rely on both wind and thermal buoyancy as driving forces, but one of them will be predominant, and both the building and the ventilation system will be designed for optimal utilisation of this driving force. The dominating natural driving force has consequences for the shape and layout of the building, for the selection of ventilation elements to be utilised (e.g. a wind scoop or an atrium) and for the air paths into, out of and through the building (ventilation principle). In a hybrid ventilation system, the natural and mechanical driving forces can either be combined, which is characterised as a fan-assisted natural ventilation system, or they can be used sequentially as a natural and mechanical ventilation system with two fully autonomous systems, where the control strategy either switches between the two systems or uses one system for some tasks and the other system for other tasks.

7.2.3 Natural ventilation principle

The natural ventilation principle used to exploit the natural driving forces can be divided into three types (see Figure 7.2).

Single-sided ventilation, where ventilation opening(s) is only on one side of the room. The main driving force is thermal buoyancy in winter and wind turbulence in summer. Compared with other principles, lower ventilation rates are generated, and the ventilation air does not penetrate so far into the space.

Cross ventilation, where ventilation openings are on two or more sides of the room. The main driving force is wind-induced pressure differentials between the openings. High ventilation flow rates can be achieved, but because of large and rapid variations in wind flows, it is difficult to control. As air is crossing the room, greater room depths can be ventilated.

Stack ventilation, where ventilation openings are at both low and high levels. The main driving force is thermal buoyancy. High and steady ventilation flow rates can be achieved at moderate temperature differences. If ventilation air is crossing the room, larger room depths can be ventilated.

Typically, different ventilation principles are applied in a specific building for different rooms depending on the outdoor climate and user behaviour. Referring to Figure 7.2, the ventilation principle will be single-sided ventilation in a cellular office, when door is closed, while it will be stack or cross ventilation, when door is open. For open doors, the ventilation principle will be stack ventilation in the building in winter and intermediate season, while it will be cross ventilation in the summer season, when the temperature difference is very small.

The ventilation principle has implications for both the shape of the building (e.g. its depth) and its plan layout. Single-sided, and to some extent cross-ventilation, requires relatively narrow building plan depths, which is usually achieved with linear building forms. Stack ventilation can be used for considerably deeper building plans, and by puncturing a plan with chimneys for inlet and outlet, there are almost no limitation on building depth.

7.2.4 Ventilation element

The third aspect is the ventilation elements used to realise the natural and hybrid ventilation strategy. Each of these ventilation elements has both a set of technical and architectural consequences and possibilities linked to them, which requires close attention of both the engineer and the architect. The most important elements are openings in the façade, wind towers, wind scoops, low pressure fans, chimneys, double façades, atria, embedded ducts, low pressure intake and outlet vents, low pressure coils, etc.

7.2.5 Building geometry

Utilisation and characteristics of natural driving forces are influenced by the height of the building and lead to distinctions in the natural ventilation concepts. Wind velocity and direction are more stable and less influenced by surrounding buildings and vegetation at a distance from ground level. The vertical distance between the inlet and the outlet can also be significant for utilisation of thermal buoyancy. A tall building therefore tends to utilise other ventilation elements and principles than a low building. A logical and practical way of sorting by building height would be to distinguish between high-rise buildings (more than 10 storeys), medium-rise buildings (3–6 storeys) and low-rise buildings (1–2 storeys).

Building geometry is also closely linked to the natural ventilation principle. Distributed and linear building forms tend to use single-sided or cross-ventilation principles, while more compact buildings use stack ventilation either by utilisation of atriums or chimneys.

7.2.6 Supply and exhaust air paths

The supply and exhaust air path is the route ventilation air travels between the outside and the occupied spaces inside a building. The supply and exhaust paths can be local or central. A local supply and exhaust air path typically implies that several inlets/outlets are scattered on the building envelope. A central inlet/outlet in most cases need horizontal and/or vertical ductworks and/or chambers inside the building to distribute the ventilation air to the desired locations. Central airflow paths facilitate heat recovery, preheating and filtering, whereas this is harder to achieve with local airflow paths. Local paths offer on the other hand greater flexibility for future changes as they usually are organised in a modular manner (e.g. inlets located in narrow horizontal bands at every floor level across the width of the façade) and are not encumbered with being linked to a dedicated distribution network in the interiors. Fan-assisted systems are used in both cases, whereas in hybrid systems using natural and mechanical ventilation, the mechanical system is most often used in combination with a natural system with distributed air paths.

7.3 System solutions and characteristics

A natural or hybrid ventilation system design is often a tailored solution based on demands for indoor climate, energy use, total costs, building design, building use and user requirements as well as user expectations. In this section, typical system solutions are described and their different characteristics evaluated.

For all natural ventilation systems, the most important issues are related to the optimum use of driving forces combined with minimising pressure losses in the system. Usually, in natural systems, the pressure loss is less than 10 Pa, whereas the pressure loss in hybrid systems can be up to 30–100 Pa, dependent on the number of ventilation components at the design airflow rate.

Natural driving forces can be increased by:

- vertically spacing the intake and exhaust openings as far apart as possible to increase the impact of thermal buoyancy;
- optimum use of wind conditions at the building site, i.e. by using wind towers;
- using large room heights and room volumes to even out variations in ventilation airflow rates.

To minimise the pressure losses, the main strategy is to aim at low air velocities in the system by:

- efficient air distribution in the building to exploit the ventilation air optimally;
- controlling ventilation airflow rates according to demand;
- using as few ventilation components as possible;
- using low pressure loss ventilation components (i.e. components that exploit aerodynamic shape and/or have large dimensions);
- using components that are easy to inspect and clean;
- minimising the need for distribution channels by using air intakes through the façade and direct air transfer between rooms;
- using air paths of large dimensions and good aerodynamic design in which the air speed is less than $1\,\mathrm{m\,s^{-1}}$ at design conditions;
- control airflows by variable speed fans instead of dampers.

As maintenance is critical to secure continuing good indoor climate and air quality, air distribution paths should be accessible without the use of special equipment. This requires that:

- air distribution channels should be passable (i.e. maintenance staff should be able to walk through them). Thus, their minimum cross-sectional area should be a height of 2 m and a width of 0.8 m;
- where passable channels are not possible, they should be short and straight;
- ventilation components must be accessible from the upstream and the downstream sides.

Systems comprising large dimensions and/or containing a large number of components and long air distribution paths are space demanding and can reduce usable building area. This problem can be minimised by:

- considering low rise buildings (or sectioning the ventilation system in high-rise buildings);
- supplying air through the façade or through channels below occupied areas (e.g. through the basement);
- exhausting air through the façade or through channels above the ceiling (e.g. through the roof space);
- supplying air to rooms through the façade (or through channels) and exploiting corridors and staircases as air distribution paths for exhaust air (or vice versa).

Periods where fans are not needed can be extended by:

- utilising local wind conditions by applying wind towers or special inlet openings;
- locating intake and exhaust with the largest distance possible to increase thermal buoyancy;
- designing buildings with large room height and room volume to even out variations in airflow rates.

Energy use for heating and cooling can be reduced by:

- using demand-controlled ventilation;
- using heat recovery;
- exploiting building thermal mass;
- using embedded ducts to reduce outdoor temperature variations (see below).

7.3.1 Natural and fan-assisted natural ventilation systems

This system is designed as a natural ventilation system where one or more fans might assist the natural driving forces. The fan(s) will only need to operate for long or short periods during the year dependent on system design and the pressure losses.

For systems incorporating air intakes in the façade of each room, air is supplied either through high- or low-level openings, depending on the air distribution principle used. Air intake can also be through an atrium or a double skin façade, which, to some extent, will provide preheated air in the winter season. In such cases, it is necessary to bypass the air supply in warm periods to avoid high intake air temperatures. For systems with a central air intake, air is distributed through an embedded duct to each room and is supplied to the room either through high- or low-level openings.

The air exhaust can be located in the ceiling/roof in each room. Air can also flow through openings to corridors, common areas or atria and be exhausted through staircases, chimneys or openings in the roof. In this way, optimal use of both wind and thermal buoyancy forces is ensured. Intake and exhaust openings can be constructed to take advantage of wind direction or be controlled to open on the windward or leeward side, respectively, to ensure proper airflow direction through the building. For hybrid ventilation systems, the assisting fans are usually integrated in the exhaust opening.

An embedded duct can be used for the passive heating and cooling of intake air. This is achieved by taking advantage of a fairly uniform ground temperature to preheat or pre-cool the supply air. In addition, the settling of large particles in the air will take place. Filtration, heat recovery, preheating and/or cooling coils can also be installed in such ducts, if necessary, for the thermal conditioning and cleaning of the air.

The pressure loss in systems with openings in the façade is usually very low. For systems with embedded ducts, careful design of the system is necessary to minimise system pressure losses. This is especially the case if filtration, heat recovery, heating coils and/or cooling coils are included in the design. Alternatively, the system should be equipped with assisting fans for use when natural driving forces are too low. In the case of multi-storey buildings, care must be taken in both design and control operation with regard to the distribution of opening areas in the building to ensure inflow of air to all floors.

Preheating is often used in cases with low-level openings in the façade to avoid draught and the preheating is usually controlled to supply air at a constant inlet temperature. Openings, fans and thereby the airflow rate are usually controlled depending on room air temperature or CO_2-level. Examples of fan-assisted natural ventilation are given in Figure 7.3.

Driving forces and system pressure loss

Systems without filtration and/or preheating will have a very low pressure loss often below 10 Pa, and they can often be ventilated using natural driving forces. The main natural driving force is typically thermal buoyancy, which is relatively straightforward to predict and control and will depend on the building height. Air intakes through double skin façades are often located on each floor, even if this reduces thermal buoyancy, to avoid too high intake temperatures on the top floors.

Central exhaust openings are relatively easy to control dependent on the wind direction in order to always ensure an opening location on the leeward side of the building. Wind pressure on openings distributed on a façade is more difficult to predict and control, but more and more products such as grilles and dampers are available on the market that can control the opening area and airflow rate independent of the driving force.

When designing hybrid ventilation systems, it must be decided if the air should flow through the fan when not operating or should bypass the fan

Figure 7.3 Built examples with fan assisted natural ventilation. Above: B&O office building, Struer, Denmark. Air intake in the facade and exhaust through staircases to exhaust hood in the roof (Hendriksen et al. 2002). Below: Waterland School, The Netherlands. Air intake through grilles below windows and exhaust through hood in each class room (van der Aa, 2002b).

to avoid pressure loss in the fan. If the first choice is made, it is necessary to select a relatively large fan dimension to increase the free area for airflow and reduce the pressure loss. Very limited information is available regarding pressure loss through stationary fans. The necessary data are normally only available for fans that are used in smoke ventilation systems. Figure 7.4 shows some examples of locating exhaust openings.

Air distribution and ventilation efficiency

For high-level openings in the façade, the aim is to supply an air jet and mixing it with room air. However, for small driving forces or pressure differences (0.2–0.4 Pa) and/or low outdoor temperatures, supply air will flow along the wall and drop towards the floor, see (1) in Figure 7.5 (Bjorn et al. 2000). Air distribution in the room will follow the displacement principle, and the draught risk will be highest along the floor (Heiselberg et al. 2001). Therefore, the opening period is often limited to a minimum in the heating season. At higher pressure differences ($\Delta p > 4 - 6$ Pa) and/or higher outdoor temperatures ($\Delta t < 5$ K), the supply airflow will act as a

Figure 7.4 Above: Exhaust hood with opening area in the top to ensure low pressure at the opening for all wind directions Axialfan with a large diameter integrated in the exhaust hood (B&O office building, Denmark, Hendriksen *et al.*, 2002). Below: Office building in Belgium with exhaust chimney and openings in all directions to ensure low pressure for all wind directions (IVEG office building, Heijmans and Wouters, 2002).

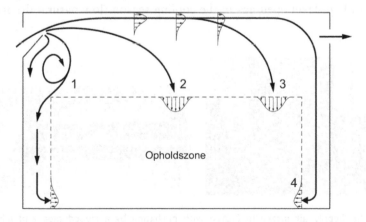

Figure 7.5 Typical air distribution conditions in room with high positioned openings.

thermal jet, see (2) and (3) in Figure 7.5, and traditional jet theory can be used to predict the airflow path and draught risk in the occupied zone (Heiselberg *et al.* 2001, 2002). For bottom-hung windows close to the ceiling, the air jet will attach to the ceiling and reach further into the room with reduced air speed and draught risk in the occupied zone as a result. In the summer situation, where the temperature difference is very small, the air will act as an isothermal jet, see (4) in Figure 7.5.

For systems with passive heating in the embedded duct, there is a risk of draught in winter during periods with low supply air temperature. Therefore, careful design of openings and their locations are needed to minimise this risk. For systems with heat recovery and/or preheating, the risk of draught can be eliminated.

For low position openings in the façade, the aim is to distribute the air to the room using the displacement principle to achieve high ventilation efficiency. Preheating with a controlled supply air temperature in the heating season is necessary to avoid draught. For a high supply air temperature (above room air temperature), the supply air will flow towards the ceiling and result in a poor ventilation efficiency, whereas a too low supply air temperature (more than 1–2° C below room air temperature) will result in draught risk at floor level. A number of 'off-the-shelf' solutions are on the market as well as a number of examples of tailored solutions as seen in Figure 7.6. The pressure loss for these systems can easily be much higher than for window openings in the façade, and therefore, a hybrid ventilation solution with a fan integrated either in the supply device or in the exhaust is often applied.

In many buildings, a combination of opening locations is used, with low-level openings in the heating season (high natural driving forces) to benefit from a high ventilation efficiency and to avoid draught by preheating the air and with high-level openings in the cooling season (low natural driving

Figure 7.6 Supply air intake in facade with prehating by a ribbed pipe and inlet opening at floor level (B&O office building, Denmark, Hendriksen *et al.* 2002).

forces) to benefit from low pressure loss, large airflow rates and night cooling of exposed ceilings.

Control strategy, automatic and/or manual control

The ventilation flow rate is often controlled by motorised openings and for hybrid ventilation systems also by frequency-controlled exhaust fans. Flow rate is typically controlled according to time, presence detection or set points for room temperature, CO_2 concentration level and/or humidity levels depending on room size and typical use. It is an advantage to control openings as part of a system as the ventilation flow rate through an opening is often dependent on other openings. It is especially important to ensure good coordination between opening areas of intake and exhaust openings to ensure proper air distribution in the building, for example inflow of air through façade openings on all floors in a multi-storey building. Alternatively, for systems with central air intake, the flow rate can be controlled by the measurement of air velocity at a characteristic point in the supply or exhaust air duct.

User control and/or manual override of automatic control is very important as it affects user acceptance of the indoor climate positively. It is

Figure 7.7 Example of facade solution with small high-positioned motorised, automatic openings and large manually openable windows close to the occupied zone for additional airing in peak load situations (Pihl & Son, office Building, Denmark).

beneficial to include openable windows in the façade, which can be controlled manually by the users depending on the need for additional airing or ventilation on warm days (see Figure 7.7). For systems with central air intake, openable windows must not affect the ventilation system negatively. In situations where the outdoor temperature increases above the indoor temperature, occupants should be advised to keep openings closed and reduce ventilation to a minimum for achieving acceptable indoor air quality.

Preheating of intake air should be controlled according to a set point temperature and must be separated from room temperature control.

Heat recovery, fan operation and energy use

For systems with air-intake openings distributed in the façade, heat recovery cannot be used and demand control becomes essential for reduction of energy use. Preheating is used for low positioned opening and is often controlled by a constant supply air temperature set point. In practice, demand control is not as ideal as in design predictions and energy use for heating as well as fan power might therefore be slightly higher than predicted.

For systems with central intake and exhaust openings, heat recovery is possible and the main challenge is to design for minimum pressure loss. Alternatively, a fan can assist the natural driving forces in periods, where heat recovery is beneficial.

The energy use by fans, including appropriate design of filtration, heat recovery and/or preheating of inlet air, is relatively low and typically is in the range of $1–3\ \mathrm{kWh\,m^{-2}}$ per annum depending on the frequency of operation.

Thermal comfort and indoor air quality

Intake of air through openings in the façade without preheating can result in draught problems at low outdoor air temperature. The problem can be reduced by minimisation of the ventilation demand and opening periods in the heating season, by room arrangement and by positioning openings at high level. For intake of air through low-level openings, it is important to ensure satisfactory preheating of intake air in winter to avoid draught problems along the floor. For systems with passive heating in an embedded duct, the periods with low supply air temperature and risk of draught in winter season are small. With heat recovery and/or preheating of air in the duct, the draught risk can be avoided.

In summer time, when indoor and outdoor temperatures are almost equal, openable windows close to the occupied zone can increase air speed around occupants and improve comfort conditions. For systems with embedded ducts, the maximum air temperature on warm days will be reduced. In Scandinavian countries, this can amount to several degrees and indoor air temperatures for well-designed buildings can be kept below 25° C on days with maximum temperatures up to 30° C. Night ventilation will often be necessary to cool the building down, which requires burglar prove intake openings, exposed thermal mass and high ventilation airflow rates ($n > 4$–$6\,h^{-1}$). In larger buildings, night ventilation control needs to be automatic. Utilisation of large room heights and air distribution by the displacement principle are also efficient to remove excess heat and avoid high air temperatures in the occupied zone.

Under the assumption that the outdoor air quality is good, this system can provide good indoor air quality, especially when the displacement principle is used, even if filtration is not possible. There are only few ventilation components and short airflow paths that can pollute the air, and it is straightforward to keep the system clean. For high-level openings without preheating, where the ventilation airflow rate is reduced to a minimum, short periods with low indoor air quality can occur, but for rooms with a large volume per person and periodic airing, the air quality could become acceptable most of the time. As the air speed is very low in the embedded duct, particles in the air (especially large particles) will settle in the duct and regular inspection and cleaning of the embedded duct are necessary. If well constructed, no problems with moisture in embedded ducts are experienced in Scandinavian countries, even if condensation occurs in shorter periods in the summer.

Complexity, robustness and maintenance

As the airflow path in fan-assisted system is clear and simple and the number of components and movable part are few, the principle should be relative robust and require a minimum of maintenance. There can be extra costs for cleaning of systems where intake air is not filtered.

If the system consists of a large number of openings in the façade, an automatic control will require a number of motors and control points, which can be expensive and, from experience, requires a long period of tuning before the system is commissioned to achieve the desired design conditions for all seasons.

In the case of a primary user controlled system, periods with low IAQ can be expected, especially for systems without preheating, but also periods with excess ventilation can occur. Automatic control of intake and exhaust openings supplemented by the option of user override usually results in a more robust solution.

Heat recovery and preheating coils (and filters) should be designed for easy access, maintenance and cleaning. For large buildings with openings in the façade, inspection, cleaning and maintenance cover many openings, which is clearly a disadvantage.

System limitations

On locations exposed to varying wind speeds and directions, problems with excess ventilation and draught can be experienced with systems having openings in the façade, because of rapidly changing pressure differences across openings. In buildings with large internal heat gains and/or solar gains, the fan extract system will only have a limited cooling capacity, even with exposed thermal mass and night cooling, which may result in high internal temperatures in the summer period. This is also the case for systems with embedded ducts and the reduction in maximum supply air temperatures that can be achieved on warm days. Careful design and reduction of heat loads are therefore essential for successful application of this system.

With air intakes in the façade, the system cannot be used in deep-plan buildings. Systems with central air intake are difficult to implement and occupy a large space in buildings with many room or several stories.

7.3.2 Natural and mechanical ventilation

This system consists of two separate ventilation systems – a natural ventilation system and a balanced mechanical ventilation system, which may include a mechanical cooling system as well. The system in use at a given time depends on the natural driving forces available, the outdoor climate, the need for outdoor air and the need for cooling, as the system that can provide satisfactory indoor climate at the lowest energy use is selected.

This type of system is used in situations where natural ventilation only can be applied part of the year. Typically, the balanced mechanical system will be in use in the heating season, when heat recovery is beneficial and/or preheating of supply air is necessary for thermal comfort reasons. The mechanical system can also be in use in periods where outdoor noise, pollution or high temperatures prohibit the use of natural ventilation. Finally, mechanical ventilation (with mechanical cooling) will also be needed in buildings where the internal heat gains and/or solar gains present are high as in peak summer periods. In cases of good building design and internal heat load control, the need for mechanical cooling is limited, and the mechanical ventilation system is usually designed to provide the basic outdoor air change for satisfactory indoor air quality.

The natural ventilation system is usually used in intermediate seasons with moderate outdoor climate conditions prevailing, for passive cooling of the building in the summer period and/or outside the periods of occupation as in passive night cooling.

Driving forces and system pressure loss

The possibility to exploit the natural driving forces and the pressure loss in both the natural ventilation system and the balanced mechanical system influences the operating period of each system. For the natural ventilation system, it will determine the duration this system will be able to fulfil the demands for indoor air quality and passive cooling. In the optimal situation, the mechanical system will only be working in the mid winter season, where heat recovery and/or preheating is favourable and maybe in the peak summer season, if mechanical cooling is needed.

The pressure loss in the mechanical system will determine the length of the period where heat recovery will be beneficial (recovered heat versus use of electricity). To reduce energy use for air circulation, the mechanical system is often designed as a low pressure system ($\Delta p < 100$–$200\,\text{Pa}$), as the whole or parts of the building are used as pathway for the airflow.

Air distribution and ventilation efficiency

In the mechanical ventilation system, the air distribution methods used are based on the mixing or displacement principle, and ventilation openings are designed in the usual way considering the risk of draught and air distribution efficiency in the space.

High positioned openings are used in the natural ventilation system in the intermediate season, supplemented by large openings close to the occupied zone in the summer period. For night cooling often the ceiling is the most important part of the exposed thermal mass and high-level bottom-hung window are at an advantage to ensure high air velocities and low air temperatures along the ceiling surface.

Control strategy, automatic and/or manual control

The most important challenge in the control of the system is the determination of the criteria for change over between the two systems – the natural and the mechanical. These criteria include both indoor climate and energy use. The criteria can, for example, be:

- outdoor temperature limit, where the risk of draught by using natural ventilation becomes too high;
- outdoor temperature limit, where heat recovery is beneficial;
- indoor/outdoor temperature limit, where mechanical cooling is needed;
- CO_2 level, where indoor air quality is no longer acceptable.

In addition, it is necessary to define criteria for the conditions at which the natural system can be used in conjunction with the mechanical system, especially in the intermediate and the summer period. It should also be evaluated if and to what extent the limits of acceptable indoor climate can be exceeded for shorter or longer periods to reduce the risk of cyclic variation of the system (the mechanical system is started and stopped too frequently) or that the mechanical system will be in use for disproportionately long times.

The mechanical system will often be controlled by indoor air quality requirements and in situations requiring mechanical cooling dictated by the indoor air temperature. It has to be ensured that the mechanical system is beneficial with the use of the natural ventilation system and/or window airing.

The natural ventilation system will be used outside the heating (and cooling) season, which means that any over-ventilation will not lead to an increase in the energy use in the building. The control of the natural ventilation system is then less critical, and manual control is often satisfactory. However, for larger buildings, night cooling should be automatically controlled, as occupants cannot control it properly.

Heat recovery, fan operation and energy use

The mechanical ventilation system can include energy efficient heat recovery in the heating season. If it is designed as a low pressure system, the energy use by the fans will be considerably lower than for traditional systems.

In situations, where mechanical cooling can be avoided, the energy use will be very low in the intermediate and the summer period, and the system should be designed so that the mechanical system is only needed in periods with extreme weather conditions.

For system with mechanical cooling, it is important to inform users about when and why the mechanical cooling is to be used to avoid window opening at high outdoor temperatures and increases energy use for cooling.

Thermal comfort and indoor air quality

Type and location of openings, arrangement and use of the room and the need for ventilation will all have a large impact on the outdoor temperature level at which the risk of draught will become unacceptable when using the natural ventilation system and a change to the mechanical ventilation system will be necessary. Typically, the outdoor temperature range for this change will be 12–15° C during occupied periods.

In the winter period, the indoor climatic conditions will correspond to the conditions for a traditional balanced mechanical ventilation system, and both air quality and temperature conditions can be maintained within tight limits. In the summer period, systems without mechanical cooling will function as natural ventilation systems. The mechanical system is often designed for a relatively low airflow rate to maintain acceptable indoor air quality conditions and will not have the airflow capacity needed for passive cooling during occupied hours or night cooling.

If mechanical cooling is used in the summer period, the indoor temperature can be kept within acceptable limits. The natural ventilation system should be used as long as the outdoor temperature is lower than the indoor temperature and could, if needed, be used in combination with the mechanical cooling system. Night cooling by natural ventilation (supplemented by the mechanical system) should be used as much as possible.

Complexity, robustness and maintenance

Application of the two different systems results in large challenges regarding the development of an appropriate control strategy for optimal use, i.e. optimisation of both indoor climate and energy use that is transparent and understandable for users. This is particularly true for the intermediate season and the peak summer periods with mechanical cooling.

System limitations

Systems without mechanical cooling has the same limitations as natural ventilation systems with regard to cooling capacity, thermal mass and the necessity for night cooling.

The principle cannot be used in deep plan buildings, as the intake of supply air for a large part of the year is located in the façade.

7.4 Ventilation components

Natural and hybrid ventilation components can be tailored components fully or partly integrated with the building design as well as off-the-shelf components or traditional ventilation components adjusted for use in natural and hybrid ventilation.

Building integrated components play a very important role. Apart from strongly characterising the building architecture, these components, if well designed, allow supply, distribution and extraction of air from the building at very low pressure drops. Most natural and hybrid ventilation systems apply a combination of components. However, to handle the specific conditions of the site and the building, such as draught control, security, air preheating, outdoor air pollution and noise, fire regulations, etc., some adaptation of the traditional components are essential for successful performance. These issues are discussed in this section.

7.4.1 Intake openings

Intake openings for air intake must have the following characteristics:

- have a low pressure drop (this is especially important for natural ventilation systems);
- have a proper location to ensure a clean and cool air intake and efficient air distribution in the building;
- prevent rain, snow, pollen, insects, etc. from penetrating the building;
- provide air without unnecessary draught risk in the occupied zone.

Application of window openings in the façade is the most simple solution for air intake. They have a very low pressure loss, but limitations for their use include rain, sound attenuation and draught. Improved performance can be achieved by designing the façade with windows of different sizes and positions with adjustable windowpanes were needed to provide occupants with opportunities for adjusting air volume flow and thermal comfort conditions. For cross- or stack-ventilation strategy, the high-level bottom-hung window (see Figure 7.8) is the best choice in winter because the air travels the largest distance and mixes with room air before it reaches the occupied zone (Heiselberg *et al.* 2001; Heiselberg and Bjorn, 2002; Wildeboer *et al.* 2002). This is also the preferred location and type for night cooling. In the

Figure 7.8 Example of office building with high-level bottom-hung windows and larger window openings close to the occupied zone (Magistrenes Hus, Copenhagen).

summer, openings close to the occupied zone are preferable to increase the air movement around the occupants.

In cases where these openings are not suitable – that is, because of the need for preheating, air filtration, sound attenuation or security – it is important that low pressure intake vents and exhaust terminal devices with low pressure drop dampers are used.

Trickle ventilators are used in both natural and hybrid ventilation systems, and with the development of both passive and electronic self-regulating trickle ventilators with direction-sensitive flow sensors, they offer good opportunities for controlling both airflow and comfort. They can also provide sound attenuation and filtration if needed and prevent penetration of rain and snow.

Low-level adjustable intake grilles have the same advantages as trickle ventilators and are often used in cases where preheating is needed, either as standalone units or in combination with perimeter radiators, to ensure an efficient air distribution (see Figure 7.9). However, if preheating and filtration are necessary, the pressure loss increases quite considerably and the natural driving forces will have to be supplemented by fan assistance.

Figure 7.9 Left: The window sill in a classroom, where the outdoor air enters and is preheated by a convector (Tånga School Sweden, Blomsterberg *et al.* 2002b). Right: Air intake in façade with preheating by convector in kindergarten (Kindergarten Solsikken, Århus, Denmark).

7.4.2 Exhaust openings

In order to optimise the natural driving forces, the exhaust opening has to be positioned as high as possible as higher wind speeds increase the driving force by wind. It is important that the wind always creates low pressure at the exhaust opening. The opening can either be designed to be independent of the wind direction or have openings in different directions that can be controlled according to the dominating wind direction to ensure that only openings on the leeward side are active (see Figure 7.10).

A high positioned exhaust opening also increases the distance to the intake openings and thereby the thermal buoyancy force which can be achieved by the application of an exhaust tower or chimney (see Figure 7.11). The exhaust chimney also decreases the useful part of the built volume, which is at a higher pressure relative to the outside. This arrangement reduces the risk of moisture damage in cold climates. In the summer period, the buoyancy force can be increased by solar heating of the air in the exhaust chimney, i.e. using a solar chimney.

In low rise buildings, skylights are often combined with exhaust openings. These rooftop openings are located relatively close to the occupied zone

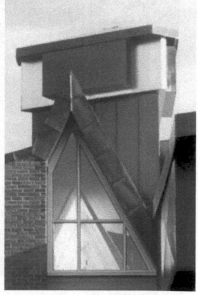

Figure 7.10 Example of exhaust openings that can be controlled according to the dominating wind directions. Above: Mediå School, Grong, Norway (Tjelfaat, 2002). Below: Søndervang School, Kolding Denmark.

Figure 7.11 Solar chimney on Tånga School, Sweden (Blomsterberg *et al.* 2002).

and should not be located close to work places as lumps of cold air can drop down in the winter season and create draughts.

7.4.3 Building integrated channels and internal openings

For natural and hybrid ventilation systems with distributed openings in the façade, the air typically flows from the room to the corridors, an atrium or stairwells from where the air is exhausted. In these systems, the building space is used for air transportation and air is transported from one room to the next, either through an open connection between the rooms or through internal openings. Internal openings are used in situations where a separation of the room is needed either because of unwanted transportation of noise, requirements for privacy, fire spread, air distribution control, possibility for manual control of ventilation, etc. The challenge is to provide

Figure 7.12 Above: Building integrated exhaust and intake duct, Mediå School, Grong Norway (Tjelflaat, 2002). Below: Several intake ducts below building to improve preheating. Straight ducts between two chamber to allow for inspection and cleaning, Klaus-Weiler-Fraxern School, Austria.

the necessary opening characteristics with as low a pressure loss as possible. Depending on requirements, the internal opening can be a simple flap integrated in the wall construction (see Figure 7.12).

For ventilation systems with central intake and/or exhaust, the air typically flows from the intake to the room and/or from the room to the exhaust through a building integrated ventilation duct. The ventilation duct can contain different ventilation components such as filtration, heat recovery, low pressure fan, etc.

From the intake to the room, the duct is often an underground duct or a concrete corridor in the basement. High construction quality is necessary to avoid problems with penetration of radon and water and ensure easy cleaning of surfaces. There is a risk of condensation in the summer period and thereby a risk of microbial growth. This risk can be reduced by regular inspection and cleaning, which means that these ducts should be passable or short and straight and they should be lit for easy inspection and cleaning.

In an underground duct with very low air velocities, sedimentation of large particles such as pollen will occur. By having a sedimentation area just after the intake, this effect can be optimised and the need for cleaning of the duct system reduced.

Underground and building integrated ducts will have a considerable thermal mass that can be used for passive cooling or reduce the need for mechanical cooling. In the winter, some preheating of intake air will be possible and defrosting of heat recovery avoided.

It is important that the duct is designed with an 'air lock' to avoid warm preheated air to flow backwards in the duct in periods where the ventilation system is stopped. If the duct is integrated in the basement, the additional cost is minimal and by establishing vertical shafts for air distribution, horizontal ducts can be avoided in the building. It is also advisable to have adjustable dampers in the duct for easy inspection.

From the room to the exhaust, the duct can be a vertical chimney or a horizontal duct above the corridor or in the attic.

7.4.4 Heating/cooling coils, heat recovery, filtration, low pressure fans

For buildings located in northern Europe, preheating of supply air can be necessary to avoid draught in the occupied zone.

In natural ventilation systems, it is very difficult to apply preheating in a traditional way as the pressure loss through the heating coil will be too high. Therefore, this is usually only used in hybrid systems. In natural ventilation systems preheating can be achieved by a calorifier in the intake chamber (Figure 7.13) as seen in Figure 7.14.

Figure 7.13 Preheating of air in intake chamber with a calorifier (Kvartershuset, Kolding, Denmark).

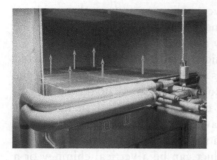
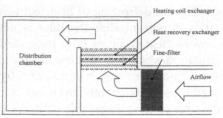

Figure 7.14 Examples of preheating of intake air with horizontal coils to ensure high efficiency for low velocity air flow (Tjelflaat, 2002).

For hybrid systems with distributed air intakes, preheating of supply air is done by means of specific components such as convectors or fan coils or by means of radiators located below windows (see Figures 7.9 and 7.14).

To avoid draught risk in buildings with central intake, air is preheated in the intake duct and supplied to rooms by low-level low-velocity diffusers. In order to minimise the pressure drop across the heating coil, the face area of the coil is relatively large, and as a result, the face velocity is relatively low. Combined with the fact that the cold intake air mainly flows as a gravity current along the bottom of the intake duct, the airflow through vertically positioned heating coils can be very unevenly distributed resulting in an inefficient preheating. This can be avoided by positioning the heating coils horizontally in the intake duct (see Figure 7.14).

In well-insulated and tight buildings with high internal and/or external heat loads, the number of hours during the period of occupation when heating is required can be limited. Therefore, to ensure preheating of intake air, it is necessary to separate this function from general room heating or at least to have a separate control of the inlet air temperature.

The possibility of heat recovery depends on the ventilation principle. For natural ventilation systems, heat recovery is very difficult to apply and not very efficient. For hybrid systems based on the natural and mechanical ventilation principles, mechanical ventilation systems are often equipped with heat recovery because this is frequently the main reason for choosing such systems, especially in cold climates. For other hybrid ventilation strategies, heat recovery can only be used in systems with central intake and exhaust of air, and a typical solution involves coupled heat exchangers with water circulation.

Passive cooling, including automatically controlled solar-shading devices as well as the use of underground ducts, culverts or plenums to precondition the supply air have been adopted in many buildings (Narita and Kato, 2002;

Schild, 2002; Tjelflaat, 2002). Underground culverts can be quite effective for preheating intake air in winter and pre-cooling in summer, especially with culverts in the basement of buildings, and the performance is often better than expected from predictions (Wachenfeldt, 2003).

Filtering of intake air is not very common in existing buildings with natural ventilation systems but are frequently used in hybrid ventilation systems. Traditional back filters are often used, and the pressure loss across the filter is reduced by a large face area (Tjelflaat, 2002), which is not always possible to implement.

Underground culverts in air intake ducts have measurable filtration effect because of settlement of particles, especially large particles. In a 60-m culvert, the number of particles with sizes above 0.3 µm was reduced by 85 per cent and for particles above 10 µm by 95 per cent (Schild, 2001).

In hybrid ventilation systems, low pressure axial fans are typically used. They have a quite different fan characteristic from fans used in mechanical ventilation systems. Figure 7.15 shows a typical fan characteristic for an axial fan used in a hybrid ventilation system. The fans can be off over shorter or longer periods and therefore provide a resistance to the flow. The relation between the airflow and the pressure difference is also shown in Figure 7.15. It can be seen that a discontinuity appears for a volume flow of about $0.4\,\mathrm{m}^3\,\mathrm{s}^{-1}$. This happens because the fan wheel starts turning for high airflow rates, whereas it is stationary for lower airflow rates.

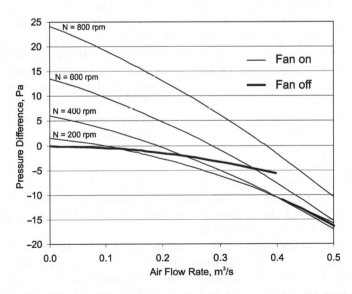

Figure 7.15 Fan characteristic for a typical low pressure fan, 4-blade axial fan, diameter 500 mm.

7.5 Control strategies

The main challenge in designing control systems for natural and hybrid ventilation is to obtain the right balance between:

- installation cost
- operating cost
- energy consumption
- indoor climate and comfort
- user satisfaction
- robustness.

The design of an 'optimal' control strategy for a given building will not only depend on technical parameters such as:

- building type and design
- ventilation system
- external noise and pollution
- solar shading
- internal heat load.

But also on other parameters such as:

- dress code
- user expectations
- user habits.

In natural and hybrid ventilation, the control system is an integrated part of the ventilation system itself. It is therefore important that the ventilation system and the control system are designed together in one process. The purpose of ventilation control is primarily to adapt the airflow rate to the actual demand, which is a requisite for a successful ventilation solution.

Hybrid ventilation systems can be made quite complex. However, it is very important to develop a control strategy and to design a control system that are easy for users to understand and can be operated by the maintenance staff. Therefore, simplicity and transparency of the user/system interface are of the utmost importance.

Control system designers need to recognise that most users are not technically literate and are not interested in learning complex operations to suit varying outdoor conditions. They want a system that responds to their needs unobtrusively and allows them to change a condition if it is perceived as unsatisfactory, with rapid feedback.

Demand control is very important in natural and hybrid ventilation systems, and in many cases, demand control also proves to be very energy efficient. However, one of the main problems encountered in automatic control of indoor air quality is the cost and reliability of CO_2 sensors used to control the ventilation demand (Willems and van der Aa, 2002). If a sensor is needed in each building zone, it can become expensive both in initial cost and in regular calibration. In some cases, the ventilation demand is controlled by infrared detection. The major advantage of this system is its relatively low cost (compared to CO_2 sensors) and its autonomy (it can work on a long-life battery; no wiring is required). The major disadvantage is that the airflow is only indirectly correlated to the demand. Sometimes the airflow can be too low or too high. Presence detection can be a good way of controlling the ventilation demand in rooms with low occupancy variation, such as cellular offices. In some cases, it has also been successfully applied in school classrooms. For rooms such as conference rooms, a CO_2 strategy is more suitable because it usually estimates the real needs more accurately. There is a strong need for reliable and cheap CO_2 sensors to be developed.

7.5.1   Typical strategies

The control strategy for a building should at least include a winter control strategy, where IAQ is normally the main parameter of concern, and a summer control strategy, where the maximum room temperature is the main concern. It should also include a control strategy to be used in the interval between winter and summer, where there might occasionally be a heating demand, as well as excess heat in the building. Both the ventilation and the control strategy are significantly influenced by the general climate in the region where the building is located. In cold climates, the control strategy should focus on minimising the ventilation energy needed to achieve good IAQ and on achieving a good indoor climate in summer and spring without mechanical cooling. In warm climates, the control strategy should focus mainly on reducing the energy consumption for mechanical cooling during summer.

If the inlet air is preheated, the best solution is to have separate control of the inlet temperature because preheating the inlet air may be needed even when there is excess heat in the room. This is especially important to consider in cases where openings are below windows and perimeter radiators are used for preheating intake air. There is a risk that the inlet temperature setpoint will be raised to compensate for insufficient room heating. This is very critical for systems with low-level low-velocity inlet diffusers because the displacement air distribution principle can be destroyed, with very low ventilation efficiency as a result. The control of outlets and night ventilation seems less problematic.

In landscape offices, automatic control is needed; but it can be difficult to find an acceptable strategy for window control that satisfies all occupants. If windows are operated automatically during occupied hours, and the external temperature is more than a few degrees lower than the room temperature, there is a great risk of user dissatisfaction because of the sensation of draught. Therefore, it is important that occupants have the opportunity to override the control for openings close to their workstation.

In classrooms, a simpler control strategy and control system is often installed, with manual control and supplementary window airing in breaks. With the high density of occupancy, the CO_2 level quickly exceeds the limit in winter, if assisting fans or mechanical ventilation systems are not operating. In schools where the occupants are responsible for indoor air quality control, there is a great risk of high CO_2 concentrations for some of the time. This risk also applies in classrooms where the inlet air is preheated. Automatic control or a combined manual and automatic control strategy is therefore advisable in school buildings, with the possibility of manual override.

In buildings that also have active mechanical cooling, where the operation mode is automatically switched between hybrid ventilation and mechanical cooling depending on the temperature or enthalpy difference between external and internal air, there is a risk that once activated, the system will stay in active cooling mode.

7.5.2 Automatic and/or user control

One of the advantages of natural ventilation systems is higher user satisfaction because of individual control of windows and indoor environmental conditions (Rowe, 2003). If possible, this feature should be maintained in the control of a hybrid ventilation system even if it could conflict with the possibility of guaranteeing a specific level of indoor thermal comfort or air quality in the rooms. Unfortunately, the relationship between the indoor climate and user acceptance in user-controlled rooms is not well known. Recent research indicates that users are more tolerant of deviations in the indoor thermal climate if the system is controlled by themselves.

Occupants want to be able to alter conditions quickly in response to unpredictable events (such as glare, draughts or outside noises). If conflicting or unsatisfactory conditions occur, occupants want to decide for themselves how to resolve the conflicts by overriding default settings rather than having conditions chosen for them. Occupants demonstrate a tendency to use supplementary mechanical cooling/heating equipment sparingly and in an energy-efficient way (Rowe, 2002) and prefer to use operable windows and other adaptive behaviours to modify conditions. Most occupants do, however, appear to have an upper 'tolerance' limit, when active intervention will be applied if the opportunity is available. This upper tolerance limit will be very individual and can be different from day to day.

Even though users should have the maximum possibility of controlling their own environment, automatic control is needed to support the users in achieving a comfortable indoor climate and to take over during non-occupied hours. In rooms for several people (e.g. open-plan offices) and in rooms occupied by different people (e.g. meeting rooms), a higher degree of automation is needed. Automatic control is also needed during non-occupied hours to reduce energy use and to precondition rooms for occupation – that is, to provide and control night cooling.

It is also very important to carefully consider how user interaction is integrated within the control system, both with regard to the type of functions that can be overruled and how and when the automatic control regains control after being overruled by the occupant. For systems with presence detection, the automatic control system usually takes over when the occupants leave the room. For other systems, it can take over after the normal occupation period has ended or after a certain time period, which can be adjusted as a part of the commissioning of the hybrid ventilation system.

7.5.3   Control tasks and parameters

The control strategy should include both time and rate control and reflect the demands of the building owner and the needs of the users.

Indoor air quality

The control of ventilation for indoor air quality can either be manual, by the occupants, simple timer control, motion detection (occupants present), based on direct measurement of indoor air quality, or a combination of these. For direct measurement of indoor air quality, CO_2 concentration is a useful indicator if occupants are the only or dominating pollutant source. If other significant sources influencing indoor air quality are present – for example, pollutants from materials and cleaning – then the CO_2 concentration in the room may be a less satisfactory indicator.

In small rooms for one or a few people – for example, cellular offices – it can normally be expected that the occupants will be able to control indoor air quality to their own satisfaction if the ventilation system provides them with the necessary facilities (such as user-controlled windows and vents of different sizes and positions).

In large rooms for many people (for instance, landscape offices) and in rooms occasionally occupied by different people (such as meeting rooms), automatic control of ventilation for indoor air quality is normally needed. The purpose of the control in this case is to reduce the ventilation energy consumption by limiting the operating hours and ventilation rate according to the occupancy pattern. The optimum strategy should have both a good user control to allow occupants to adjust conditions locally at their work station and an automatic back up.

Even during non-occupied hours, there might be a need for indoor air quality-controlled ventilation, especially in tight buildings. This includes ventilation after the end of the occupancy period to remove built-up pollution, ventilation during non-occupied periods to remove pollution from materials, and cleaning and ventilation before occupancy to start the occupancy period with fresh air in the building.

Thermal comfort and draught

Room temperature control during occupied hours in summer can be either manual or automatic. Occupants do have a very clear sense of their own thermal comfort, but typically they react too late, when the temperature already is above the acceptable temperature limit.

Automatic control of openings, as well as solar shading, can be beneficial because it ensures action as soon as the indoor temperature begins to increase. The need for direct automatic control of room temperature during occupancy is mainly related to large rooms catering for many people and to rooms occasionally occupied by different people. Direct automatic control of room temperature is also necessary if comfort is achieved by mechanical means – for example, mechanical cooling or additional mechanical fan-forced airflow.

In temperate climates during the summer, the normally small difference between indoor and external air temperature on warm summer days has limited potential to reduce room temperature, even if the flow rate is high. In many cases, the body cooling potential of air movement because of open windows might be the most important in relation to thermal comfort. If the external air temperature is higher than the indoor air temperature, external airflow will increase room temperature. This will often be the situation for buildings with efficient night cooling, mechanical cooling and/or efficient solar shading. In such cases, this can be handled through an automatic control system by changing the control mode from temperature control to indoor air quality control.

To avoid sensations of draught, it might be necessary to preheat incoming external air; this might also be necessary even if cooling is needed in the room. Coils or radiators for preheating the supply air should normally be controlled based on the temperature of the inlet air.

It is important that occupants are carefully instructed on how to operate windows when outdoor air temperature is high or when mechanical cooling is on.

Night ventilation during summer

The control of night ventilation is of great importance to achieve acceptable thermal comfort during hot summer days in buildings without mechanical cooling and to reduce energy consumption for mechanical cooling.

Building structures should be as cold as possible without creating thermal discomfort in the morning. The control of night ventilation should normally be automatic; but it is possible to have night ventilation with manual user-controlled windows or hatches in the individual rooms. Manual control by occupants requires clear and easy-to-understand instructions. Automatic control can be local per room or central for the building or a section of the building. Local control is normally only relevant in larger rooms and especially if local fan assistance is used. Central control must normally be based on measured temperatures in representative rooms. The selection of the representative rooms is of great importance.

The actual night ventilation strategy depends on the system. If fans are included, it is preferable to have a few degrees of cooling potential available from the external air before fans are started because of fan power consumption. Night ventilation must continue until the building is sufficiently cooled or occupied again. If the building structures are cooled to low temperatures, it might be necessary to interrupt night ventilation before the end of the non-occupied period in order to regain acceptable surface temperatures before the start of building occupation.

Natural and mechanical mode switch

One of the main characteristics of hybrid ventilation system is the ability to switch automatically between natural and mechanical modes in order to optimise the balance between indoor environmental quality and energy use. This challenge differs between the different hybrid ventilation principles.

For a fan-assisted system, the fan can be controlled by the temperature or by the indoor air quality in the rooms, by the pressure in the supply or exhaust ducts, or by the airflow rate through the fan. If the fan is in the natural ventilation flow path, the control can be either on or off, stepped or continuous, depending on the natural driving forces. If the fan is in parallel to the natural ventilation flow path and uses part of the same flow path, it is difficult to have continuous control and to determine when the conditions allow the fan to be switched off again.

Alternating natural and mechanical ventilation must normally be controlled based on the external temperature and humidity. Alternatively, it can be controlled by a time schedule. Good information for the occupants is needed about the actual mode of the ventilation system.

7.6   Examples

7.6.1   NCC headquarters, Denmark

The headquarters of NCC Contractors in Denmark is an office building of five floors with a central atrium (see Figure 7.16). The building was completed in 2000. In the first phases of the design, the design team, consisting of architects, consulting engineers and NCC as contractor, was supported by a specialist team whose main purpose was to improve indoor

Figure 7.16 NCC headquarters, Copenhagen, Denmark. External view and atrium.

environmental quality and to apply passive technologies such as daylighting, passive cooling and natural ventilation in the building.

The building is located at the waterfront in the centre of Copenhagen with surrounding buildings of similar height. It is a square building, and each floor has three landscape (open plan) office rooms with east, north and west orientation (see Figure 7.17). Larger meeting rooms, the canteen and a glazed buffer zone are located towards the south. These non-office rooms are air-conditioned and are designed to prevent solar radiation on the south façade from heating the office zones.

Building design and construction

Each open plan 'landscape' office incorporates separate zones, which consist of small meeting rooms, printer rooms and service rooms containing mechanical exhaust plant. Zones with higher heat or contaminant loads (such as larger meeting rooms, toilets, kitchens, etc.) are located in the corners of the building and are separately treated with mechanical ventilation or air-conditioning. This reduces the heat and contaminant loads of the landscape offices and thereby reduces the need for ventilation.

The room depth of the office zones is 15 m, the floor-to-floor height is 4 m and the room height to ceiling is 3 m. As the ratio between room depth and room height is 5, cross-ventilation of the office spaces is necessary to achieve acceptable conditions. This requires that the landscape offices are open to the square atrium in the middle of the building, and a stack ventilation principle is applied with air intakes in the façades and air exhaust in the atrium. The building has five storeys and hence can create large thermal buoyancy forces in the lower floors with progressively lower values in the upper floors where the vertical distances between the intake and the roof level exhaust opening become much smaller.

The building has fully glazed south and north façades. For all façades, it is possible to establish small, high-level openings as well as larger openings

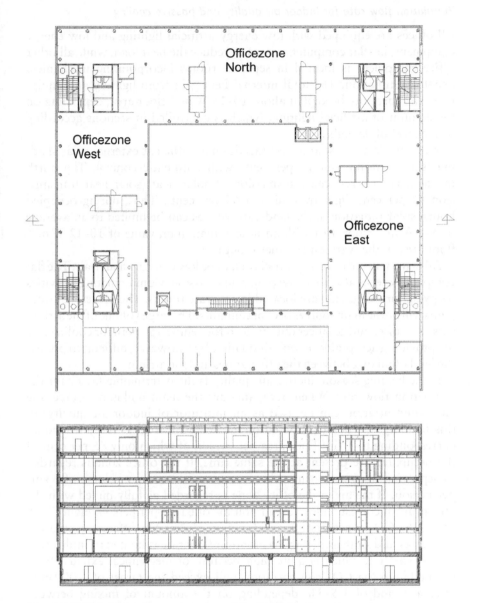

Figure 7.17 Third floor plan and section of the NCC headquarters.

close to the occupied zone. The landscape offices have suspended ceilings and wooden floors. Only the walls in the staircases in the corners of the building have high thermal mass. Therefore, the building cannot be considered to have significant thermal mass.

Ventilation flow rate for indoor air quality and passive cooling

All offices are equipped with low energy artificial lighting and low energy equipment, i.e. flat computer screens to reduce the heat load, while all other office equipment is located in separate rooms incorporating mechanical exhaust ventilation. The total internal heat gain from lighting and equipment can therefore be kept at about 10–18 $W m^{-2}$ floor area, depending on the amount of artificial lighting, which is controlled in sections according to the level of daylight.

Windows in the east and west façades have efficient external solar shading (solar transmission 19 per cent) with automatic control. The north façade glass has reduced transmission of solar heat (solar heat transmission 41 per cent, light transmission 65 per cent). Thus, during occupied hours, solar radiation in the landscape offices can be limited to an average of 5–8 $W m^{-2}$ floor area with an hourly maximum value of 10–12 $W m^{-2}$ floor area in the warmest summer month.

All materials and coverings are chosen to be low emitting materials. The flat computer screens also have low pollutant emission. Contaminating activities (copying, printing, etc.) are located in separate rooms. Therefore, the occupants will be the main sources of contaminants. The large volume of the atrium acts as a buffer and balances out the need for ventilation both for cooling and for providing acceptable indoor air quality. It is, however, difficult to predict the level of mixing between the office zones and the atrium.

In the heating season, indoor air quality is the determining factor for the ventilation flow rate. When occupants are the main pollutant source, the CO_2 concentration can be used as an indicator of indoor air quality. In this building, the design is based on an occupant load in the office zones corresponding to $10 m^2$ floor area per person of which only 85 per cent of the occupants are present at the same time. If the office zone is regarded as separated from the atrium, this corresponds to an air volume of $35 m^3$ per person. If the air in the office zone is regarded as fully mixed with the atrium air, it corresponds to an air volume of $61 m^3$ per person. Figure 7.18 shows the CO_2 concentration as a function of time for different ventilation flow rates and strategies. If the ventilation flow rate is limited to infiltration ($n = 0.3 m^3 h^{-1}$, which is quite high because of mechanical exhaust from the building) the CO_2 concentration will exceed the target of 1,000 ppm after a period of 1.5–3 h, depending on the amount of mixing between offices and atrium. If the infiltration flow rate is supplemented by periodic airing ($n = 6 h^{-1}$) by window opening for periods of 12 min, an average concentration level of 1,000 ppm can be achieved with an airing periodicity of between 2–4 h, depending on the level of mixing.

In the summer period (cooling season), the ventilation flow rate will be determined by the need for passive cooling. The necessary ventilation flow rate in the summer period for passive cooling of the whole building is quite small

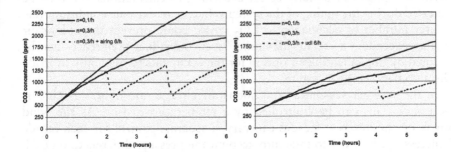

Figure 7.18 Calculation of CO2 concentration in office zone – left: without mixing of air between office zones and atrium; right: with complete mixing of air in the building.

$(n = 1 - 2\,h^{-1})$, because of the very low heat loads in the buildings, compared with typical air exchange rates of $n = 4 - 6\,h^{-1}$ in Danish office buildings.

Air distribution principles and opening locations

The ventilation principle applied to this building is stack ventilation with air intake in the façade, airflow across the landscape offices and exhaust through openings in the roof of the atrium. As no preheating is included, small openings located far from the occupied zone will result in the best thermal comfort conditions. Therefore, openings are located close to the ceiling. Narrow openings along the whole length of the façade are used as they can be opened during the night too. The air distribution principle is mixing.

All openings can be controlled by the occupant. High-positioned openings are automatically controlled with manual override, while larger openings, close to the occupied zone, can only be controlled manually. The openings in each landscape office are divided into three zones to allow fine tuning of the system according to the local wind and surface pressure conditions as well as user preferences.

Control strategy, automatic and user control, thermal comfort and indoor air quality

In the heating season, acceptable indoor air quality can be obtained by infiltration and periodic airing by window opening. During the 2–3 short airing periods each day, air velocities and low temperatures will cause draughts, and this solution can only be used if this is accepted by the occupants. Alternatively, a solution with preheating needs to be established.

The opening period and time between airing periods can be controlled by a timer or by a CO_2 sensor. A timer is preferable because then the airing periods are known to the occupants, and they can act accordingly. The time between airings can initially be set according to calculations and then be adjusted during the tuning of the systems. In case of extreme conditions (i.e. a large number of occupants), a CO_2 sensor could prevent the indoor air quality exceeding a certain maximum level. Individual preferences can be ensured by implementing an option for the occupants to manually override the automatic control system.

In the summer season, a continuous air change rate of $n = 1-2\,h^{-1}$ is necessary. This can be controlled by a temperature sensor. As the wind forces on the different façades can vary, a separate control for each façade is necessary. In order to take into account the pressure differences on each façade, the control could even be divided into several zones per façade. In the present case, three zones per façade were chosen. In situations with extreme weather conditions, additional openings that can be controlled manually by the occupants are useful. In periods where outdoor temperatures are above indoor temperatures, the control strategy should minimise ventilation. Indoor air quality should be the determining parameter, and occupants should be advised to keep manually openable windows closed.

Night ventilation is necessary in the building to cool it to an acceptable starting temperature in the morning.

In some buildings, the intermediate season requires an additional control strategy. In this case, it is not needed, as the necessary ventilation flow rate in this period is very small. The ventilation flow rate needed for cooling corresponds to that needed for acceptable indoor air quality which, for an airflow rate of $10\,l/s$ person, will be $n = 1\,h^{-1}$.

7.6.2 Bang & Olufsen headquarters, Denmark

Bang & Olufsen required an office building of high quality and a minimum of technical installations, which should be simple and hidden. The entire office building is U-shaped with three connected wings (Building 1–3, see Figure 7.19). The building constructed in 1998 and is situated in the outskirts of the town of Struer, facing an open land without significant noise or air pollution. Building 3 includes meeting rooms, auditoria and a canteen and is ventilated by a balanced mechanical ventilation system. Buildings 1 and 2 include offices and are ventilated by a hybrid system. The office layout of Building 1 is a combination of open plan areas between the two staircases and some cellular offices. The building has three similar floors with a heated gross floor area of $1680\,m^2$.

The building is specifically designed for natural ventilation. In the design stage of the ventilation system, the architects and engineers took into account both the thermally generated pressures as well as the wind-induced pressures. The design team, the client and the main contractor had a thorough co-operation to optimise the initial costs of the building.

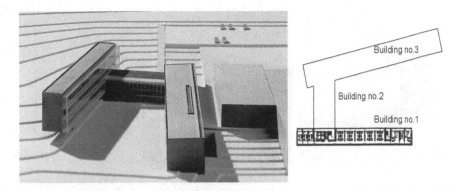

Figure 7.19 Model of B&O office building and plan of one of the floors in Building I.

The present case study description is based on the work performed during the international project IEA-ECBCS Annex 35 "Hybrid Ventilation" and reported by Hendriksen *et al.* 2002.

Building design and construction

The building has pre-cast concrete planks, exposed ceilings, a south façade of concrete inside and bricks outside and a fully glazed north façade. The net ceiling height is 3.1 m and the gross height is 3.4 m. Depth of floor is 7.5 m.

The main façades are facing north and south. Figure 7.20 illustrates the north façade, which is fully glazed with narrow windows in the horizontal divisions. These serve as inlet openings for the hybrid ventilation system with ribbed heat pipes for preheating of incoming air. The south façade has an insulated double-brick wall and a moderate window area to allow daylight. These windows may also be opened directly by the users and are automatically controlled, during nighttime, for passive cooling of the building in the summer period. Ceilings consist of exposed concrete slabs. The vertical partitions of each office section are for noise reduction.

Air distribution principles and opening locations

The principle of the hybrid ventilation system is shown in Figure 7.21. The ventilation principle is stack- and wind-driven natural ventilation with fan assistance, and the air distribution principle is displacement ventilation.

Outdoor air is supplied through automatic low positioned, narrow hatches (windows) in front of the floor slab in the north façade (see Figure 7.3). Supply air is preheated to a temperature of 18° C, using ribbed heat pipes located between the inlet opening and the floor-mounted grille, to prevent draught (see Figure 7.6). After leaving the floor-mounted inlet

Figure 7.20 Glazed north facade and brick wall south façade as well as cross section of the B&O building.

Figure 7.21 Visualisation of air distribution principle of the B&O building (Birch & Krogboe A/S, Consultants and Planners).

grille, inlet air is assumed to act according to the displacement principle. Two central stairwells serve as 'extract channels' for the three storeys. In each stairwell, air is extracted through a specially designed cowl on the roof. These have fans integrated into them to provide assistance, when the natural driving forces are insufficient (see Figure 7.4). Windows in the south façade are used for supplementary ventilation during summertime.

Control strategy, automatic and user control, thermal comfort and indoor air quality

The hybrid ventilation system is demand controlled both by using metabolic CO_2 concentration, as an indicator of IAQ, and by room temperature in case of overheating. The control system is based on centralised components divided into local zones with sensors connected in a BEMS-system. This system gives first priority to natural driving forces and second priority to mechanical fan support.

When ventilation is needed, the hatches and the dampers in the extract cowl are kept open. If the necessary ventilation airflow rate is not achieved by natural means, the fan speed is controlled. It is active according to a certain time schedule or to a selected control mode, i.e:

- *Constant mode based on time schedule.* In periods with insufficient natural driving forces, the assisting fans are turned on. The control of the assisting fans is based on a mechanical principle of constant airflow, which means that the fans are turned on, if a certain design air change rate is not obtained by natural driving forces. The design air change rate is $1.5\,h^{-1}$ during daytime and $3\,h^{-1}$ during night cooling. The air change rate is measured indirectly by air velocity sensors located in a cross-section in each of the two extracts.
- *Demand control.* CO_2 concentration is used as indicator for IAQ, and room air temperature is used as an indicator of thermal comfort. There are two sensors on each of the three floors. Demand control is active from 7 a.m. to midnight. It is activated if the set value of CO_2 exceeded (1,000 ppm) or if a pre-set room temperature is exceeded (e.g. $t_r >$ 25° C). This typically occurs during summer days.
- *Night cooling mode.* Night cooling with fan boost can be active according to a time schedule from midnight to 7 a.m. It comes into operation when the average room temperature on all floors exceeds 25°C and, at the same time, the outdoor temperature is below 20° C. This condition commonly occurs at night during summer in Denmark.

Inlet temperature sensors control the ribbed heat pipes, at the supply openings, in order to fulfil the requirements for the supply air temperature of the displacement ventilation system. Besides controlling the flow in the heat pipes, the position of the heat valve is used to balance the airflow rate through each inlet. This is accomplished by the valve adjusting the position of each inlet opening on the north façade, according to variations in wind pressures at different positions along the façade.

When the outdoor temperature is below 5°C, the hybrid ventilation system turns off to prevent inlets from being stuck in an open position by the freezing of condensed water. This temperature condition occurs for approximately 25 per cent of the total number of hours each year. During these periods, occupants can use the south facing windows for airing. The ventilation system is also shut down during periods of rain or strong winds (wind speed $> 11\,\mathrm{m\,s^{-1}}$, which occur approximately 1 per cent of the time in a year). During these periods, the north façade inlet openings and the extract dampers are closed, and the fans are turned off.

Room temperature sensors are located at a height of 1.6 m above the floor, and CO_2 sensors are located at a height of 2.2 m. External sensors are located on the roof and on a mast located next to the neighbouring building.

Building performance

Lessons learned from this building are based on experiences obtained from the facility management staff of Bang & Olufsen and from experiences gained during monitoring of the building (Hendriksen *et al.* 2002; Brohus *et al.* 2003).

Building ventilation and system design. In periods with strong winds from the west direction, some unforeseen air exchange between building 1 and building 2 was encountered. This has resulted in periods with too large air change as well as draught perceived by occupants. Glazed doors were installed in the hallways to eliminate this high air change.

Cellular offices are situated at each end of the building. These rooms do not have direct inlet openings, and the doors need to be open for ventilation of the small office rooms.

Occupants in the open space office have complained about cold draught from the glazed north façade. Hereafter, the ribbed heat pipes in the air inlet were used as convectors to prevent cold draught in periods when the hybrid ventilation system was inactive.

When a ventilation strategy based on a constant airflow principle is used, assisting fans tend to start immediately if natural driving forces are insufficient. This results in undue running hours of assisting fans. Also the system tends to cycle between natural and mechanical mode and the result that the full potential of natural driving forces are not used, i.e. the fans are running in 2/3 of operating hours of the hybrid ventilation system.

Hybrid ventilation components. The hybrid ventilation system has many distributed components and some centralised components. The components are standard products supplied from either natural or mechanical ventilation manufacturers. The amount of distributed components has resulted in time-consuming fault detection and the need to replace some components during the first years of operation. The operable windows, which serve as inlets, showed some variations in the degree to which they were closed by the actuators. This resulted in either air leakage through some vents and too much compression in others, which, in some cases, resulted in cracks in the glazing.

Inlet air was found to bypass the ribbed heating pipes by finding a low-resistance path close to the connecting pipes. This resulted in insufficient heating of the supply air. The problem was eliminated by sealing the route with insulation plates.

Extract dampers in the roof cowl have been found to stick in various positions from open to closed, which in some cases have lead to air leakage. The noise level from assisting fans was found to be unacceptable. This was overcome by reducing the maximum rotational speed of the fans during occupied periods.

Energy performance. The displacement air distribution principle works quite well, resulting in a high ventilation efficiency. Energy use of the assisting fans is very low ($1.7 \text{ kWhm}^{-2}\text{year}^{-1}$) and accounts for only about 3 per cent of the electrical energy use. The measured indoor air quality (CO_2 used as an indicator) was very high. The energy demand for heating was remarkably higher than expected. This can be related to the large areas for transmission heat loss, a very large glazed area towards the north, and an infiltration rate that is larger than expected.

Night cooling of the building proved to be effective and resulted in lower room temperatures during daytime. It was active for 4 per cent of the year, which corresponds to 350 h or at least 50 days.

7.6.3 Mediå School, Norway

Mediå School is a single-storey school building for 6–13-year-old children located in a small town. The hybrid ventilation system is fully integrated with the building design. To achieve the design goals of acceptable indoor climate at the lowest possible energy consumption, a number of elements related to the ventilation system design were defined (Tjelflaat, 2002):

- easy inspection and cleaning of ventilation airways.
- ventilation intake designed to be protected from precipitation and to utilise wind forces.
- underground ventilation culvert with large thermal mass to reduce daily temperature swings.
- demand-controlled displacement ventilation.
- ventilation airflow driven by buoyancy and wind forces combined with a supply and an extract fan.
- heat recovery from exhaust air by 'run around system'.

The present case study description is based on the work performed during the international project IEA-ECBCS Annex 35 'Hybrid Ventilation' and reported by Tjelflaat (2002).

Building design and construction

The single-storey building has a gross floor area of $1,000\,m^2$ with 8 class-rooms for 30 students each, wardrobes, service rooms and small offices. A distribution culvert, made of concrete and insulated, supports the inner longitudinal walls of the building. The culvert has an inner height of 2.0 m. It is also used for piping of the tap water and of the hydronic heating for spaces and for heating of supply air (see Figures 7.22 and 7.13).

An intake culvert connects the intake tower on the North side of the building with the distribution culvert in the basement. The intake culvert is 2.0 m high and 1.5 m wide. The length of this culvert is 15 m, and it has a slope away from the building of about 1.5 m (see Figure 7.22).

Brick walls are used on the two classroom clusters along the corridors and on the end-walls (on the outside). Otherwise, gypsum plates and glazing are used for interior walls. Some walls to classrooms are installed as folding walls. The floor is concreted with a vinyl cover.

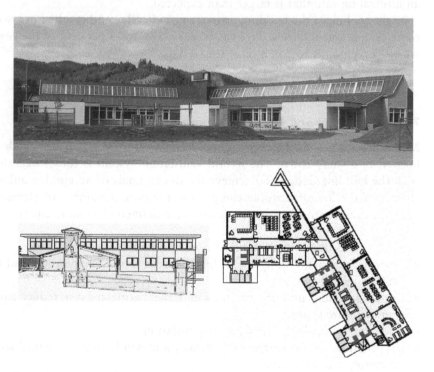

Figure 7.22 Picture, plan and section of Mediå School, Grong, Norway (Tjelflaat, 2002).

Classrooms have a sloping ceiling with heights between 2.5 and 4.5 m. The attic is used for daylighting of classrooms. Design (maximum) occupancy is 224 people and typical occupancy is 175 people. Normal school hours are 0830 till 1400 during Monday through Friday.

Ventilation flow rate for indoor air quality and passive cooling

The design capacity of the ventilation system is about $8,000\,\mathrm{m^3h^{-1}}$ for the heating mode and $10,000\,\mathrm{m^3h^{-1}}$ for the cooling mode (when the heat exchangers on the supplyside are bypassed). It is designed to maintain a ventilation airflow rate that is necessary to keep the set air quality or temperature level in rooms. The ventilation is never shut off in the building. In periods of no occupancy, the ventilation works only in 'leakage' mode that gives a low airflow rate to remove contaminants. The airflow rate is dependent on natural driving forces. All exhaust dampers in rooms are set to leave a 20-mm gap when closing in order to allow the leakage flow.

Air distribution principles and opening locations

This is stack- and wind-supported, which is a balanced low-pressure mechanical system with both air supply and extract in the classrooms. Air is taken from an inlet tower at some distance from the building that utilises wind forces where the inlet fan is located. Airflows through an underground culvert with a large thermal mass to reduce daily temperature swings and is distributed via a purpose-made basement corridor to low-positioned supply air terminal devices in classrooms to increase the ventilation efficiency (see Figure 7.23).

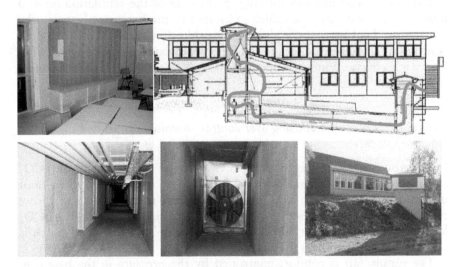

Figure 7.23 Air supply system of the Mediå School (Tjelflaat, 2002).

Figure 7.24 Air exhaust system (Tjelflaat, 2002).

Air is extracted from classrooms through a high-positioned hatch into a purpose-made lightwell corridor and exhausted through a roof tower with outlet valves that ensure suction by wind from any direction. A heat recovery unit and a low-pressure exhaust fan are located in the tower (see Figure 7.24).

The system also includes filtering, preheating of the ventilation air and heat recovery with bypass, which is located in the basement between the underground culvert and the basement corridor (see Figure 7.14). The flow is driven by low-pressure fans in the supply and extract, supported by wind and stack effects. Window opening is possible and the ventilation system will normally adapt to it.

Control strategy, automatic and user control, thermal comfort and indoor air quality

Ventilation is demand controlled by a CO_2-sensor in each classroom. CO_2 and temperature sensors are located 0.9 and 0.8 m, respectively, above floor in most rooms (see Figure 7.25), where the picture to the right shows, from the bottom, temperature sensor, CO_2-sensor and light switches. If the CO_2 level or the room temperature exceeds the set-point, the extract hatch is opened and adjusted by a motor.

The supply fan is comfort controlled by the pressure in the basement supply corridor to 2 Pa over-pressure compared with the external. The

Figure 7.25 Location of sensors for temperature and CO2 level in the Mediå School (Tjelflaat, 2002).

extract fan is controlled to maintain a 5 Pa pressure drop between the basement supply corridor and the lightwell extract corridor to avoid over-pressure in the building. Both fans are frequency controlled. The ventilation is controlled by the building management system.

Supply air temperature set point: 19° C during the heating season. For warm periods, like when the school starts in August, preheating of air is shut off, and the supply air temperature is allowed to swing freely to take full advantage of the thermal mass in the distribution culvert. Such adjustments must be done manually by decreasing the set point for the supply air tempeature according to the weather and the actual air supply temperature. During hot days, a supply temperature down to 17° C has not resulted in complaints.

In hot periods, forced ventilation must be used to utilise the thermal mass of the intake culvert for cooling purposes.

Building performance

There are some main conclusions that can be drawn from the experience with the design and the monitoring of the school.

The building proved to perform very well with respect to indoor air quality and thermal comfort. The hybrid ventilation system has ended up with a higher pressure drop than designed. This causes lower maximum ventilation airflow, but it is sufficient for 95 per cent of the occupancy periods. The extra fan energy is not noticeable. The naturally driven leakage ventilation is found to work sufficiently well outside occupancy hours. The ventilation system is very easy to inspect and maintain. The cooling effect of the underground culvert and the basement corridor is higher than expected. The energy use corresponds to the reference consumption in Norway, but is higher than predicted. The reasons for this are initial failures in the control system (now solved) and under-prediction of the energy loss by cold bridges and energy use by BEMS, pumps, etc.

Experience shows that the design of the building and the integration with the design of the ventilation system must be given enough resources early in the design process, and the basis for design must be well established. It is important that all aspects of energy use are evaluated simultaneously, and a good simulation tool with realistic input data is very valuable.

7.6.4 Tånga School, Sweden

Tånga School is a two-storey school building located in a residential area in the city of Falkenberg on the west coast of Sweden, 100 km south of Göteborg (Figure 7.26). The school is located in a mostly residential area.

The school was built in 1968 and is a complete school with 20 classrooms, 10 workshops, dining hall, kitchen, gymnasium and offices with a total floor area of 9350 m². The school was due for renovation, and the

Figure 7.26 Tånga School building B (Blomsterberg *et al.* 2002).

owners were interested in implementing energy saving features and hybrid ventilation.

The present case study description is based on the work performed during the international project IEA-ECBCS Annex 35 'Hybrid Ventilation' and reported by Blomsterberg *et al.* (2002).

Building design and construction

The school consists of four buildings (A, B, C and D), all of them with two storeys. Building A is with auditorium, dining hall, kitchen and offices. The building is almost rectangular and has a flat roof. Building B, which is mainly classrooms, is E-shaped and has a flat roof. Building C with mainly workshops is rectangular and has a flat roof. Building D is a gym. These buildings and the HVAC systems were rather typical for the period 1961–1975. Some improvements were carried out in 1989 (new windows) and 1991 (added thermal insulation). The school needed a general renovation because of wear and tear.

Buildings A, B and C were retrofitted. A hybrid ventilation system was installed in building B. The objective was to reduce the overall use of electricity for ventilation by installing a demand control hybrid ventilation system combining natural (passive stack and solar chimney) and mechanical (fan assistance) driving forces, instead of the existing balanced ventilation system without heat recovery. The height of building B is 7.5 m, the gross floor area $3672\,m^2$ and the room height 3.3 m. Every room is heated by radiators or convectors with thermostatic valves.

Ventilation flow rate for indoor air quality and passive cooling

The national requirements for minimum ventilation airflow rates are $7\,ls^{-1}person^{-1}$ during periods of occupancy and $0.35\,ls^{-1}m^{-2}$ during periods of non-occupancy. The design airflow rate for the actual hybrid ventilation system is however only $4.5\,ls^{-1}person^{-1}$ based on maximum occupancy in the classrooms. The arguments for this design value is that there seldom is maximum occupancy, children have lower metabolism than adults and that for shorter periods of time, one can allow a higher CO_2 level than 1,000 ppm. However, if the hybrid ventilation system for some reason does not give an acceptable indoor air quality, it should always be possible to manually change to a third Constant Air Volume (CAV) operation mode with the fans running to ensure an airflow rate of $7\,ls^{-1}person^{-1}$ based on maximum occupancy in the whole building.

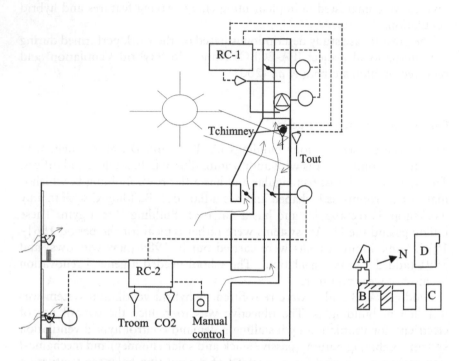

Figure 7.27 The principle of the hybrid ventilation system with supply air through convectors in the facade and exhaust through the passive stack of the Tånga
Building (Blomsterberg *et al.* 2002).

Air distribution principles and opening locations

The main principle of ventilation of school building B is fan-assisted natural
ventilation (see Figure 7.27). The outdoor air is distributed to the rooms
through three air intakes below the windows in the exterior walls into a
stub duct from where it is distributed to the room (see Figure 7.9). Air is
preheated by convectors under the stub duct and is distributed by the mixing
principle. Air is exhausted through air terminal devices below the ceiling on
the opposite side of the room and evacuated through a vertical ventilation
duct. Local dampers are mounted both in the air intakes and in the exhaust
duct from each room to allow individual control of the flow rate.

To increase the stack effect, a 6-m high passive stack and solar chimney
have been installed on the roof with assisting fan and a central damper
mounted in parallel (see Figures 7.11 and 7.27). In addition to extending
the length of the exhaust ducts, the solar chimney consist of a flat plate
solar air collector that heats the air in the chimney and increases the stack
effect in the last 6 m of the exhaust ducts. There are in total three solar
chimneys, each one serving a separate part of the building. It is desirable

to achieve equal stack effects on both floors and when needed having the exhaust fans working simultaneously. The cross-section area of the exhaust ducts from the first floor is reduced to achieve this.

The pressure drop in the air distribution system is very low. Low-pressure vents in the façade and low-pressure exhaust air terminal devices are used. There are no filters, large ventilation ducts and no heat recovery. Window airing is possible at any time. Night cooling can be used.

Control strategy, automatic and user control, thermal comfort and indoor air quality

The control system is a combination of individual and central control. The ventilation system should supply a basic ventilation rate during the lessons, and during breaks, the ventilation can if necessary be forced. The users are able to override the automatic control of the ventilation system.

A CO_2 sensor in each room controls the local inlet and outlet dampers. At a CO_2 level of 1,000 ppm or less the local dampers are set to a minimum open position, which can be varied as a function of the outdoor temperature and are controlled by a timer between 6^{00} and 18^{00} weekdays. At low outdoor temperatures and/or high wind speeds, the airflow rate is therefore automatically limited to prevent excessive energy use and problems with dry indoor air. If the CO_2 level exceeds 1,000 ppm, this is indicated by a signal lamp in the classroom. At CO_2 levels above 1,500 ppm, the local dampers open 100 per cent. The teacher can always override the local control system and manually change the position of the local dampers between 50 and 100 per cent.

When the stack effect is no longer sufficient to maintain the design airflow rates, the fan is started and the central damper is closed. To avoid a high frequency of starting and stopping of the fan, the dampers are opened and the fan is stopped at a somewhat higher temperature difference. When running, the exhaust fan is controlled by the pressure difference across the fan. The exhaust fan increases the pressure difference continuously as the temperature difference decreases.

Window airing is possible at any time. In summertime, the stack effect can also be utilised for night cooling of the building.

Building performance

The results of the Tånga School show that it is possible to control the indoor climate in a school building with a hybrid ventilation system with a fairly simple automatic and manual control system. Occupants perceived the indoor climate as rather good. The CO_2 concentration is mostly around 1,000 ppm or lower and only for short periods (10–20 min) is higher, but very seldom above 1,500 ppm. The personnel appreciate that the system can be operated manually and do so fairly often. However, at times, it would probably make sense if the automatic control of the hybrid ventilation was influenced not only by the CO_2 content in the classroom but also by the air temperature of the classroom.

For building B, the savings is 30 per cent in energy use for space heating, and the overall reduction in use of electricity for ventilation is 55 per cent.

References

Aggerholm, S., Heiselberg, P. and Bergse, N. C. (2006). Hybrid ventilation i kontorer og institutioner. SBi Dokumentation (in Danish).

Andersen, K.T., Heiselberg, P. and Aggerholm, S. (2002). Naturlig ventilation i erhvervsbygninger – beregning og dimensionering. By og Byg Anvisning 202, By og Byg (in Danish).

Bjorn, E., Jensen, J., Larsen, J., Nielsen, P. V. and Heiselberg, P. (2000). Improvement of thermal comfort in a naturally ventilated office. Proceedings of the 21st AIVC Conference, Den Haag, The Netherlands, September 22–25 (CP21). Air Infiltration and Ventilation Centre, AIVC.

Blomsterberg, Å., Wahlström, Å. and Sandberg, M. (2002). 'Tånga School': Case Study Report in Principles of Hybrid Ventilation (Ed. Heiselberg) Hybrid Ventilation Centre, Aalborg University, Aalborg, ISSN 1395-7953 R0207, Hybrid Ventilation Centre, Aalborg University, Aalborg.

Brohus, H., Frier, C. and Heiselberg, P. (2003). Measurements of hybrid ventilation performance in an office building. The International Journal of Ventilation, 1, 77–88 (Special Edition).

Heijmans, N. and Wouters, P. (2002). IVEG office building: case study report in Principles of Hybrid Ventilation (Ed. Heiselberg) Hybrid Ventilation Centre, Aalborg University, Aalborg, ISSN 1395-7953 R0207.

Heiselberg, P. (ed.) (2002). Principles of Hybrid Ventilation. IEA ECBCS-Annex 35final report. Hybrid Ventilation Centre, Aalborg University, ISSN 1395-7953 R0207. The booklet, technical and case study reports of the project are available for download at http://hybvent.civil.auc.dk.

Heiselberg, P. and Bjørn, E. (2000). Impact of open windows on room air flow and thermal comfort. International Journal of Ventilation, 1(2), 91–100.

Heiselberg, P., Svidt, K. and Nielsen, P. V. (2001). Characteristics of airflow from open windows. Building and Environment, 36, 859–869.

Hendriksen, O., Brohus, H., Frier, C. and Heiselberg, P. (2002). Bang and Olufsen Headquarters: case study report in Principles of Hybrid Ventilation (Ed. Heiselberg) Hybrid Ventilation Centre, Aalborg University, Aalborg, ISSN 1395-7953 R0207.

Hummelgaard, J., Juhl, P., Sæbjörnsson, K. O., Clausen, G., Toftum, J. and Langkilde, G. (2005). IAQ and occupant satisfaction in five mechanically and four naturally ventilated open-plan office buildings. Indoor Air, International konference, Beijing, September 2005.

Kleiven, T. (2003). Natural Ventilation in Buildings – Architectural Concepts, consequences and possibilities. Norweigian University of Science and Technology, NTNU, Trondheim. Ph.D. thesis, 13, Department of Architectural Design, History and Technology, ISBN 82-471-5560-5.

Narita, S. and Kato, S. (2002). Fujita Technology Center: Case Study Report in *Principles of Hybrid Ventilation* (Ed. Heiselberg) Hybrid Ventilation Centre, Aalborg University, Aalborg, ISSN 1395-7953 R0207.

Rowe, D. (2002). Wilkinson Building The University of Sidney: Case study report in *Principles of Hybrid Ventilation*, Hybrid Ventilation Centre, Aalborg University, ISSN 1395-7953 R0207.

Rowe, D. (2003). A study of a mixed mode environment in 25 cellular offices at the University of Sydney. *International Journal of Ventilation*, 1, 53–64 (Hybvent – Hybrid Ventilation Special Edition).

Schild, P. G. (2001). An overview of Norwegian buildings with hybrid ventilation, in Proceedings of the Second International Forum on Hybrid Ventilation, HybVent Forum 01, Delft, The Netherlands, May 2001.

Schild, P. G. (2002). Jaer School: Case Study Report in *Principles of Hybrid Ventilation* (Ed. Heiselberg) Hybrid Ventilation Centre, Aalborg University, Aalborg, ISSN 1395-7953 R0207.

Seppänen, O. and Fisk, W. J. (2002) Association of ventilation system type with SBS symptoms in office workers. *Indoor Air*, 12(12), 98–112.

Tjelflaat, P. O. (2002). Mediå School: Case Study Report in *Principles of Hybrid Ventilation* (Ed. Heiselberg) Hybrid Ventilation Centre, Aalborg University, Aalborg, ISSN 1395-7953 R0207.

van der Aa, A. (2002a). Cost of Hybrid Ventilation Systems: Technical Report in *Principles of Hybrid Ventilation* (Ed. Heiselberg) Hybrid Ventilation Centre, Aalborg University, Aalborg, ISSN 1395-7953 R0207

van der Aa, A. (2002b). Hybrid Ventilation Waterland School Building, The Netherlands – First Results of the Monitoring Phase: Case Study Report in *Principles of Hybrid Ventilation* (Ed. Heiselberg) Hybrid Ventilation Centre, Aalborg University, Aalborg, ISSN 1395-7953 R0207.

Vik, T. A. (2003). Life Cycle Cost Assessment of Natural Ventilation Systems, PhD thesis, Norwegian University of Science and Technology, NTNU, Trondheim

Wachenfeldt, B. J. (2003). Natural Ventilation in Buildings – Detailed Prediction of Energy Performance, PhD Thesis, Norwegian University of Science and Technology, NTNU, Trondheim.

Wildeboer, J. and Fitzner, K. (2002) Influence of the height of the supply air opening for natural ventilation on thermal comfort. Proceedings of ROOMVENT2002, 8th International Conference on Air Distribution in Rooms, Copenhagen, September 2002, pp. 545–548.

Willems, E. M. M. and van der Aa, A. (2002). CO_2-Sensors for Indoor Air Quality – A High-Tech Instrument is Becoming a Mass Product: Market Survey and Near Future Developments: Case Study Report in *Principles of Hybrid Ventilation*, Hybrid Ventilation Centre, Aalborg University, Aalborg.

Chapter 8

Measurement and visualization of air movements

Claude-Alain Roulet

8.1 Introduction

The air is 'lazy' and always follows the easiest path. It is also fully transparent and airflow patterns are invisible. These may be felt when they are too strong, causing draughts, or smelt when the air drives smelly pollutants at people. This means that the airflow does not necessarily follow expected patterns and it is only noticed when things go wrong.

Therefore, in ventilation, measurements may be useful to check whether the airflow patterns are as expected. Such measurements allow checking that:

- the air change efficiency is as high as possible
- clean air is supplied at right places, etc.

The airflow patterns should, in principle, be organized so that new air reaches as close as possible to the head of the occupants and that vitiated air be evacuated as quickly as possible before being mixed with the new air. The air change efficiency (see Section 8.5.1) expresses this quality. This should be larger than 50 per cent and preferably close to 100 per cent if piston ventilation is used.

Rather high air change efficiencies, indicating piston-type ventilation, can be seen in rooms ventilated by units 7 to 10, which are high auditoriums (Figure 8.1). The normal-height rooms ventilated by units 2 and 3 represent a complete mixing, whereas room one shows poor air change efficiency, partly explained by the fact that supply and exhaust are both at the ceiling in this room.

These few examples clearly illustrate the usefulness of measurements to detect malfunction.

8.2 Air velocity measurement

Anemometers measure the air velocity in one point at a time. Mapping this quantity then requires multiple measurements at points distributed in

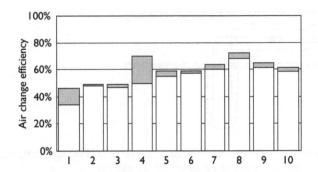

Figure 8.1 Air change efficiencies in some ventilated areas. Dark bands are uncertainty bars. Note that in one unit the efficiency is below 50 per cent, indicating short-circuits and dead-air zones.

a two- or three-dimension grid. Measurements take along time because the airflow in buildings is normally characterized by low velocities, high turbulence levels and in most cases also unsteady. Particle-image velocimetry (PIV) allows assessing air velocity at several places on one plane at a time.

Because airflow patterns in buildings are in most cases turbulent, the velocity measured at a particular location varies with time. If the anemometer is fast enough to measure instantaneous values (response time less than about 100 ms), the following statistical quantities can be calculated from recorded velocity with time, $v(t)$

the mean velocity

$$\bar{v} = \frac{1}{(t_2 - t_1)} \int_{t_1}^{t_2} v(t)\, dt \qquad (8.1)$$

the standard deviation

$$\sigma = \frac{1}{(t_2 - t_1)} \sqrt{\int_{t_1}^{t_2} [v(t) - \bar{v}]^2\, dt} \qquad (8.2)$$

and the turbulence intensity

$$Tu = \frac{\sigma}{\bar{v}} \qquad (8.3)$$

Different measuring techniques such as thermal anemometers, laser Doppler anemometers (LDAs), particle-image anemometers or ultrasonic anemometers are available and these will be described in the following sections.

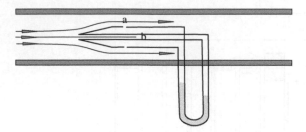

Figure 8.2 Schematics of a Pitot tube installed in an air duct. Pressure difference between points a and b depends on the air velocity.

8.2.1 Pitot tube

Pitot tubes use the pressure difference between dynamic and static pressure to measure air velocity. This instrument is directional and measures the velocity component parallel to the tube axis (see Figure 8.2).

Advantages

This is a very simple, cheap and robust instrument that allows measuring velocity with a manometer, but pressure transducers are often used for diagnosing air-handling units (AHUs). This instrument is commonly used in mechanical ventilation systems.

Disadvantages

The Pitot tube is a point-measuring technique and not very sensitive to low velocities and relatively slow. Measuring velocities of $1\,\mathrm{ms}^{-1}$ requires a manometer able to measure $0.6\,\mathrm{Pa}$, which is not common. Therefore, low velocities found in draughts and turbulence cannot be measured with this instrument. The velocity is evaluated at a single point and in one direction.

8.2.2 Thermal anemometers

Basic principle

A thin electrically heated element (film or wire) is cooled by the airflow to be measured. The heat transfer from the element is a function of the air velocity, the surface and the air temperature. With regard to control, there are three basic types of anemometers, the constant-current anemometer (CCA), the constant temperature anemometer (CTA) and the constant temperature difference anemometer (CTDA). To take into account the effects of the probe geometry, natural convection induced by a heated surface and the air turbulence, one normally carries out calibration in specially designed calibration rig. Standard probes for airflow measurements in ventilation are of spherical or cylindrical shape with a diameter of 2–3 mm. The velocity range is from 0.10 to 20 ms^{-1}.

Advantages

The thermal anemometer technique is widely used in ventilation. The measuring system is relatively simple and easy to operate. There is a wide variety of systems ranging from standard versions for measuring the mean velocity and turbulence properties to the more sophisticated versions for comfort measurements including probes for temperature, radiation and humidity measurements.

Disadvantages

The thermal anemometer technique is also a point-measuring technique. The information for mean velocity, turbulence and, in most cases, temperature is evaluated at a single point within the room. A typical measuring time for one point is about 2–10 min. For a complete survey of the airflow pattern within a large enclosure, this method is very time consuming. Measurements in enclosures with unsteady airflow pattern, for example due to solar radiation effects, cannot be realized even when several probes at different positions are used simultaneously.

The probes, the probe-mounting tubes, etc., disturb the airflow pattern at the measuring point, especially in highly three-dimensional airflow patterns with low velocities as it is the case in buildings. At very low air velocities, the natural convection effects are also important. At a velocity of 0.10 ms^{-1}, the heat transfer by forced convection and the heat transfer by natural convection due to heated surface of the probe at say 60°C are of the same order of magnitude. In particular, the natural convection effects cannot be estimated for turbulence measurements with three-dimensional velocity fluctuations.

8.2.3   Laser Doppler anemometer

Basic principle

Two laser beams intersect within a measurement volume resulting in a fringe pattern. If a small tracer particle floating in the airflow crosses the fringe pattern, light is scattered. From the detected frequency of the scattered light during the flight of the particle across the fringe pattern, one component of the particle velocity can be evaluated from straightforward geometrical

Figure 8.3 Laser Doppler anemometer used to measure airflow patterns around a manikin (www.ie.dtu.dk).

considerations. The system can be extended to two and three component measurements (see Figure 8.3). On the basis of the optics and the data acquisition system, a large range of velocities from a few mms^{-1} up to several 100 ms^{-1} can be measured.

Advantages

The LDA measuring techniques do not need calibration. If the system is operated correctly, the results are free from errors. The size of the measurement volume depends on the optical configuration and is, in any case, small enough for applications in ventilation. The technique is non-intrusive. The distance between the optical probe, which may be connected to the laser light source by an optical fibre, and the measuring point depends on the size of the measurement volume to be realized. Typical distances could range from 0.2 to 5 m. The frequency response depends on the size of the tracer particles used and is uncritical for ventilation applications.

Disadvantages

The LDA technique is also a point-measuring technique, with all the disadvantages mentioned above. In large enclosures, it is often a problem to provide tracer particles such as oil droplets, water droplets and solid particles in appropriate quantities.

The most important disadvantages are the complexity, the cost and weight of the equipment. The set-up of the optical system and the evaluation of the data cannot be carried out simply following a manual but need a skilled operator. Especially for applications in large enclosures, lasers with relatively high powers ($>$ 1 W) including an effective cooling water system have to be provided. LDA systems are also expensive compared with other systems.

8.2.4 Particle-image anemometer

Basic principle

The principle of the PIV, which is derived from the well-known qualitative light-sheet technique, is the measurement of the displacement of markers that travel with the airflow either by traces on a single frame taken with longer exposure times or by traces on multiple frames taken with short exposure times within subsequent time intervals. The displacement of the markers (particles, bubbles or even balloons) must be small enough to represent a good approximation of the actual velocity. Commercial systems are available that record the particle images in a plane of the flow field, which is illuminated either by white light or by a laser system. Two velocity components are evaluated by image processing software, but to obtain the third velocity component, one needs several additional techniques, such as the stereo photogrammetry or an extended laser sheet technique using laser beams with different colours (see Figure 8.4).

Advantages

The PIV technique is a plane-measuring technique that allows the simultaneous evaluation of the two velocity components in one plane of the flow field. New methods to measure the third velocity component within this plane are currently under development. This would allow the measurement of all three velocity components in a given plane simultaneously. No calibration is needed. An accuracy of 5–10 per cent is expected for the two-component technique. The PIV technique is non-intrusive.

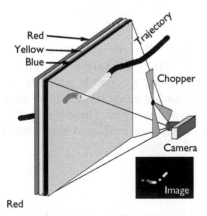

Figure 8.4 One way to get 3-D information on velocity of particles.

Disadvantages

The two-component PIV technique has been mostly used up-to-now in small-scale applications. For laboratory-based ventilation studies, measuring areas of several square metres have been achieved. However, for measurements within large enclosures, new systems have to be developed. The maximum allowable measuring area depends on the size of the tracers (e.g. helium-filled bubbles), on the light intensity (such as laser power) and on the resolution of the detecting medium (photo-camera, CCD-chip, etc.). Furthermore, bubbles are difficult to concentrate within an illuminated plane of a large enclosure and the control of neutral buoyancy for the bubbles is critical.

Only snapshots of velocity fields can be evaluated. The evaluation of mean velocities and turbulent quantities, which has to be based on large numbers of such snapshots, is beyond the capabilities of standard computer systems. The price of the equipment strongly varies with its sophistication and resolution in space and time, but remains high.

8.2.5   Ultrasonic anemometers

Basic principle

The technique is based on the measurement of the travelling time of a sound wave between an emitter and a receiver, which is affected by the air velocity. This measuring principle is well known in meteorology and has been introduced in ventilation only recently. The velocity is a mean value for the measurement volume of a typical dimension of about 50 mm and ranges from a few mms^{-1} up to 10 ms^{-1}.

Advantages

Commercial systems are able to measure all three velocity components simultaneously and to evaluate mean velocities and turbulence intensities. The measuring system is relatively simple, easy to operate and, once properly calibrated at the factory, needs no further calibration. Unlike thermal anemometers, the probes are linear, even at low speed. There is no influence of temperature. The air temperature can also be evaluated simultaneously from the same signal.

Disadvantages

The ultrasonic anemometer is an intrusive point-measuring technique, the disadvantages of which have been discussed with the thermal anemometers.

8.3 Measuring air tightness

8.3.1 Why check air tightness?

Having the correct airflow rate to the appropriate locations are imperative for obtaining good indoor air quality. Leakage, which allows air to follow inappropriate ways, should therefore be avoided. This requires an airtight building envelope and an airtight ductwork.

Figure 8.5 shows the infiltration ratios (i.e. the part of outdoor air that is not supplied by the AHU) for ten buildings.

Exfiltration has a negative effect on heat recovery (if any), and may, in some climates, create serious condensation problems at leakage locations. Infiltration has a negative effect on indoor air quality, because infiltrated air is not filtered, dried, cooled or heated.

A study performed in France and Belgium (Carrie *et al.* 1997) has found that, on average, 40 per cent of the supplied air is lost through ductwork leakage before reaching the user. This either reduces the effective ventilation rate or requires an increase of supply air, leading to energy loss.

8.3.2 Methods of determining the air leakage coefficients

Principle

The air tightness is actually measured by the air permeability of the envelope of the measured object. This type of measurement is standardized at an international level (ISO, 1998). For this, a fan maintains a pressure differential between the interior and the exterior of the object, and the airflow rate needed to maintain this pressure differential is measured. The pressure differential is measured with a manometer (range 0–100 Pa), and the airflow rate through the fan is measured using one of the following methods:

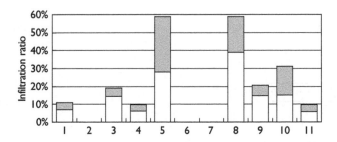

Figure 8.5 Infiltration ratios in different buildings, shown with uncertainty band. Of these 10 buildings, 7 have an infiltration rate significantly different from zero!

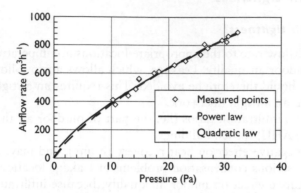

Figure 8.6 Airflow rates and pressure differences as measured in a real test, together with power law and quadratic fits.

1   The fan is calibrated so that the airflow rate can be determined from pressure differential and rotation speed (blower door).
2   A nozzle or another suitable flow meter is installed in the airflow circuit.
3   Tracer gas dilution technique (see Section 6.6).

The measurements are repeated for various pressure differentials, ranging from twice the natural pressure differential to about 60 Pa, or even higher for some cases (Figure 8.6).

The following general models are used for the characterization of air permeability.

The power law:

$$\dot{V} = K\Delta p^n \tag{8.4}$$

and the quadratic law, proposed by Etheridge (Etheridge and Sandberg, 1996):

$$\Delta p = a\dot{V}^2 + b\dot{V} \tag{8.5}$$

or:

$$\dot{V} = \frac{-b + \sqrt{b^2 - 4a\Delta p}}{2a} \tag{8.6}$$

where

$\dot{V}$ is the volume airflow rate through the leakage area (m³s⁻¹);
Δp is the pressure difference across the leakage area (Pa);

n is the flow exponent $(0.5 < n < 1)$;

K is the airflow coefficient $(\mathrm{m}^3\mathrm{s}^{-1}\mathrm{Pa}^{-n})$ and

a and b are coefficients representing the turbulent and laminar parts of the quadratic law, respectively $(\mathrm{Pasm}^{-3}$ and $\mathrm{Pas}^2\mathrm{m}^{-6})$.

The purpose of quantitative pressurization measurements is to determine these coefficients and exponents of either of the above models describing the airflow through the envelope or component.

The power law is fully empirical and reflects the fact that leakage is made of various cracks arranged in series and parallel. The quadratic law expresses that the flow is a mix of laminar and turbulent flow arranged in parallel. This law is not in common use because it is less practical.

Getting the airflow coefficients

If measurements are performed at two pressures only, for example at the lowest accurately measurable pressure differential and at the maximum acceptable one, the results of the measurements are Δp_1, $\dot{V}_1$ and Δp_2, $\dot{V}_2$. The coefficients of the power law are then

$$n = \frac{\log \dot{V}_1 - \log \dot{V}_2}{\log \Delta p_1 - \log \Delta p_2} \quad \text{and} \quad K = \frac{\dot{V}_1}{\Delta p_1^n} = \frac{\dot{V}_2}{\Delta p_2^n} \tag{8.7}$$

The coefficients of the quadratic law are

$$a = \frac{\Delta p_1 \dot{V}_2 - \Delta p_2 \dot{V}_1}{\dot{V}_1 \dot{V}_2 \left(\dot{V}_2 - \dot{V}_1\right)} \quad \text{and} \quad b = \frac{\Delta p_1 \dot{V}_2^2 - \Delta p_2 \dot{V}_1^2}{\dot{V}_1 \dot{V}_2 \left(\dot{V}_2 - \dot{V}_1\right)} \tag{8.8}$$

More than two measurements may be useful for testing the fitness of the model and to increase the accuracy of results. In this case, the least square fit method can be applied to obtain the coefficients. For this, Equation (8.4) can be liberalized by taking the logarithm of both sides:

$$\log \dot{V} = \log K + n \log \Delta p \tag{8.9}$$

which expresses a linear relationship between $\log \dot{V}$ and $\log \Delta p$ (Figure 8.7).

The measurement points can also be interpreted using the inverse problem theory (Tarantola, 1987), taking into account a priori knowledge such that the exponent n is between 0.5 and 1. This leads to a probabilistic relationship between the parameters n and K or a and b, giving the most probable pair and a clear view of their variability (Fürbringer et al. 1994).

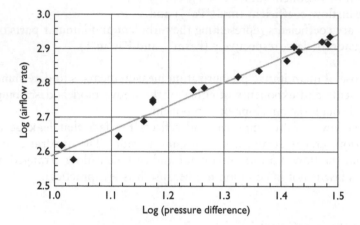

Figure 8.7 Logarithmic plot of airflow rates and pressure differences. The slope of the best-fit line is an estimate of *n* and its ordinate is an estimate of log (*K*). Natural logarithm is used here.

Ways of expressing the air tightness

In addition to the coefficients of the flow equations (i.e *K* and *n* or *a* and *b*), one of the following ways is used to represent the permeability as a single value only:

1   The airflow rate at a given pressure, calculated from Equations (8.4) or (8.5), depending on which parameters are available. The conventional pressures used in various countries are 1, 4, 10 or 50 Pa. The pressure 50 Pa corresponds to a pressure differential commonly used for measurements, whereas 4 Pa better represents a typical pressure differential present across building envelopes.

2   This airflow rate divided by the internal volume of the tested enclosure. This is a theoretical air change rate at that pressure. For this figure, 50 Pa is the most widely used pressure difference, and the number is then called $n_{50}(h^{-1})$. This value is less than $1\,h^{-1}$ in airtight buildings but, depending on the climate, buildings may have figures larger than $10\,h^{-1}$.

3   This airflow rate divided by the area of the envelope of the tested enclosure. This gives a figure specific to this envelope. For such application, the most common pressure differential is 4 Pa, and this parameter is then v_4, or specific leakage rate at 4 Pa, and is expressed in $m^3h^{-1}m^{-2}$. This figure is less than one, even $0.5\,m^3h^{-1}m^{-2}$ for airtight envelopes.

4   An equivalent leakage area, i.e. the area of a circular hole with sharp edges that would have the same airflow rate at a given pressure differential. This area is

$$A_L = K\sqrt{\frac{\rho}{2}}\Delta p^{\left(n-\frac{1}{2}\right)} \tag{8.10}$$

The standard deviation of the leakage area is

$$\delta A_L = \sqrt{A_L^2\left[\left(\frac{\delta K}{K}\right)^2 + (\delta n \ln \Delta p)^2\right]} \tag{8.11}$$

5   The ratio of the equivalent leakage area to the area of the envelope of the tested enclosure.

8.3.3   Neutral height method

Another, simpler method that offers, in some cases, a good estimate of the leakage area and determines whether specifications are met or exceeded is the neutral height method. It is based on the determination of the neutral height and the equipment necessary is reduced to simply an airflow direction detector (small smoke generator such as a cigarette or an incense stick, small flame, etc.) and a yardstick.

The neutral height is the plane in the building or part of it where the indoor–outdoor pressure differential is zero. This height is dependent on the size and position of the ventilation and leakage openings so that the airflows going in and out of the building are balanced.

The measurement should be performed without mechanical ventilation system (with the system switched off) on a preferably cold day without wind and when the building is heated. In this case, the airflow through the building results from buoyancy (stack effect) only.

To determine the position of the neutral height of an opening, one should stand by the outside opening when it is open. The airflow direction detector is then moved from bottom to top of the opening to observe the flow direction. The neutral level is located between the ingoing and outgoing flow directions. Sensitivity can be increased by reducing the width of the opening.

In an airtight building with a single opening, the neutral height will be located at mid-height (Figure 8.8 left). Cold air enters the building through the lower half of the opening, and warm air leaves the building through the upper part.

If there is another opening or leakage area above the test opening, part of the incoming air will leave the building through it and will not, therefore, pass through the upper part of the test opening. The neutral height will

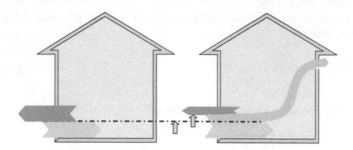

Figure 8.8 Principle of the neutral height method for assessing leakage area. Left: no leak (except the test opening), right: leakage above the test opening.

rise above the lower opening to balance the two airflows, and the opening area between this neutral level and the mid-height of the opening is equal to the equivalent leakage area (Figure 8.8 right). If the leakage is small, the sensitivity can be increased by reducing the width of the opening.

If the neutral height is below the mid-height of the test opening, the leakage area is below the opening. If the neutral height is not within the test opening, even when this is wide open, it is because the other leakages or openings are larger than the test opening. In this case, a walk through the building is necessary to identify and if possible close these large openings.

The equivalent area measured this way is the difference between the equivalent areas of the openings or leakage areas located above and below the opening. Therefore, it is useful to make this measurement at two test openings located at different heights in the building.

8.3.4 *Measurement of air tightness of single-duct or duct network*

To ensure the integrity of the supply air quality and to avoid energy waste when the air is either heated or cooled, there should be negligible leaks from the transport ducts and connections. Significant energy may be wasted, for example, where leaky ductwork passes through an unheated space such as an attic, basement or crawl space. As an example, it was found that the ductwork is the most significant source of leakage in western U.S. houses, together with damperless fireplaces (Dickerhoff *et al.* 1982). This fact is confirmed by a more recent study (Modera, 1989), but some houses, which likely have more airtight ductwork, were found acceptable (Palmiter and Bond, 1991).

Various techniques exist to check the air tightness of ductwork. In some countries, the air tightness of the ventilation system has to be checked when commissioning the system (NBCF, 1987), but in most countries measurements are seldom carried out. In addition, there is little guidance available, except an ASTM standard (ASTM, 2003). Appropriate methods of measuring duct air leakage are presented below.

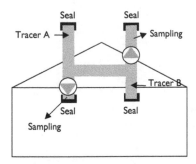

Figure 8.9 Location of tracer injection and sampling tubes for the measurement of leakage airflow rates in a ventilation system.

8.3.5 Pressurization method

The principle of this method is the same as those described in Section 8.3.2. All intakes, supply terminals, and exhaust and extract terminals connected to the system must be carefully sealed, e.g. using plastic sheet and adhesive tape.

Tracer gas injection and air-sampling tubes may be installed at appropriate points in the system to quantify any residual flow rate resulting from leakage. Suitable locations are the main supply or exhaust ducts (Figure 8.9).

The system fans (or a fan added at one register if required) are used to pressurize the supply side and depressurize the exhaust side of the network. The flow, Q_L, through the fan(s) is due to leakage and is measured as described in Section 8.2 (airflow meters) or 6.6 (with a tracer gas) together with the pressure difference, Δp, between the inside and outside of the ducts. The flow rate is the sum of all leaks downstream of the measurement point in the pressurized ducts and upstream for the depressurized ducts.

A series of measurements are made at different fan speeds, and the coefficients of Equations (8.4) or (8.5) are determined, and the relationship is subsequently used to calculate the leakage rate at the service pressure difference.

8.3.6 Flow rate difference method

If a duct is very leaky, the leakage can be obtained by measuring the difference between the flow rates at two locations along the flow. Because additional pressure drop should be avoided, it is recommended that tracers be used to measure the airflow rates. One tracer should be injected at a point upstream of the first location and a second tracer injected at the first location. The concentrations of each tracer are measured after the second, downstream location, at a distance where a good mixing is achieved (see Section 6.6). If steady flows can be assumed, two sequential measurements using a single tracer at each point may be used instead.

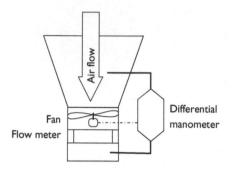

Figure 8.10 Schematics of a compensated flow meter. The differential manometer adjusts the fan so that there is no pressure drop through the flow meter.

For depressurized ducts, only one tracer is necessary. It is injected at the upstream end of the duct, and its concentration is measured at both ends to give the flow rate at each.

The leakage of the whole supply or exhaust network may be determined by measuring the difference between the airflow rate in the main duct (near the fan) and the sum of all the flow rates at the individual inlet or extract terminals. For this purpose, the main airflow rate can be measured with a tracer, in the vicinity of the fan, and the flow rates at the terminals may be determined with a compensated flow meter (Figure 8.10).

Because the result (i.e. the leakage flow rate) is a small difference of two large numbers, this method can give rise to large uncertainties for tight or only slightly leaky ducts. Because of its ease of use, it can nevertheless be used for diagnostic purposes, to detect whether the ductwork is very leaky or not.

8.3.7 Differential building pressurization

The methods described above measure the leakage of the whole duct system. From the point of view of avoiding energy waste alone, it may be useful to measure the leaks to or from outside only, and not those between the system and the interior of the building.

For this purpose, the duct system is considered as a part of the envelope and the indirect component testing method can be used. In a first test, the closed building is pressurized after sealing the outdoor air intake and exhaust of the building, with all the registers and returns open (Figure 8.11 left). In a second test, all registers and returns are sealed (Figure 8.11 right). The difference in airflow rate between the two tests, for each pressure, is caused by duct leakage to the outside.

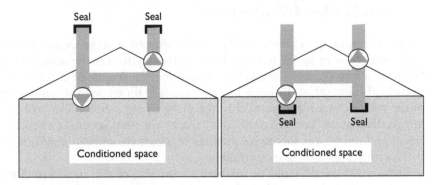

Figure 8.11 The difference between two measurements providing the duct leakage to outside of the conditioned space.

The major advantage of this method is the minimal requirement for equipment as additional equipment for envelope leakage measurements is not necessary. However, it is prone to inaccuracy, because the duct leakage is obtained as the difference between the measurements of two large airflow rates.

8.4 Visualization of air movement

8.4.1 Quantitative flow visualization

Methods applicable to room airflows are briefly described below.
    Required materials are:

- lighting system – standard or laser
- tracer particles diffusing the light coming from the lighting system – metaldehyde flakes, helium-filled soap bubbles or balloons
- a recording system – photographic camera, video recording.

Three dimensional approaches are:

- stereophotogrammetry using two or more cameras (Scholzen *et al.* 1994)
- illumination with gradually coloured light sheet.

Velocity can be measured using various techniques:

- tracking of particles or balloons in time (Maas *et al.* 1993; Alexander *et al.* 1994) (Lagragian approach)
- streak photography (Eulerian approach) (Scholzen *et al.* 1994)

- correlation techniques on one double exposed frame or two consecutive frames (standard PIV techniques).

All these techniques follow solid particles suspended in the air and moving with them. Measuring the distance between two pictures of the same particle taken at a known time interval or measuring the length of a track of a particle photographed during a known exposure time gives the component of the velocity in a plane parallel to the plane of the picture. The problems are to distinguish between individual tracks on photographs (when they cross) and to connect successive pictures of a particle without confounding these with pictures of other particles.

These techniques should be adapted to the size of the enclosure. Small particles or bubbles provide a dense velocity information, but in a small and thin field of a few square metres. Balloons allow the coverage of large rooms but provide only a few tracks.

Two-dimensional PIV system is now commercially available measuring equipment, generally used for small areas. Its use for measurements in large enclosures with low air velocities is more difficult (impossible without major changes of the commercial systems) as lighting, particle seeding and recording systems have to be adapted to their large dimensions (see also Section 8.2.4).

8.4.2   Visualization in scale models

Scale models are often used for the visualization of airflow patterns in large enclosures. In principle, scale models should be built and used in such a way that most important non-dimensional numbers (Reynolds, Froude, etc) should be conserved. In practice, however, it is not always possible to conserve all non-dimensional numbers, especially when thermal gradients and thermal exchanges are taken into account. It is then up to the modeller to make the best choice! More information on modeling techniques is also found in the work of Awbi (Awbi, 2003).

Scale models in water are often used. Density gradients may be obtained either by heating or by adding salt to water. Temperature fields can be visualized by adding tiny spheres containing cholesteric liquid crystals to the water. Density fields are made visible by adding a colour together with the salt. These methods are useful to study flow patterns but cannot model correctly the thermal exchanges with walls (i.e. only adiabatic conditions). Moreover, water models are in general less turbulent than the real airflows they simulate.

The ammonia absorption method (Krückels, 1969) is a measuring technique in which the mass transfer results in a chemical reaction that causes a colour change on the body surface. This measuring technique is used to research problems of heat and mass transfer involving air under the conditions of forced convection. In this case, the wall is coated with a thin wet layer (filter paper or gel) containing an aqueous solution of $MnCl_2$ (manganese chloride) and H_2O (water). The reacting gas NH_3 is added in short pulses and at low concentrations to the main air stream. The air stream and ammonia must be well mixed. Ammonia is absorbed by the wet layer based on the local partial concentration differences, and a reaction takes place in which MnO_2 (manganese dioxide) is one product. While the original coating is bright, MnO_2 is dark; thus the observed colour intensity of MnO_2 is a measure of the local mass transfer rate. As a result of the absorption of ammonia by the moistened foils, the ammonia concentration on the body surface is zero. This means that in the analogue temperature field, the surface temperature is constant (T_{wall} = constant). A photometric evaluation is performed to obtain the local distribution of heat and mass transfer coefficient. There are two possibilities for the photometric evaluation: the transmission method or the reflectance method. It is necessary to perform a calibration to obtain a correlation between the colour (transmission coefficient) and the mass transfer.

An example for a measured distribution of the heat transfer coefficient is found in Figure 8.12. The heat transfer coefficients are related to the temperature difference between the temperature at the intake air and the wall temperature.

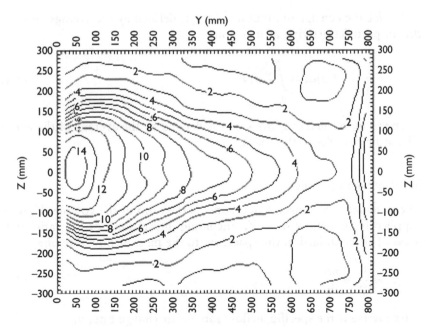

Figure 8.12 Distribution of the heat transfer coefficient ($W\,m^{-2}k^{-1}$) at the ceiling of a model room for an air change rate of $75\,h^{-1}$.

8.5   Age of air and air change efficiency

8.5.1   Definitions

The quantities defined below are explained in greater detail elsewhere (Sandberg and Sjöberg, 1984; Sutcliffe, 1990; Roulet and Vandaele, 1991) and are only briefly described here.

Age of the air

The fresh air particles coming from outside or from the ventilation system arrive at a given location r in a room after a time τ_r, which will vary from one particle to the other. τ_r is called the residence time of the particle in the room, or its age, as if it were born when entering the room. Because there is a large number of air particles, we may define a probability density $f(\tau_r)$ that the age of particles arriving at a given location is between τ and $\tau + d\tau$ and a probability $F(\tau_r)$ that this age is larger than τ. The following relationships always hold between these two functions:

$$f(\tau_r) = -\frac{dF(\tau_r)}{d\tau} \quad \text{and} \quad F(\tau_r) = 1 - \int_0^\tau f(t_r)dt \tag{8.12}$$

The *local mean age of air* at a point $\underline{r}$ is defined by the average age of all the air particles arriving at that point:

$$\bar{\tau}_r = \int_0^\infty t f_r(t)dt = \int_0^\infty F_r(t)dt \tag{8.13}$$

The *room mean age of air* $< \tau >$ is defined by the average of the ages of all the air particles in the room.

Nominal time constant

The nominal time constant of a zone, is the ratio of the volume, V, to the supplied fresh airflow rate, $\dot{V}$, (including infiltration), or the ratio of the mass of air contained in the space, m, to the mass airflow rate, $\dot{m}$:

$$\tau_n = \frac{V}{\dot{V}} = \frac{m}{\dot{m}} \tag{8.14}$$

Its inverse is the specific airflow rate or air change rate, n.

These parameters can be measured using a tracer gas, assuming a complete mixing of the tracer. However, even in case of poor mixing, it is shown that the nominal time constant is equal to the mean age of air at the exhaust (Sandberg, 1984):

$$\tau_n = \bar{\tau}_e \tag{8.15}$$

Air exchange efficiency

This expresses how the fresh air is distributed in the room. The time, τ_a, required on average to replace the air present in the space is given by the following expression (Sandberg and Sjöberg, 1984):

$$\tau_a = 2 < \tau > \tag{8.16}$$

where $< \tau >$ is the room mean age of air. At a given flow rate and space volume, the shortest time required to replace the air within the space is given by the nominal time constant.

The air exchange efficiency η_a is calculated using the relation:

$$\eta_a = \frac{\tau_n}{2 \langle \tau \rangle} \tag{8.17}$$

The air exchange efficiency is equal to one for piston-type ventilation whereas for complete mixing it is equal to 0.5. Short-circuiting of air will give rise to an efficiency that is lower than 0.5.

Examples

Figure 8.13 shows some typical probability curves for the age of the air particles at the exhaust, with the air exchange efficiency shown as a parameter, whereas Figure 8.14 shows the corresponding probability density curves. In this example, the nominal time constant is one h.

Note that these curves are theoretical, and for illustration. Some of them, in particular those for very high efficiency, η, are not likely to be found in practice.

100 per cent air exchange efficiency occurs when all air particles reach a given location in the room at the same time: the air is displaced like a piston. The exhaust is reached at a time corresponding exactly to the nominal time constant. At 99 per cent air exchange efficiency, there is already some spreading of the ages around an average still equal to the nominal time constant. When the probability density function spreads out and the probability function smoothens, the mean age at the exhaust remains the same, but there are more 'young' air particles and more 'aged' ones. Consequently, the most probable age is reduced. Both functions gradually change into an exponential, which corresponds to 50 per cent efficiency or complete mixing. With this distribution, the most probable age is zero, but the mean age

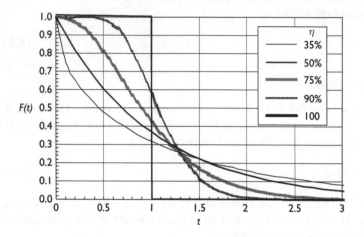

Figure 8.13 Typical probability curves for the age of the air (η is the air exchange efficiency).

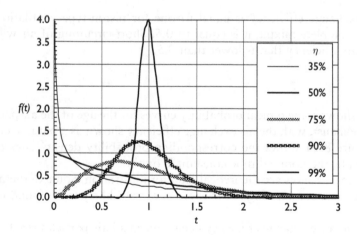

Figure 8.14 Typical probability density curves for the age of air (η is the air exchange efficiency).

is still equal to the nominal time constant. The last curve, with 35 per cent efficiency, represents a situation with a short-circuit, where there are many very young particles reaching the exhaust, but there are also more very old particles, because the short-circuiting induces dead zones (Table 8.1).

Table 8.1 Nominal time constant and the room mean age of air corresponding to the probability curves shown in Figure 8.13 and Figure 8.14

Air exchange efficiency	η_a	35%	50%	75%	90%	99%
Mean age at exhaust or nominal time constant	$\tau_e = \tau_n$	1.00	1.00	1.00	1.00	1.00
Room mean age of the air	$<\tau>$	1.44	1.00	0.67	0.55	0.50

8.5.2   Measurement method

The basic principle is to mark the air to be traced with a gas (the tracer gas), according to a known schedule, and to follow the concentration of that tracer gas at the location of interest. This technique is based on the assumption that the tracer gas behaves the same as the air: no adsorption and of equal buoyancy. It can be readily seen that if the air is marked at the inlet by a short pulse of tracer gas, and if the tracer molecules follow the air molecules, they will arrive at a given location at the same time as the air molecules. In fact, the pulse technique is not the only one and the probability functions and the local mean ages can be measured by recording the time history of the net tracer concentration, $C_r(t)$, at any point, r, by either of three strategies as follows:

1   *step down*: uniform concentration of tracer is achieved at the beginning of the test, when the injection is stopped,
2   *step-up*: the tracer is injected at air inlet, at a constant rate from the starting time throughout the test,
3   *pulse*: a short pulse of tracer is released in the air inlet at the starting time.

It was shown, however, that for rooms with a single air inlet and a single air outlet, the step up method is to be preferred, because it is the easiest to perform for that case and gives the best accuracy (Roulet and Cretton, 1992). Therefore, in the following, we will consider only this method.

For a step-up technique, tracer gas is injected into the supply air in the outside air duct at a constant rate, starting at a known time t_0. It is assumed that the tracer and the air are fully mixed in the supply duct to produce a steady concentration, C_3, at the inlet. If C_3 cannot be measured, the equilibrium concentration within the enclosure, C_4, may be used instead.

Tracer gas concentration at the locations where the age of air is required is recorded. The sampling time interval should be short enough to record the transient evolution of the concentration. It should then be much shorter than the expected age of air.

One important location is in the exhaust duct, where $C_e = C_6$ is measured. This measurement provides both the nominal time constant and the mean age of air in the ventilated space. There, the recording time interval shall

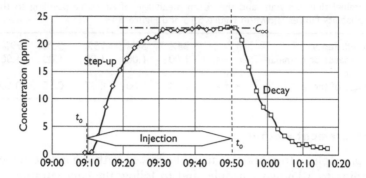

Figure 8.15 Record of tracer gas concentration in the exhaust duct during measurement of the age of air.

not be longer than the fifth of the nominal time constant. Injection rate is maintained constant until a steady state is obtained. An example of such a record is given in Figure 8.15.

When the concentration stabilizes, the step-up experiment is ended. However, it is recommended that decay experiment is performed and recording the concentration continued after having stopped the tracer gas injection. Also note the time when injection is ceased: this time is the starting time of the decay experiment.

To interpret the recorded tracer gas concentrations and obtain the age of air, the background (or supply) concentration should first be subtracted from all measurements, and the elapsed time should be calculated by subtracting the starting time from all time values. In the following formulae, the net concentration, C_r, is the difference between the concentration measured at location r and the concentration in the outdoor air.

Note that after some time, the concentration at all points tends towards a simple exponential value of the form:

$$\text{Step-up}: C_e\left(t\right) = C_e\left(\infty\right)\left(1 - e^{-t/\tau_c}\right) \qquad \text{Decay}: C_e\left(t\right) = C_e\left(0\right)e^{-t/\tau_c}$$

$$(8.18)$$

This characteristic is used to estimate the integrals involved in the calculation of air change efficiency.

Once $C_e(\infty)$ is known, the probability function of the age of air can be calculated from the concentration ratio:

$$\text{Step-up case } F(\tau) = 1 - \frac{C_e\left(t - t_0\right)}{C_e(\infty)} \qquad \text{Decay case } F(\tau) = \frac{C_e\left(t - t_0\right)}{C_e(t_0)}$$

$$(8.19)$$

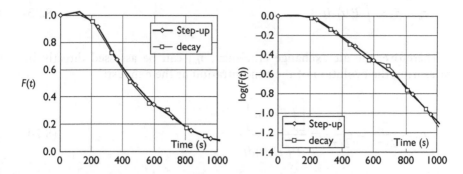

Figure 8.16 Probability functions of the age of air obtained from the recorded concentration illustrated in Figure 8.15 (left: linear scale, right: logarithmic scale, showing an exponential decay after 400 s).

Figure 8.16 shows the concentration ratio calculated from the recorded concentration illustrated in Figure 8.15.

The mean age of air in the ventilated space and the ventilation effectiveness is then found as follows. The local mean age of air at any location is the integral of the probability distribution:

$$\tau_r = \int_0^\infty F_r(t)\, dt \tag{8.20}$$

And the room mean age of the air is the average of all ages at all locations in the room:

$$\langle \tau \rangle = \frac{1}{V} \int_V \bar{\tau}_r dr \tag{8.21}$$

When there is only a single exhaust, the room mean age of air can be deduced from the tracer concentration measurements in the exhaust duct, $C_e(t)$:

$$\langle \tau \rangle = \frac{\mu_1}{\mu_0} = \frac{\int_0^\infty t F_e(t)\, dt}{\int_0^\infty F_e(t)\, dt} \tag{8.22}$$

In this case, the nominal time constant of the ventilated space, τ_n, which is the ratio of the space volume and the volumetric airflow rate, is equal to the mean age of air at the exhaust:

$$\tau_n = \bar{\tau}_e = \int_0^\infty F_e(t)\, dt \tag{8.23}$$

Therefore, the air exchange efficiency, η_a, can be assessed directly by measuring the evolution of the concentration at the exhaust:

$$\eta_a = \frac{\tau_n}{2\langle\tau\rangle} = \frac{\bar{\tau}_e}{2\langle\tau\rangle} = \frac{\left[\int_0^\infty F_e(t)dt\right]^2}{2\int_0^\infty tF_e(t)dt} \tag{8.24}$$

8.5.3 Interpretation

In practice, the various moments in the above formulae are calculated numerically, on the bases of discrete recorded values of concentration and time. The following section describes a simple method to calculate these moments, using the trapezium method, whose general formulation is

$$\int_0^{t_N} f(t)dt \cong \sum_{j=0}^{N-1} \frac{1}{2}(f_j + f_{j+1})\Delta t \tag{8.25}$$

where f_j is represents $f(t_j)$ and Δt represents $t_{j+1} - t_j$.

Assuming a linear variation of the concentration in each time step, we obtain, for the two integrals defined above:

$$\mu_0 = \int_0^\infty F_e(t)dt = \left(\frac{F_0 + F_N}{2} + \sum_{j=1}^{N-1} F_j\right)\Delta t + \varepsilon_0(N, \tau_d) \tag{8.26}$$

$$\mu_1 = \int_0^\infty tF_e(t)dt = \left(\frac{NF_N}{2} + \sum_{j=1}^{N-1} jF_j\right)\Delta t^2 + \varepsilon_1(N, \tau_d) \tag{8.27}$$

where

F_j is the probability distribution at time $t = j\Delta t$.

$$\text{Step-up case } F_j = 1 - \frac{C_e(t_0 + j\Delta t)}{C_e(\infty)} \qquad \text{Decay case } F_j = \frac{C_e(t_0 + j\Delta t)}{C_e(t_0)} \tag{8.28}$$

N is the last measurement integrated using the trapezium method.

$\varepsilon_n(N, \tau_d)$ is the rest of the integral, evaluated using an exponential fit on the last measurements (see below).

The number of measurements, N, could be large enough to ensure that the sum of the terms for $j > N$ is negligible, or, in other words, that C_N is very close to the steady-state value. In this case, the remaining parts, $\varepsilon_n(N, \tau_d)$, are negligible. In practice, however, the measurement can be stopped before reaching the steady-state condition. In this case, the tail in the integral of the moments is not measured but is estimated.

As shown in Section 8.5.2, this tail is, in most cases, exponential. Therefore, for a time larger than $t_N = N\Delta t$, it can be assumed that

$$F(t > t_N) = F_N \cdot \exp\left(\frac{t_N - t}{\tau_d}\right) \tag{8.29}$$

where τ_d is a time constant determined by a curve fit on the last measurements in the exponential part. The time required for reaching an exponential decay depends not only on the nominal time constant of the room but also on the ventilation system. In case of complete mixing, the decay will be exponential from the beginning of the test. In the case of perfect displacement ventilation, the decay will be very sharp after a time equal to the age of air.

If N is chosen in such a way that $F(t > t_N)$ is exponential, the remaining part, $\varepsilon_n(N, \tau_d)$, of the moments can be calculated analytically from

$$\varepsilon_0(N, \tau_d) = \int_{t_N}^{\infty} F_e(t)dt = F_N \int_{t_N}^{\infty} \exp\left(\frac{t_N - t}{\tau_d}\right) dt = F_N \tau_d$$

$$\varepsilon_1(N, \tau_d) = \int_{t_N}^{\infty} t F_e(t)dt = F_N \tau_d (t_N + \tau_d) \tag{8.30}$$

8.5.4 Error analysis

Assuming that random and independent errors spoil the measurements of tracer gas concentration, the confidence interval of any result, y, is

$$[y - \delta y; \ y + \delta y] \quad \text{with} \quad \delta y(x_i) = T(P, \infty) \sqrt{\sum_i \left(\frac{\partial y}{\partial x_i}\right)^2 \delta x_i^2} \tag{8.31}$$

where

$T(P, \infty)$ is the Student coefficient for having the actual value within the confidence interval with probability $1 - P$.

x_i is any variable on which the airflow rate $\dot{m}$ depends.

δx_i is the standard deviation of the variable x_i, assumed to be a random variable of mean x_i and normal distribution.

The confidence interval of $F(\tau)$ is

$$[F_j - \delta F_j; F_j + \delta F_j] \quad \text{with}$$

$$\delta F_j = T(P, \infty) \sqrt{\left[\frac{\delta C_e(j\Delta t + t_0)}{C_e(j\Delta t + t_0)}\right]^2 + \left[\frac{C_e(j\Delta t + t_0)}{C_e^2(\infty)} \delta C_e(\infty)\right]^2} \tag{8.32}$$

Then, the confidence intervals of the moments are

$$\delta \mu_0 = T(P, \infty)\sqrt{f_{\mu_0}} \quad \text{and} \quad \delta \mu_1 = T(P, \infty)\sqrt{f_{\mu_1}} \tag{8.33}$$

with

$$f_{\mu_0} = \left(\frac{\Delta t}{2}\right)^2 [(\delta F_0)^2 + (\delta F_N)^2]$$

$$+ (\Delta t)^2 \sum_{J=1}^{N-1} \delta F_j^2 + \left(\frac{F_0 + F_N}{2} + \sum_{j=1}^{N-1} F_j\right)^2 (\delta \Delta t)^2 + (\delta \varepsilon_0)^2 \tag{8.34}$$

and

$$f_{\mu_1} = \left(\frac{\Delta t}{2}\right)^2 [(\delta F_0)^2 + (\delta F_N)^2]$$

$$+ (\Delta t)^2 \sum_{J=1}^{N-1} j^2 \delta F_j^2 + \left(\frac{NF_N}{2} + \sum_{j=1}^{N-1} jF_j\right)^2 (\delta \Delta t)^2 + (\delta \varepsilon_1)^2 \tag{8.35}$$

in which

$$\delta \varepsilon 0_1 = T(P, \infty)\sqrt{\tau_d^2 \delta F_N^2 + F_N^2 (\delta \tau_d)^2} \tag{8.36}$$

and

$$\delta \varepsilon_1 = T(P, \infty)\sqrt{(t_N + \tau_d)^2 \tau_d^2 \delta F_N^2 + (t_N + 2\tau_d)^2 F_N^2 (\delta \tau_d)^2} \tag{8.37}$$

where $\delta \tau_d$ is the confidence interval resulting form the exponential fit.
Finally we get

$$\delta \tau_n = \delta \tau_e = \delta \mu_0(F_e) \tag{8.38}$$

$$\delta \langle \tau \rangle = \left\{\frac{1}{[\mu_0(F_e)]^2}\right\} \sqrt{[\mu_1(F_e)]^2 [\delta \mu_0(F_e)]^2 + [\delta \mu_1(F_e)]^2} \tag{8.39}$$

and

$$\delta \eta_a = \frac{\delta \mu_1(F_e)}{[2\mu_0(F_e)]^2} \tag{8.40}$$

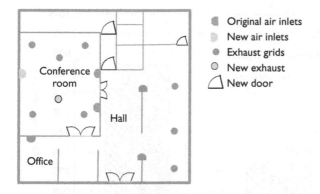

Figure 8.17 Arrangement of the conference room.

8.5.5 Case study

Age of air measurements were performed in a conference room as shown in Figure 8.17, in order to check whether the airflow pattern in the room was as expected, i.e. a piston flow from bottom to top.

The supply air, slightly colder than room, is introduced at low speed close to the ground through three inlets. This cold air spreads onto the floor like a lake and should go up faster where there are heat sources such as occupants. The exhaust grilles are placed on the ceiling.

These measurements were performed as a first step in the room as it was. It was found that the air was poorly distributed in the room, because there were only two air inlets to the right of the room, and not enough exhaust grilles. In addition the room was found leaky, and much air from the ventilation system was leaving the room quickly after entering it, thus reducing the purging effect. The indoor air quality was however good, because the ventilation rate was high (250 s nominal time constant, or 15 outdoor air change per hour!).

On the basis of these results, improvements were made to the system. The leaks from the room were sealed, and new inlet and exhaust grilles were installed. A second measurement campaign was performed, showing a significant improvement of the air change efficiency (Figure 8.18). The air change efficiency was doubled; and the mean age of air was maintained despite a reduction of the ventilation rate by a factor the two.

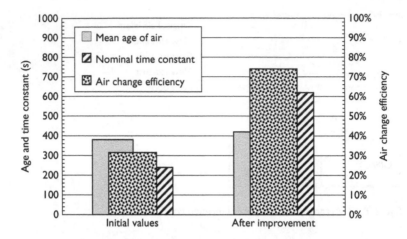

Figure 8.18 Room ventilation characteristics before and after improvement.

8.6   Mapping the age of air in rooms

Mapping either the contaminant concentration or the age of air in a room can be of great advantage in studying the contaminant or airflow pattern and their effects on occupants. Such maps have already been calculated using computer codes (Davidson and Olsson, 1987) and some qualitative representations have been drawn from measurements (Valton, 1989).

The purpose of this chapter is to propose a systematic way to obtain a preliminary map of contaminant concentration or of the age of air from measurements in a room. Such a map allows one to identify poorly ventilated locations or dead zones within the measured room, under the conditions prevailing during the experiments.

8.6.1   Minimum number of measurements

Measuring the variable at each node of a network and interpolating between these nodes basically provides a map of any scalar variable y in a three-dimensional room. Such measurements are however very expensive and may be unfeasible; if only 5 values are taken on each axis, at least 125 measurements are required. Therefore, it makes sense to look for methods needing a minimum number of measurements points.

The minimum number of measurements depends on the objective of the mapping experiment, or more precisely on the empirical model that is chosen to represent the map of the variable v.

Any infinitely derivable function (as y is assumed to be) can be developed in a Taylor series around a given point. This gives a polynomial, which can be approximated by its $k+1$ first terms, k being the degree of the polynomial. In the following, models of degree 1 and 2 will be considered. If a *linear* model is adopted (degree 1), such as

$$y = a + \Sigma_i b_i x_i \qquad (8.41)$$

where x_i are the three coordinates of the measured point, only four measurements are needed to obtain a set of coefficients $\{a, b_i\}$. If more measurements are made, the coefficients may be obtained by a least square fit procedure provided there is no (or negligible) uncertainty on the coordinates. If their coordinates differ for the other points, these supplementary measurement points give information on the validity of the used model.

If the linear model does not appear to be valid, higher degree models may be used. For example, a quadratic model

$$y = a + \Sigma_i b_i x_i + \Sigma_{i \neq j} b_{ij} x_i x_j + \Sigma_i b_{ii} x_i^2 \qquad (8.42)$$

which contains ten coefficients, can be chosen. Such a model may already fit many practical situations and present minimal and maximal value(s). To determine these coefficients, measurements at ten locations is the minimum.

An intermediate model is the *interactions* model:

$$y = a + \Sigma_i b_i x_i + \Sigma_{i \neq j} b_{ij} x_i x_j \qquad (8.43)$$

for which seven coefficients must be determined. Table 8.2 summarizes the minimum number of measurements needed.

8.6.2 Location of the measurement points

The next problem is: where should we locate the measurement points? There are numerous possible experimental designs, but they do not give the expected results with the same accuracy. For example, it is obvious that, to fit a linear model of one dimension only ($y = ax + b$), the location of the two measurement points (the minimum number) which gives the best accuracy on a and b is at the ends of the experimental domain. If the model is more sophisticated or is for a larger number of dimensions, the choice is not so

Table 8.2 Minimum number of measurements needed to obtain the coefficients of a kth degree polynomial empirical model representing a variable in a two and three-dimensional space

Model Dimensions	Linear	Interaction	Quadratic	Cubic	4th degree
2	3	4	6	10	15
3	4	7	10	20	35

obvious. However, several tools exist for planning such experiments, which may be found in the literature (Fedorov, 1972; Box *et al.* 1978; Bandemer and Bellmann, 1979; Feneuille *et al.* 1983) and are applied below.

However, because points close to the walls do not represent the inner volume, the sampling points should not be located near the walls or in the corners of the room, in practice not nearer the wall than 0.1 times the characteristic enclosure dimension. In the followings, the 'room' or the 'experimental domain' is a volume that is smaller than the actual measured space by about 20 per cent in each direction.

Let us take a coordinate system in such a rectangular volume that as a unit for each direction is half-length of that domain in that direction. Three numbers, included in the interval $[-1, +1]$ locate any point in the 'room'.

The experimental design can be represented by a rectangular matrix with three columns (one for each coordinate) and as many lines as measurement points. A general condition is that, to obtain the coefficients of a polynomial of degree k, each of the variables x, y and z shall take at least $k+1$ values in the experimental design, which should have at least $k+1$ levels on each axis.

The coefficients a, b_i, b_{ij}, etc. of the model are obtained from the ages measured at all points of the experimental design by solving a system of equations. A good experimental design should reduce the variance (or increase the accuracy) of the results.

Because the measured data are not known with infinite accuracy, the system of equations should be well conditioned, i.e. its solutions should not dramatically change when the input data are slightly modified. The condition number gives an estimate of the ratio of the confidence intervals of the solutions to the confidence intervals of the inputs. The smallest this number is, the best conditioned the system of equation is.

Mathematical methods to assess the performance of experimental design are found in the literature mentioned above.

8.6.3   Examples of experimental designs

Several experimental designs were examined with the aim to map a rectangular volume or a rectangular area (Roulet *et al.* 1991). The tested models were the linear, the interaction and the quadratic models. Several of these designs were found to be unusable (singular matrix or too large a condition number for the quadratic model). Some useable examples are given below.

Factorial designs

A k-dimensional, l-level factorial design is obtained by dividing the experimental domain (e.g. the interval $[-1, 1]$) on each axis into l equidistant levels. The complete factorial design contains all the points obtained by the l^k combination of the l possible values of the k coordinates.

The number of points in a full-factorial design is l^k. If l and k are greater than 2, the full-factorial designs often have many more points than the minimum required and are therefore seldom used. However, partial factorial designs can be obtained by selecting the required number of measurement points from the full design. Some examples are given below.

Two-dimensional designs

The two-dimensional, two-level full-factorial design (Table 8.3) is optimal for a linear model, providing the coefficients of that model with the best accuracy. If, for economical reasons, one point is omitted, the confidence intervals of the coefficients are twice that based on four measurement points.

Table 8.3 Two-dimensional, two level full-factorial design

No	x	y	
1	−1	−1	
2	1	−1	
3	−1	1	
4	1	1	

It is very important to note that the very often-used design consisting of changing one variable at a time (Table 8.4) is less accurate than the former.

Table 8.4 Two-dimensional design changing one variable at a time

No	x	y	
1	1	0	
2	0	1	
3	−1	0	
4	0	−1	

Adding a fifth point at the centre (0,0) of the two-dimensional, two-level level full-factorial design allows assessing the coefficient of the interaction term b_{12}, without loss of accuracy.

The following two points

Number	x	y
6	−1	0
7	1	0

can be added to obtain a minimum design for a quadratic model, which has a condition number of 6.3 (Figure 8.19).

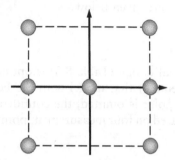

Figure 8.19 Minimum design for a 2-D quadratic model.

The two-dimensional full-factorial design with three levels shown in Table 8.5 has a better condition number (4.4) for a quadratic model.

Table 8.5 Two-dimensional full-factorial design with three levels

No	x	y	No	x	y
1	−1	−1	6	1	0
2	0	−1	7	−1	1
3	1	−1	8	0	1
4	−1	0	9	1	1
5	0	0			

Three-dimensional designs

In three dimensions, the four-point design of Table 8.6 is perfect for a linear model.

Table 8.6 Minimum three dimensional design for assessing the coefficients of a linear model

No	x	y	z
1	−1	−1	1
2	1	−1	−1
3	−1	1	−1
4	1	1	1

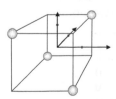

Table 8.7 Full-factorial design for assessing the coefficients of a linear
model with interactions

No	x	y	z
1	−1	−1	−1
2	1	−1	−1
3	−1	1	−1
4	1	1	−1
5	−1	−1	1
6	1	−1	1
7	−1	1	1
8	1	1	1

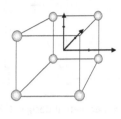

Table 8.8 Three-dimensional centered star design

No	x	y	z
9	1	0	0
10	0	1	0
11	0	0	1
12	−1	0	0
13	0	−1	0
14	0	0	−1
15	0	0	0

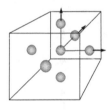

It can be expanded to a full-factorial design (Table 8.7) which is good when used with interaction model.

Here again, the star design shown in Table 8.8 is less accurate and needs more work than the minimum design of Table 8.6. However, combining the centred star design with the full-factorial design of Table 8.7 gives a so-called composite centred design, suitable for a quadratic model, having a condition number of 4.4. If fewer points are wanted, the points 8, 5 and 2 can be deleted (in that order) giving finally a design having twelve points and a condition number of 4.8. Finally, deleting two more points (3 and 15) gives the design C3, which has six points in the centre of the faces and four points at opposite corners (Figure 8.20).

The condition number of these designs and three models are given in Table 8.9.

There are numerous other possibilities that can be imagined or found in the referred literature. If such a design is planned to be used, it is advisable

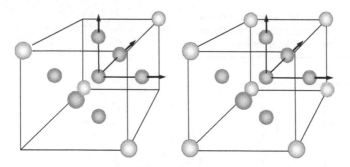

Figure 8.20 Experimental designs C3 (left) and composite centered (right).

Table 8.9 Condition number for some experimental designs and three models

Experimental design	Number of points	Quadratic model	Interactions model	Linear model
2-D Designs				
2-level part factorial	3	–	–	2.0
2-level full factorial	4	–	–	1.0
Centered 2-level factorial	5	–	1.0	1.0
Minimum for quadratic	6	6.3	1.0	1.0
3-level full factorial	9	4.4	1.0	1.0
3-D Designs				
2-level half factorial	4	–	–	1.0
2 level full factorial	8	–	1.0	1.0
C 3	10	4.3	3.2	1.0
Composite centered	15	4.4	1.0	1.0

No figure indicates that this design cannot be used with the corresponding model (not enough experiments).

to first compute the condition number and values of the variance at several locations in the room.

8.6.4 Application example

After improvements, a map of the age of air was measured in the conference room shown in Figure 8.17. These measurements were performed in a plane located at the head level of sitting persons, using a 9-sampling-points full-factorial design. The results are shown in Figure 8.21. In the unoccupied room, the air is older at the middle left, where there is only one air inlet. When the room is occupied, the air is younger in the middle of the room, where there are occupants.

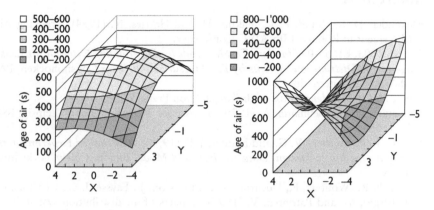

Figure 8.21 Map of the age of the air at head level. Left: unoccupied room, right: room occupied by 10 persons sitting around the conference table.

8.7 Summary

Vizualization and assessment of airflow patterns may be useful to check whether the reality is actually as designed, or to look for reasons of dysfunctions. Several tools are available that allow assessing various characteristics of the airflow patterns in buildings.

Several types of anemometers allow assessing air velocities within a wide range, from the detection of draughts up to air speed found in ducts.

The control of airflows requires that the air passes through the control systems, therefore that the building envelope, partition walls, ducts, casings of the air handing unit, etc. are reasonably airtight. The air permeability of enclosures and ductwork can be quantified using various methods, from a very simple technique providing a rough estimate of the permeability of the building envelope, up to sophisticated techniques used to check ductwork.

The airflow patterns themselves can be either visualized using visible tracers such as fumes, flakes or balloons, or quantified using laser Doppler or particle imaging velocimetry. The age of the air, measured at any place using tracer gas techniques, provides information on the air distribution efficiency in the room.

Mapping the age of the air – or any other variable related to airflow pattern such as contaminant concentrations – provides more detailed information but requires more measurement efforts. These efforts can be reduced by an appropriate planning of the experiments.

References

Alexander, D. K., Jones, P. J., Jenkins, H. and Harries, N. (1994), Tracking Air Movement in Rooms: 15th AIVC Conference.

ASTM, 2003, E1554-03 *Standard Test Methods for Determining External Air Leakage of Air Distribution Systems by Fan Pressurization*. West Conshohocken, PA, USA: ASTM International.

Awbi, H. (2003), *Ventilation of Buildings*. Spon Press, 522 p.

Bandemer, H. and Bellmann, A. (1979), *Statistische Versuchsplanung*. Leipsig: BSB G. Teubner Verlag.

Box, G. E. P., Hunter, W. G. and Hunter, J. S. (1978), *Statistics for Experimenters, an Introduction to Design, Data Analysis and Model Building*. New York: John Wiley.

Carrie, F. R., Wouters, P., Ducarme, D., Andersson, J., Faysse, J. C., Chaffois, P., Kilberger, M. and Patriarca, V. (1997), Impacts of air distribution system leakage in Europe: the SAVE duct European programme. 18th AIVC Conference, pp. 651–660.

Davidson, L. and Olsson, E. (1987), Calculation of age and local purging flow rate in rooms. *Building and Environment*, 22, 111–127.

Dickerhoff, D. J., Grimsrud, D. T. and Lipschutz, R. D. (1982), *Components Leakage Testing in Residential Buildings*. Summer Study in Energy Efficient Buildings, Santa Cruz CA Lawrence Berkeley Lab.

Etheridge, D. and Sandberg, M. (1996), *Building Ventilation, Theory and Measurement Techniques*. Hoboke, NJ, USA; Chichester, UK: Wiley & Sons, 709 p.

Fedorov (1972), *Theory of Optimal Experiments*. New York: Academic Press.

Feneuille, D., Mathieu, D. and Phan-Tan-Luu, R. (1983), *Méthodologie de la recherche expéri-mentale*. Marseille Cedex: Cours IPSOI, R. H. Poincaré.

Fürbringer, J.-M., Foradini, F. and Roulet, C.-A. (1994), Bayesian method for estimating air tightness coefficients from pressurisation measurements. *Building and Environment*, 29, 151–157.

ISO, 1998, ISO 9972 – *Thermal Insulation – Assessment of the Airtightness of Buildings – Fan Pressurisation Method*. Genève: ISO.

Krückels, W. (1969), Eine Methode zur photometrischen Bestimmung örtlicher Stoffübergangszahlen mit Hilfe chemischer Nachweisreaktionen. *Chemie-Ingenieur-Technik*, 41, 427–433.

Maas, H. G., Grün, A. and Papantoniou, D. (1993), Particle tracking velocimetry in three-dimensional flows. *Experiments in Fluids*, 15, 133–146.

Modera, M. P. (1989), residential duct system leakage: magnitude, impacts, and potential for reduction. *ASHRAE Transactions*, 95(Pt 2), 561–569.

NBCF, 1987, *Indoor Climate and Ventilation in Buildings. Regulations and Guidelines 1987*. Helsinki: The Finnish Ministry of the Environment.

Palmiter, L. and Bond, T. (1991), *Modeled and Measured Infiltration: A Detailed Case Study of Four Electrically Heated Homes*. Seattle, WA: Ecotope Inc.

Roulet, C.-A. and Vandaele, L. (1991), Airflow patterns within buildings - measurement techniques. *AIVC Technical Note*, 34, 265 p.

Roulet, C.-A. and Cretton, P. (1992), Field comparison of age of air measurement techniques. *Roomvent 92 Proceedings*, 3, 213–229.

Roulet, C.-A., Compagnon, R. and Jakob, M. (1991), A simple method using tracer gas to identify the main airflow and contaminant paths within a room. *Indoor Air conference*, 3, 311–322.

Sandberg, M. (1984), The multi-chamber theory reconsidered from the viewpoint of air quality studies. *Building and Environment*, 19, 221–233.

Sandberg, M. and Sjöberg, M. (1984), The use of moments for assessing air quality in ventilated rooms. *Building and Environment*, 18, 181–197.

Scholzen, F., Moser, A. and Suter, P. (1994), Particle – Streak-Velocimetry for Room Air Flows, 15th AIVC Conference.

Sutcliffe, H. C. (1990), A guide to air changes efficency. *AIVC Technical Note*, 28.

Tarantola, A. (1987), *Inverse Problem Theory, Method for Data Fitting and Model Parameter Estimation*. Amsterdam: Elsevier.

Valton, P. (1989), Renouvellement d'air dans les bâtiments. *PROMOCLIM E*, 18, 279–297.

Nomenclature

Symbols and units

A	area, cross section	m^2
A	vector containing coefficients	
C	concentration	JK^{-1}
f	probability density function	
F	probability function	
K	leakage coefficient in the power law	$m^3(h, Pan)^{-1}$
m	mass	kg
$\dot{m}$	mass flow rate	kgs^{-1}
M	model matrix	
n	exponent in the power law	–
N	number	
p	pressure	Pa
P	probability	
r	position vector, containing the coordinates of a point	
R	recirculation ratio in an AHU	–
t	time	s
$T(P, \nu)$	Student distribution function	
Tu	turbulence intensity	–
v	velocity	ms^{-1}
V	volume	m^3
$\dot{V}$	volume flow rate	m^3s^{-1}
x	coordinate in a room	m
y	variable	
Y	vector containing variables	
Δp	pressure differential	Pa
Δt	time interval	s
ε	effectiveness	–

$\varepsilon_n(N, \tau_d)$	is the rest of an integral	
γ	ratio	–
η	efficiency	–
μ_i	statistical moment of order i	
ρ	density	kgm^{-3}
σ	standard deviation	
τ	time constant, age or the air	s

Subscripts

0	zero, reference	r	room
a	air	t	tracer gas
e	exhaust	x	extract
i	internal	k	numbering the tracer gas
i	infiltration	i, j	numbering sampling points
L	leakage	n	nominal
o	outdoor		

Index

Milton Keynes UK
Ingram Content Group UK Ltd.
UKHW021855071024
449327UK00021B/1579